Methods in Molecular Biology

Series Editor
John M. Walker
School of Life and Medical Sciences
University of Hertfordshire
Hatfield, Hertfordshire, UK

For further volumes:
http://www.springer.com/series/7651

For over 35 years, biological scientists have come to rely on the research protocols and methodologies in the critically acclaimed *Methods in Molecular Biology* series. The series was the first to introduce the step-by-step protocols approach that has become the standard in all biomedical protocol publishing. Each protocol is provided in readily-reproducible step-by-step fashion, opening with an introductory overview, a list of the materials and reagents needed to complete the experiment, and followed by a detailed procedure that is supported with a helpful notes section offering tips and tricks of the trade as well as troubleshooting advice. These hallmark features were introduced by series editor Dr. John Walker and constitute the key ingredient in each and every volume of the *Methods in Molecular Biology* series. Tested and trusted, comprehensive and reliable, all protocols from the series are indexed in PubMed.

Targeting Enzymes for Pharmaceutical Development

Methods and Protocols

Edited by

Nikolaos E. Labrou

Department of Biotechnology, Agricultural University of Athens, Athens, Greece

Humana Press

Editor
Nikolaos E. Labrou
Department of Biotechnology
Agricultural University of Athens
Athens, Greece

ISSN 1064-3745 ISSN 1940-6029 (electronic)
Methods in Molecular Biology
ISBN 978-1-0716-0162-4 ISBN 978-1-0716-0163-1 (eBook)
https://doi.org/10.1007/978-1-0716-0163-1

This Humana imprint is published by the registered company Springer Science+Business Media, LLC, part of Springer Nature.
The registered company address is: 233 Spring Street, New York, NY 10013, U.S.A.

Preface

Drug development is the process of bringing a new pharmaceutical drug to the market once a lead compound has been identified through the process of drug discovery. Enzymes offer unique opportunities for drug discovery, as they are one of the most important groups of drug targets. Many of the significant biochemical processes in the cell are enzyme-catalyzed reactions (biosynthesis and catabolism) or specific cellular signaling pathways that enzymes are involved with. Several important diseases are often associated with the elevation or repression of certain enzymes. The attractiveness of enzymes as targets for drug discovery stems from their high levels of disease association (target validation) and druggability (target tractability), which typically characterize this class of proteins. In general, drug discovery starts by manipulating the target enzyme with a compound that usually leads to inhibition or activation of its catalytic activity. Today, there are seventy-one human, bacterial, viral, and fungal enzymes that have been used successfully for the development of currently marketed clinically utilized drugs. All these drugs act as reversible or irreversible enzyme inhibitors.

This volume contains methods and detailed experimental protocols on the evaluation of the effect of a compound or a mixture of compounds on the action of enzymes that are significant targets in pharmaceutical industry. It examines the most reliable and robust methods for both bench and R&D scientists and sets a standard for best practices in the field. This volume consists of three different sections, each of which deals with different steps in the process from target selection and compound design to inhibitor evaluation. The first section presents biocomputing and bioinformatics protocols that have been developed recently. It underlines the progress in this area and points out the advantages that enzymologists and medicinal chemists can exploit for new target selection, druggability assessment, and structure-based design. The next section contains a selection of the state-of-the-art modern biophysical, electrophoretic, and chromatographic methods and high-throughput screening approaches that have been developed and are currently used for the assessment of enzyme/inhibitor interaction. The subsequent section provides detailed protocols and examples of the inhibition analysis and evaluation of selected enzymes. It contains critical information on enzyme structure-function relationships as well as mechanistic aspects on how these enzymes are inhibited.

This volume has been written by international scientists, who are active in biochemical and biomedical research, with expertise in chemistry, protein biochemistry, enzymology, molecular biology, and genetics. While it is not possible to detail and include every possible method and protocol related to enzyme inhibition, the present volume attempts to provide working tips with examples and analysis relevant to a wide range of more important enzyme targets and commonly available enzyme inhibition techniques and protocols.

The present book would definitely be an ideal source of scientific information for advanced students, junior researchers, and scientists involved in health sciences, cellular and molecular biology, biochemistry, biotechnology, cosmetology, and other related areas in academia. It is also aimed at professionals including academic faculty members, industrial scientists, and anyone working in the pharmaceutical, food, and cosmetics industries.

I sincerely hope that the reader will enjoy the information provided in this book and find its contents interesting and scientifically stimulating. I also hope that I have established a successful compilation of chapters within the exciting area of enzymes as drug targets. I would like to thank all the contributing authors for their enthusiasm and for the time they spent preparing the chapters for this book. I would also like to thank Dr. John Walker, the series editor, for his help and encouragement, and everybody at Springer for their helpful advice and support. I would especially like to thank my family for their understanding and patience during the editing and organization of the book chapters.

Athens, Greece *Nikolaos E. Labrou*

Contents

Contributors

ANNA MARIA ALMERICO • *Department of Biological, Chemical and Pharmaceutical Sciences and Technologies (STEBICEF), University di Palermo, Palermo, Italy*

BOLORMAA BALJINNYAM • *National Center for Advancing Translational Sciences, National Institutes of Health, Rockville, MD, USA*

EVANGELIA G. CHRONOPOULOU • *Laboratory of Enzyme Technology, Department of Biotechnology, School of Applied Biology and Biotechnology, Agricultural University of Athens, Athens, Greece*

RICARDO L. E. FURLAN • *Facultad de Ciencias Bioquímicas, Universidad Nacional de Rosario-CONICET, Rosario, Argentina*

NIKOLAOS GEORGAKIS • *Laboratory of Enzyme Technology, Department of Biotechnology, School of Applied Biology and Biotechnology, Agricultural University of Athens, Athens, Greece*

ZDENĚK GLATZ • *Department of Biochemistry, Faculty of Science, Masaryk University, Brno, Czech Republic*

BASKARAN GUNASEKARAN • *Faculty of Biotechnology and Biomolecular Sciences, Department of Biochemistry, Universiti Putra Malaysia, Serdang, Selangor, Malaysia*

CHARLOTTE S. HUGHES • *Diamond Light Source Ltd., Chilton, Didcot, UK*

ROHANAH HUSSAIN • *Diamond Light Source Ltd., Chilton, Didcot, UK*

ELISAVET IOANNOU • *Laboratory of Enzyme Technology, Department of Biotechnology, School of Applied Biology and Biotechnology, Agricultural University of Athens, Athens, Greece*

PRANAV JOSHI • *Department of Chemical and Biomedical Engineering, Cleveland State University, Cleveland, OH, USA*

SOO-YEON KANG • *Department of Chemical and Biomedical Engineering, Cleveland State University, Cleveland, OH, USA*

JANEZ KONC • *Faculty of Chemistry and Chemical Technology, University of Maribor, Maribor, Slovenia; National Institute of Chemistry, Ljubljana, Slovenia; Faculty of Mathematics, Natural Sciences and Information Technologies, University of Primorska, Koper, Slovenia; Faculty of Pharmacy, University of Ljubljana, Ljubljana, Slovenia*

PENMETCHA K. R. KUMAR • *Biomedical Research Institute, National Institute of Advanced Industrial Science and Technology, Tsukuba City, Ibaraki, Japan*

NIKOLAOS E. LABROU • *Laboratory of Enzyme Technology, Department of Biotechnology, School of Applied Biology and Biotechnology, Agricultural University of Athens, Athens, Greece*

MOO-YEAL LEE • *Department of Chemical and Biomedical Engineering, Cleveland State University, Cleveland, OH, USA*

SAMO LEŠNIK • *Faculty of Chemistry and Chemical Technology, University of Maribor, Maribor, Slovenia*

XIAO-YUAN MAO • *Department of Clinical Pharmacology, Xiangya Hospital, Central South University, Changsha, People's Republic of China; Institute of Clinical Pharmacology, Central South University, Hunan Key Laboratory of Pharmacogenetics, Changsha, People's Republic of China; Engineering Research Center of Applied Technology of Pharmacogenomics, Ministry of Education, Changsha, People's Republic of China;*

National Clinical Research Center for Geriatric Disorders, Changsha, Hunan, People's Republic of China
RUMANA MOWLA • *School of Pharmacy and Medical Sciences, University of South Australia, Adelaide, SA, Australia*
ABRAHAM OCHOA • *Facultad de Química, Laboratorio de Biofísicoquímica, Universidad Nacional Autónoma de México (UNAM), Ciudad de México, Mexico*
LUIS F. OLGUÍN • *Facultad de Química, Laboratorio de Biofísicoquímica, Universidad Nacional Autónoma de México (UNAM), Ciudad de México, Mexico*
STEVEN W. POLYAK • *School of Pharmacy and Medical Sciences, University of South Australia, Adelaide, SA, Australia*
GEORGIOS PREMETIS • *Laboratory of Enzyme Technology, Department of Biotechnology, School of Applied Biology and Biotechnology, Agricultural University of Athens, Athens, Greece*
I. AYELEN RAMALLO • *Facultad de Ciencias Bioquímicas, Universidad Nacional de Rosario-CONICET, Rosario, Argentina*
ROMAN ŘEMÍNEK • *Faculty of Science, Department of Biochemistry, Masaryk University, Brno, Czech Republic*
MICHAEL RONZETTI • *National Center for Advancing Translational Sciences, National Institutes of Health, Rockville, MD, USA*
MARIO O. SALAZAR • *Facultad de Ciencias Bioquímicas, Universidad Nacional de Rosario-CONICET, Rosario, Argentina*
PATRICK M. SCHAEFFER • *Molecular and Cell Biology, College of Public Health, Medical and Veterinary Sciences, James Cook University, Douglas, QLD, Australia*
JAN SCHEJBAL • *Faculty of Science, Department of Biochemistry, Masaryk University, Brno, Czech Republic*
PIYOOSH SHARMA • *Pharmaceutical Chemistry Research Laboratory, Department of Pharmaceutical Engineering and Technology, Indian Institute of Technology (Banaras Hindu University), Varanasi, India*
SUSHANT KUMAR SHRIVASTAVA • *Pharmaceutical Chemistry Research Laboratory, Department of Pharmaceutical Engineering and Technology, Indian Institute of Technology (Banaras Hindu University), Varanasi, India*
MOHD YUNUS SHUKOR • *Faculty of Biotechnology and Biomolecular Sciences, Department of Biochemistry, Universiti Putra Malaysia, Serdang, Selangor, Malaysia*
GIULIANO SILIGARDI • *Diamond Light Source Ltd., Chilton, Didcot, UK*
ANTON SIMEONOV • *National Center for Advancing Translational Sciences, National Institutes of Health, Rockville, MD, USA*
ALANNA E. SORENSON • *Molecular and Cell Biology, College of Public Health, Medical and Veterinary Sciences, James Cook University, Douglas, QLD, Australia*
WOLFGANG THORMANN • *Clinical Pharmacology Laboratory, Institute for Infectious Diseases, University of Bern, Bern, Switzerland*
FRIDA TREJO • *Facultad de Química, Laboratorio de Biofísicoquímica, Universidad Nacional Autónoma de México (UNAM), Ciudad de México, Mexico*
MANISH KUMAR TRIPATHI • *Pharmaceutical Chemistry Research Laboratory, Department of Pharmaceutical Engineering and Technology, Indian Institute of Technology (Banaras Hindu University), Varanasi, India*
MARCO TUTONE • *Department of Biological, Chemical and Pharmaceutical Sciences and Technologies (STEBICEF), University di Palermo, Palermo, Italy*

CHRISTINA VAROTSOU • *Laboratory of Enzyme Technology, Department of Biotechnology, School of Applied Biology and Biotechnology, Agricultural University of Athens, Athens, Greece*

HENRIETTA VENTER • *School of Pharmacy and Medical Sciences, University of South Australia, Adelaide, SA, Australia*

ADAM YASGAR • *National Center for Advancing Translational Sciences, National Institutes of Health, Rockville, MD, USA*

KYEONG-NAM YU • *Department of Chemical and Biomedical Engineering, Cleveland State University, Cleveland, OH, USA*

Chapter 1

In Silico Laboratory: Tools for Similarity-Based Drug Discovery

Samo Lešnik and Janez Konc

Abstract

Computational methods that predict and evaluate binding of ligands to receptors implicated in different pathologies have become crucial in modern drug design and discovery. Here, we describe protocols for using the recently developed package of computational tools for similarity-based drug discovery. The ProBiS stand-alone program and web server allow superimposition of protein structures against large protein databases and predict ligands based on detected binding site similarities. GenProBiS allows mapping of human somatic missense mutations related to cancer and non-synonymous single nucleotide polymorphisms and subsequent visual exploration of specific interactions in connection to these mutations. We describe protocols for using LiSiCA, a fast ligand-based virtual screening software that enables easy screening of large databases containing billions of small molecules. Finally, we show the use of BoBER, a web interface that enables user-friendly access to a large database of bioisosteric and scaffold hopping replacements.

Key words Drug discovery, Binding sites, Enzyme binding, Virtual screening, Ligand homology modeling, Protein sequence variants, Bioisosterism, Maximum clique algorithm, Scaffold hopping, ProBiS, ProBiS-CHARMMing, GenProBiS, LiSiCA, BoBER

1 Introduction

Computational methods that predict protein-ligand binding have become the cornerstone in the early steps of the drug discovery process [1, 2]. In recent years, we developed a number of such tools that can each be used separately or one complementing the other in the drug discovery pipeline. All of the tools are based on an efficient maximum clique algorithm [3], which enables fast detection of similarities between several different molecule descriptors. This enables mining of large databases, such as the Protein Data Bank (PDB) [4], for the discovery of novel protein functions [5], for identifying novel binding sites on proteins [6], for the study of effects of genetic variation on ligand binding and for drug

Nikolaos E. Labrou (ed.), *Targeting Enzymes for Pharmaceutical Development: Methods and Protocols*, Methods in Molecular Biology, vol. 2089, https://doi.org/10.1007/978-1-0716-0163-1_1, © Springer Science+Business Media, LLC 2020

discovery, specifically for the identification of drug repurposing opportunities [7].

First, we present the ProBiS algorithm [8–15], which enables a fast comparison of a query protein structure against the entire nonredundant PDB, and detects structures in this database that share similar three-dimensional (3D) amino acid motifs with the query protein. Next, we describe the web server implementation of the ProBiS algorithm combined with the CHARMMing tool for simulating molecular dynamics [16], the ProBiS-CHARMMing web interface [17], which additionally enables user-friendly execution of the ProBiS algorithm, identification of ligands that are able to binding to the query protein binding site and the subsequent minimization and potential energy prediction of such predicted protein-ligand complexes.

Then, we present the GenProBiS [6], an interactive web platform, which allows the mapping of human somatic missense mutations related to cancer and nonsynonymous single nucleotide polymorphisms from genome sequences of 21 species to protein binding sites from the PDB. GenProBiS enables detection of sequence variants within a protein binding site and visual exploration of specific protein-ligand interactions or, alternatively, loss of such interactions due to specific missense mutations.

Next, we present LiSiCA [18], a ligand-based virtual screening program that enables 2D and 3D screening of large databases consisting of millions of small molecules, such as the ZINC database [19], in a very short time. The program is also implemented as a PyMOL [20] plugin to enable easy setup of the virtual screening procedure and subsequent visualization of the obtained molecules [21].

Finally, we present BoBER [22], a web tool for identifying new bioisosteric and scaffold hopping replacements by mining the whole PDB [23]. The tool enables the input of a drug molecular structure after which it automatically fragments the molecule and finds possible replacements for the found fragments.

All these software tools have been widely used in various drug discovery projects by us and by other researchers. For example, binding site comparison for ligand prediction with ProBiS was successfully used to discover novel potent inhibitors of the InhA enzyme of *Mycobacterium tuberculosis* [7]. Using ProBiS-CHARMMing, we have successfully predicted the holo version of the MurA bacterial enzyme of *Escherichia coli*, which enabled virtual screening by molecular docking that led to the discovery of two novel inhibitors of this enzyme [24]. Using LiSiCA, we successfully identified new butyrylcholinesterase inhibitors with low binding-efficiency indices that could potentially lead to new treatment of Alzheimer's disease [18]. Reigada and coworkers [25] used LiSiCA to repurpose the acne drug isotretinoin against *Trypanosoma cruzi* parasite, the causative agent of Chagas disease. We also successfully used BoBER to increase the selectivity of covalent inhibitors of monoamine oxidase (MAO) enzyme, from MAO-A toward the MAO-B subtype (data not yet published).

2 Computational Tools

2.1 ProBiS

1. **ProBiS algorithm** [8–15] superimposes complete protein surfaces, surface motifs, or protein binding sites. It enables pairwise alignments of entire protein structures or selected binding sites as well as fast database searches for similar protein binding sites throughout large protein databases. The algorithm can find similar binding sites even in proteins with different folds and without prior knowledge of their location. ProBiS algorithm is able to run in parallel on multiple CPU threads. It is available as a stand-alone program for Linux operating systems at http://insilab.org/probis-algorithm/.
2. **ProBiS-CHARMMing** is a web server [17] that enables simple user-friendly setup of pairwise or database searches and superimposition of protein structures. It requires less knowledge of the command line scripting as the stand-alone version described above. This web server combines the functionality of ProBiS and CHARMMing [16]. The ProBiS part enables the prediction of ligands (small molecules, proteins, nucleic acids, or ions) that may bind to a query protein. This is achieved by comparing a protein's surface structure against a pre-prepared nonredundant database of protein structures and finding those that have binding sites similar to that of the query protein. Existing ligands found in the similar binding sites are then transposed to the query according to the generated binding site superimpositions by ProBiS. The CHARMMing part then enables the optimization of the geometry of the predicted protein-ligand complex as well as the calculation of the protein-ligand interaction energy, both using the extensively validated CHARMM force field [26, 27]. The web server is freely accessible at https://probis.nih.gov/.
3. **GenProBiS** web server [6] maps amino acid sequence variants from the UniProt Variants dataset [28] and somatic missense mutations from the COSMIC database [29] to protein structures from the PDB [4] which has been enriched with the ProBiS predicted protein–protein, protein–nucleic acid, protein–compound, and protein–metal ion binding sites. The server allows intuitive visual exploration of comprehensively mapped variants, such as human somatic missense mutations related to cancer and nonsynonymous single nucleotide polymorphisms from a variety of biological species, within the predicted binding sites regions for about 80,000 PDB protein

structures visualized in 3D using fast WebGL graphics. The GenProBiS webserver is open and free to all users and is available at http://genprobis.insilab.org/.

2.2 LiSiCA

1. **LiSiCA** is a software [18] that searches for 2D or 3D similarities between a reference compound and a database of target compounds and enables ligand-based virtual screening for drug discovery. The similarities are detected based on SYBYL atom typing. The software enables the execution on multiple processor threads, therefore enabling quick screening of very large (many millions) small molecule databases. LiSiCA can be used in Linux and Windows operating systems and can be obtained from http://insilab.org/lisica/.
2. **LiSiCA plugin for PyMOL** is a graphical user interface [21] that enables efficient use of LiSiCA functionality within the PyMOL molecular visualization program without the need to be familiar with the command line tool. Additionally, it offers the possibility of direct visualization of the highest scoring molecules and their visual comparison to the superimposed query molecules. The plugin is available at http://insilab.org/lisica-plugin/.

2.3 BoBER

BoBER web server [22] implements an interface to a database of bioisosteric and scaffold hopping replacements. The server enables user-friendly search for bioisosteric and scaffold hopping replacements, which are obtained by mining the PDB. It uses the ProBiS algorithm to search for local physicochemically similar surface functional groups in binding sites of holo proteins, i.e., proteins with bound ligands. The superimposition of similar binding sites subsequently leads to the alignment of bound ligands. Functional groups of ligands that align well within these superimposed binding sites are deposited to the web server as possible bioisosterically replaceable moieties. The web server is accessible at http://bober.insilab.org/.

3 Methods

Here, we describe the protocols for using the computational tools for similarity-based drug discovery described in the previous section. The stand-alone tools require a Linux operating system, whereas the web servers are compatible with Firefox or Chrome web browsers on any operating system. For the stand-alone methods, the commands in the following sections can be copied and pasted into the terminal to be executed directly. Each command starts with "$," which stands for the terminal prompt.

3.1 Use of the ProBiS Stand-Alone Program

We show on five different examples in a step-by-step fashion how to use the ProBiS stand-alone program as a command line tool. First, we show how to superimpose a pair of two whole protein structures based on their local surface alignments. We then show how to specify and superimpose certain binding sites within two compared proteins. Next, we show how to compare a single whole protein to a database containing many protein structures and also how to compare a single binding site to a database of whole protein structures. Finally, we demonstrate how to compare a binding site against many binding sites.

1. **Prerequisites**
   (a) Download the ProBiS precompiled program from: http://insilab.org/files/probis-algorithm/probis. The examples discussed here are also freely available for download from the ProBiS algorithm source code repository at: https://gitlab.com/janezkonc/probis.
   (b) Files for each example are in a directory "example{number}" in the downloaded repository. In a terminal "cd" to an example directory to begin. New terms, options, and modifiers, introduced in each example, are explained in more detail in Subheading 4.
2. **Example 1: Superimpose a pair of protein structures (pairwise alignment I)**

   ProBiS finds all local structural alignments of the compared protein surfaces, and superimposes the second protein's coordinates onto the first protein's coordinates according to the found alignments.

   (a) Superimpose two proteins using the command (*see* **Note 1**):

   ```
   $ ../probis -compare -super -f1 1phr.pdb -c1 A
   -f2 3jvi.pdb -c2 A
   ```

   This outputs files "*.rota.pdb" contain 3jvi's coordinates superimposed onto 1phr according to the three different alignments found in this case. The best superimposition, i.e., the one with highest z_score (*see* **Note 2**) is in file "1phrA_3jviA.0.rota.pdb." Alignment scores are in "REMARK PROBIS" lines in the ".rota.pdb" files.
3. **Example 2: Superimpose a pair of binding sites (pairwise alignment II)**
   (a) Superimpose the two binding sites defined as residues in a 3.0 Å radius around the SO4 ligands as follows (*see* **Note 3**):

   ```
   $ ../probis -compare -super -dist 3.0 -f1
   1phr.pdb -c1 A -bsite1 SO4.158.A -f2 3jvi.pdb -
   c2 A -bsite2 SO4.201.A
   ```

(b) Alternatively, you can directly select the binding site residues around SO4 using their residue numbers and chain identifier(s) with the "-motif" modifier (*see* **Note 3**):

```
$ ../probis -compare -super -motif1 "[:A and
(12-19,129-131)]" -motif2 "[:A and (7-15)]" -f1
1phr.pdb -c1 A -f2 3jvi.pdb -c2 A
```

The output file "1phrA_3jviA.0.rota.pdb" contains 3jvi's coordinates superimposed onto 1phr according to the generated alignment of the two binding sites. Alignment scores are in the "REMARK PROBIS" lines in the outputted ".rota.pdb" files.

4. **Example 3: Compare a protein against many protein structures**

(a) Convert protein files from the .pdb format to the surface . srf format, which allows for faster computations (*see* **Notes 4** and **5**):

```
$ ../probis -extract -f1 1phr.pdb -c1 A -
srffile 1phrA.srf
$ for i in $(cat proteins.txt); do ../probis -
extract -f1 ${i:0:4}.pdb -c1 ${i:4:1} -srffile
$i.srf; done
```

(b) Run ProBiS on 8 processors for query protein 1phr.A. Resulting pairwise alignments are saved in a .nosql file. The host file "hosts" is a text file with worker hosts specified one per line. Run the MPI version with (*see* **Note 6**):

```
$ mpiexec -v -hostfile hosts -np 8 ../probis -
surfdb -sfile srfs.txt -f1 1phr.pdb -c1 A -
nosql example.nosql
```

(c) Alternatively, a non-MPI version can be executed using, e.g., 8 processor threads; however, this is only possible on a single host computer:

```
$ ../probis -ncpu 8 -surfdb -sfile srfs.txt -f1
1phrA.srf -c1 A -nosql example.nosql
```

(d) Read a ".nosql" file and convert alignments to Json format (*see* **Note 7**). Output the query protein PDB with residues marked by degrees of structural conservation (*see* *.cons.pdb and **Note 8**):

```
$ ../probis -results -f1 1phr.pdb -c1 A -nosql
example.nosql -json example.json
```

(e) To get superimposed proteins as .pdb run:

```
$ for i in $(cat proteins.txt); do ../probis
align --alno 0 --nosql example.nosql --f1 1phr.
pdb
--c1 A f2 ${i:0:4}.pdb --c2 ${i:4:1}; done
```

Outputs are in the form of "*.alno.rota.pdb" files which contain translated and rotated coordinates of aligned proteins and the original coordinates of the query protein. Alignments numbers start with zero, therefore to obtain alignments number two or three use "-alno 1" or "-alno 2" options.

5. **Example 4: Compare a binding site against many protein structures**

(a) Extract the SO4 ligand's binding site and save it to a .srf file. This binding site is defined here as all surface residues in a radius of 3.0 Å around the SO4 ligand (*see* **Note 9**). Use an intact PDB file, containing the whole protein and ligands, with ProBiS. Do not cut out binding sites into a separate file by hand, because ProBiS will do this for you (*see* **Note 10**):

```
$ ../probis -extract -bsite SO4.158.A -dist 3.0
-f1 1phr.pdb -c1 A -srffile 1phrA.srf
```

(b) Alternatively, the SO4 binding site residues can also be selected by their residue numbers and chain identifiers using the "-motif" modifier:

```
$ ../probis -extract -motif "[:A and (12-19),
129-131)]" -f1 1phr.pdb -c1 A -srffile
1phrA.srf
```

(c) Convert .pdb files to .srf for faster computation:

```
$ for i in $(cat proteins.txt); do ../probis
-extract -f1 ${i:0:4}.pdb -c1 ${i:4:1} -srffile
$i.srf; done
```

(d) Run ProBiS on 8 processors for query protein 1phr.A. Resulting pairwise alignments are saved in a .nosql file. The host file "hosts" is a text file with hosts specified one per line. Run the MPI version with:

```
$ mpiexec -v -hostfile hosts -np 8 ../probis -
surfdb -local -sfile srfs.txt -f1 1phr.pdb -c1
A -nosql example.nosql
```

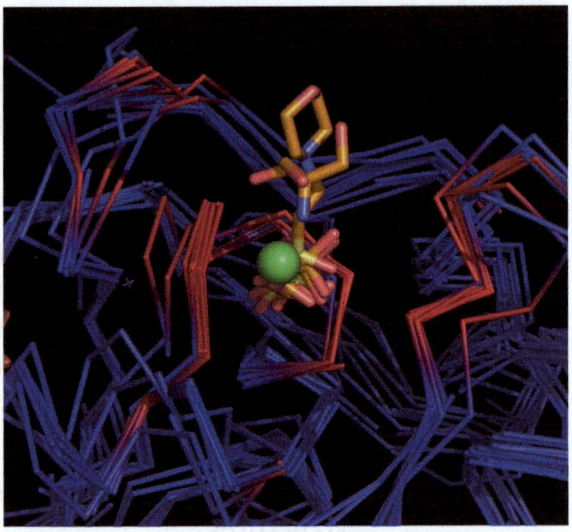

Fig. 1 ProBiS stand-alone program. Multiple superimposed binding sites (red) with co-laterally superimposed ligands that can be seen in the center

(e) Alternatively, the non-MPI version can be started using, e.g., 8 processor threads, however this is possible only on a single computer:

```
$ ../probis -ncpu 8 -surfdb -local -sfile
srfs.txt -f1 1phrA.srf -c1 A -nosql
example.nosql
```

(f) Read a .nosql file and convert alignments to JSON format. Output the query protein PDB with residues marked by degrees of structural conservation (*see* ∗.cons.pdb):

```
$ ../probis -results -f1 1phr.pdb -c1 A -nosql
example.nosql -json example.json
```

(g) To obtain the superimposed proteins (Fig. 1) as .pdb file execute:

```
$ for i in $(cat proteins.txt); do ../probis
align -alno 0 -nosql example.nosql -f1 1phr.pdb
-c1 A f2 ${i:0:4}.pdb -c2 ${i:4:1}; done
```

Outputs are in the form of "∗.alno.rota.pdb" files which contain translated and rotated coordinates of aligned proteins and the original coordinates of the query protein. Alignment numbers start from zero; therefore, to get alignments number two or three use "-alno 1" or "-alno 2" options.

6. **Example 5: Compare a binding site against many other binding sites**
   (a) Everything is the same as in Example 4, except that a file with ligand codes should be prepared. The ligand codes need to follow the same format as the "bsite" modifier and they define the binding sites deemed for comparison (see "proteins.txt" in this example's directory). Here, the command for converting .pdb to "srf" files is:

```
$ for i in $(cat proteins.txt); do ../probis -
extract -biste ${i:6} -dist 3.0 -f1
${i:0:4}.pdb -c2 ${i:4:1} -srffile
${i:0:5}.srf; done
```

3.2 Use of the ProBiS-CHARMMing Web Server

We demonstrate how the ProBiS-CHARMMing web server can be used to identify new ligands that bind to a known inhibitor's binding site on the example of the butyrylcholinesterase enzyme complexed with a nitroquinoline inhibitor. We also demonstrate how to perform energy minimization on the new predicted complex and how to calculate the energy of ligand binding.

1. Access the ProBiS-CHARMMing web server at https://probis.nih.gov/.
2. In the *Enter Query Protein* window enter "4XII" into the *PDB ID* box and to the *Chain ID(s)* box enter "A." This is an X-ray structure of the human butyrylcholinesterase complexed with a reversible inhibitor ligand (*see* **Note 10**).
3. Click on the *Select Binding Site (optional)* dropdown menu. In this example, the "40V.1001.A" represents the reversible inhibitor of the enzyme. Therefore this radio button should be selected. At the bottom, click on the *Distance (Å)* dropdown menu and select 6.0. Consequently, a binding site consisting of protein amino acid residues that are within 6.0 Å of the inhibitor ligand (*see* **Note 11**) is defined. On the right side, an interactive preview in JSmol of the protein and the location and size of the currently defined binding site can be seen.
4. The other optional parameters, that is, *Pairwise Align Two Proteins or Binding Sites (optional)* and Advanced Settings (optional), are not used in this example (*see* **Note 12**).
5. In the *Your e-mail address (optional)* box an email address can be entered. An email will be sent to this address with the link to the ProBiS results after the screening procedure is completed.
6. Click on the *Submit Job*. This will initiate a screening of the selected binding site (or whole protein structure) against the prepared nonredundant PDB (*see* **Note 13**). The screening procedure should take about 20 min to complete. ProBiS

output page will automatically open when the screening is finished.

7. The protein model is shown in a JSmol molecular viewer on the left side of the output page. The level of evolutionary conservation of the defined binding side is color coded, ranging from nonconserved (blue) to conserved (red). Hold the left mouse button and drag to rotate the view. Click the right mouse button to show the *More option* dropdown menu. Hold the left mouse button and drag up/down to zoom. Double-click the left mouse button and drag to translate the view.
8. On the right side of the output screen we see the predicted small molecule ligands shown (*see* **Note 14**)—arranged according to *Confidence* as default, which is express using *z*-score values (*see* **Note 2**) (Fig. 2).
9. Click on the *View 3D* button on the ligand from the "1eve" source protein (fifth row). This will load the predicted pose of the selected ligand into the protein's binding site, as can be

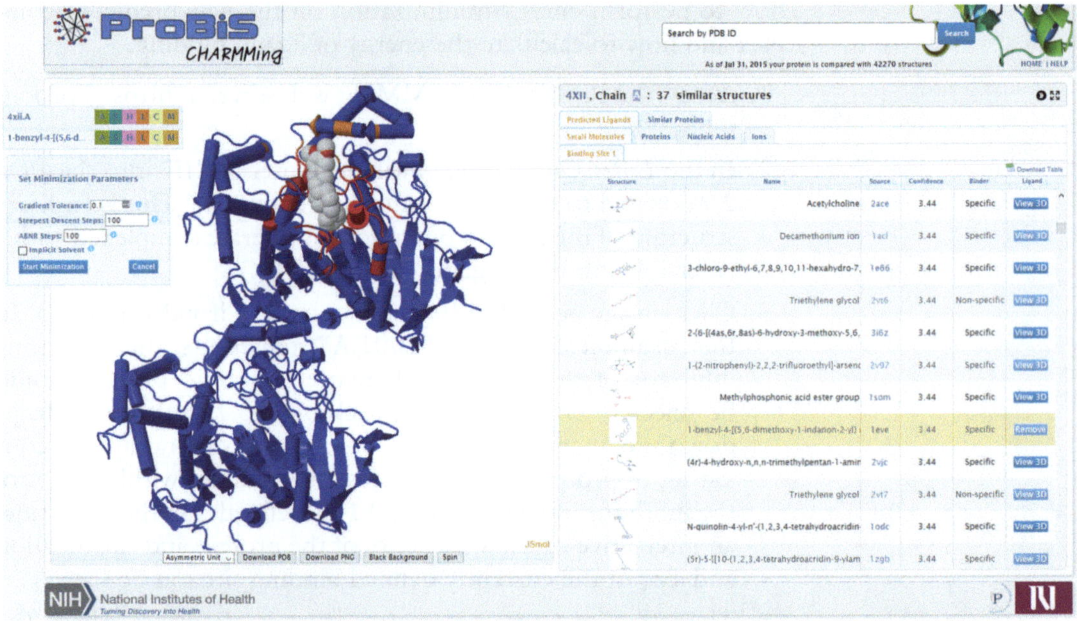

Fig. 2 ProBiS-CHARMMing web server. On the left side of the screen is a ligand (carbons as gray spheres) that was transposed from the 1eve protein structure to the predicted binding site on the butyrylcholinesterase enzyme. Butyrylcholinesterase is shown as a cartoon with rockets from blue (least conserved residues) to red (most conserved residues). On the same side of the screen also notice two "ASHLMCM" menus for the enzyme structure and for the inhibitor. Clicking on the "*M*" button and then "*Minimize*" will bring up the "*Set Minimization Parameters*" menu with the options to set the number of minimization steps, to set the tolerance gradient and to set the use of implicit water during the minimization process. The right side of the screen shows the predicted ligands visible from the available tabs. These ligands can be small molecules, proteins, nucleic acids, or ions. The functionality of the ProBiS stand-alone program can also be replicated here by clicking on the "*Similar Proteins*" tab

observed in the molecular viewer. Explore the ligand pose within the binding site using the mouse commands described in the previous step.

10. When the View 3D button was clicked in **step 9**, a new row (1-benzyl-4-[(5,6-d...) containing the "*ASHLCM*" buttons was added to the blue floating menu on the left side of the screen (by default). This new row represents the predicted ligand–protein complex. Click on the orange *M* button and then click *Minimize.* This will open the *Set Minimization Parameters* menu.
11. In the *Set Minimization Parameters* menu leave everything as default and click on the *Start Minimization* button. This will send the data to the CHARMMing server where the structure of the predicted ligand-protein complex will be energy minimized and the interaction energy of binding of the ligand to the protein will be calculated (*see* **Note 15**). This computation may take a few of minutes to complete and the calculated energy of binding should be −10.365 kcal/mol.
12. A new row in the "*ASHLCM*" window will be added, which stands for the minimized geometry of the complex. Use the mouse functions and the ASHLCM buttons to explore the optimized structure in Jsmol viewer or download the structure using the *Download PDB* button at the bottom of the page and explore it in any molecule viewer (e.g., PyMOL or UCSF Chimera).

3.3 Use of the GenProBiS Web Server

On the example of a human matrix metalloproteinase-13 complexed with a highly selective tetrazole inhibitor, we demonstrate the options available in the GenProBiS web server. First, we describe the basic mouse movement commands, and then we show how the server can be used to analyze the effects of sequence variants on ligand binding.

1. Access the GenProBiS web server at http://genprobis.insilab.org/ (Fig. 3).
2. Select the *Basic input* search window, enter "3kejA" (meaning PDB ID: 3kej, chain ID: A) and click *Submit.* After a few moments the output window will open.
3. To rotate the structure, move the mouse while holding the left mouse button. To zoom in or zoom out, either use the middle mouse button to scroll or move the mouse while holding down the right mouse button. To translate the structure, move the mouse while either holding the scroll (middle) mouse button or both right and left mouse buttons. Open the information window by clicking on a molecule you wish to be the new center of rotation. Information window contains the name, PDB, and UniProt number of the clicked amino acid and the

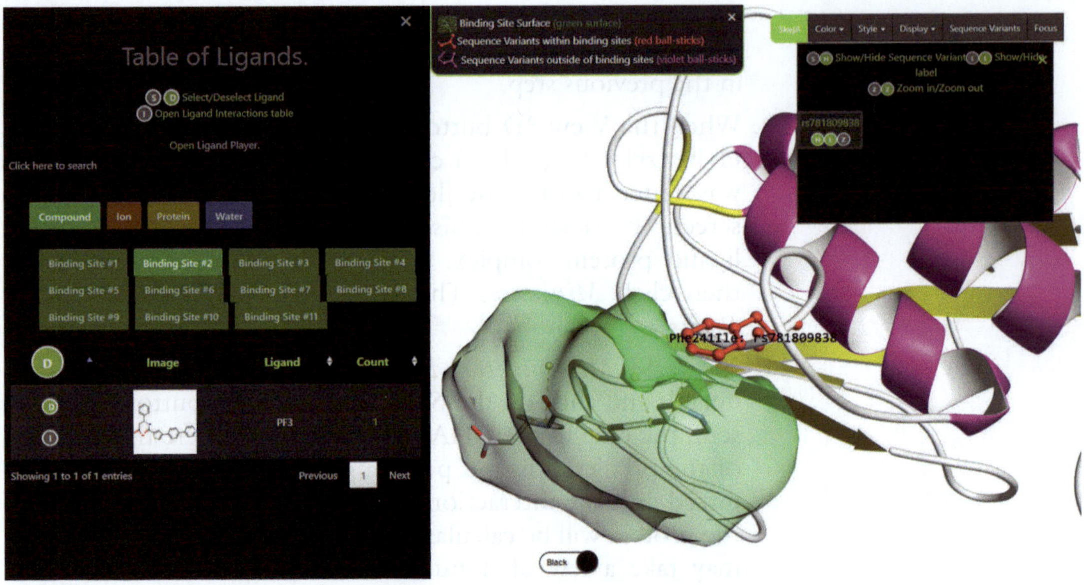

Fig. 3 GenProBiS web server. The right side of the view shows a population sequence variant *rs781809838* (red sticks) protruding into the binding site (green surface) and interacting with a ligand (CPK colored sticks) of a macrophage elastase enzyme. On the left side is a 2D structure of the ligand in the *Table of Ligands*, which shows that there are 11 compound binding sites in this enzyme, in addition to *Ion*, *Protein*, and *Water* binding sites

information if the residue is a known sequence variant or somatic missense mutation.

4. Click on the *Table of Sequence Variants*. In the table header is general information about the displayed protein and links to the genome sequence and to the species information. In the table below are listed sequence variants identified by their accession numbers, one on each row of the table. Each sequence variant is linked to the original database, where more information about the variant can be found, e.g., the frequency of occurrence in population. If the sequence variant is a part of a binding site, a circle (or more circles if it is in more binding sites) is present; the color of each circle corresponds to the binding site type (which is determined by the type of the ligand that binds to this binding site). The hash symbol "#" stands for the binding site number. Information on amino acid change is also provided. If the sequence variant has been connected with a disease or influences drug pharmacology, a link is shown to the corresponding ClinVar [30] and PharmGKB [31] databases. For each Sequence Variant there are two buttons available, the *Select* or *Deselect* button, the *Label* button and *Zoom* button (*see* **Note 16**). A click on the first selects or deselects the observed sequence variant, which causes the variant to appear or disappear from the floating

Table of Sequence Variants on the right side of the viewer and from the viewed 3D protein structure. Only one sequence variant per residue can be selected. The Label button shows or hides the label on the selected sequence variant in the 3D viewer and the Zoom button zooms to and highlights the selected sequence variant. In the header of the table there is a *Select* or *Deselect All Sequence Variants* button, which selects all the sequence variants on the protein (but only one per residue) or deselects all of them. For faster search there is a search bar customized to search sequence variants by accession number and amino acid change. Sequence variants that could not be mapped to the protein structure are listed below the sequence variants table in the No structural data available for the following Sequence Variants table.

5. Click the *Table of Binding Sites*. Here all binding sites are listed as tabs (*see* **Note 17**). The first level of tabs are the binding sites organized by the type of binding site and the second level are the binding sites of a specific type organized by their binding site numbers. Only one binding site can be viewed at one time. For each binding site, there is a Table of Sequence Variants listing the sequence variants that are within the binding site. The table is imported from the Table of Sequence Variants and offers the same functionality. All other binding site amino acids that are not sequence variants are listed in the table Other Amino Acids in This Surface Region, below the sequence variants table. Binding sites can be viewed as a surface model (default option) or as sticks, which enable to view individual amino acid residue structures.
6. Click the *Table of Ligands*. This table contains the ligands that bind to each binding site (*see* **Note 18**). The layout of the binding site tabs is the same as for the *Table of Binding Sites* (*see* previous point). Only one binding site can be viewed at one time. Under each binding site tab there is a table of ligands belonging to that binding site. Ligands are grouped by their three-character identifiers. For each ligand group, there is information about the structure-image, name and the number of the ligands in binding site with the same name. For each group, there are also two buttons: the *Select* or *Deselect* button, which shows or hides the ligands in the group in the 3D viewer bound to a protein structure, and *Ligand Interactions Table* button which opens the Table of Ligand Interactions. *Select* or *Deselect All* button shows/hides all groups of ligands simultaneously. To browse quickly through the ligands, click on the *Ligand Player*, which enables visualization of ligands bound to a protein as a movie.
7. The *Table of Ligand Interactions* contains detailed information on ligand interactions (*see* **Note 19**). A click on a row in this

table zooms on the corresponding sequence variant in the structure and shows its ligand pair and the bond between them.

8. In the *Sequence Viewer*, one can visualize the protein sequence interactively and mapped to it the sequence variations, the binding sites, and the evolutionary conservation. The evolutionary conservation is shown as default and the sequence variants are shown as magenta colored amino acid one-letter codes in the protein sequence. A specific binding site can be selected, which shows as an intermittent line underneath the protein sequence. The sequence variants that are in the binding site are colored red. The meaning of different colors is explained in the *Legend*. Every amino acid residue in the protein sequence is numbered according to the PDB and UniProt system and sequence variant accession numbers and secondary structure information is shown upon hovering over a particular amino acid residue. A click on a sequence variant in the protein sequence zooms on the selected amino acid residue in the 3D viewer. The image of the protein sequence can also be exported to one's computer by clicking the *Save as Image* button. The Sequence Viewer can also be minimized for easier interaction with the protein structure in the 3D viewer.
9. The draggable menu can be dragged across the screen. It has several submenus:
   (a) *Color*: choose the color of the protein structure.
   (b) *Style*: choose the style of the protein structure.
   (c) *Display*: toggle display of water, hydrogen, and possible ligands.
   (d) *Sequence Variants*: opens a draggable *Table of Sequence Variants* selected in the *Table of Sequence Variants* or *Table of Binding Sites*.
   (e) *Focus*: zooms in on the structure and sets the center of rotation at the geometric center of the protein structure.
10. In the draggable *Table of Sequence Variants*, the sequence variants selected in the Table of Sequence Variants or Table of Binding Sites are displayed as a grid. In each cell there is accession number of one sequence variant and three buttons:
   (a) *Show/Hide* button that shows or hides the sequence variant from the displayed structure, but it does not deselect it from the Table of Sequence Variants or the Table of Binding Sites.
   (b) *Label* button that shows and hides the label of a sequence variant (*see* **Note 18**).
   (c) *Zoom in/Zoom out* button that focuses display on the selected sequence variant and changes the center of the

display to that sequence variant. Second click on the button zooms out and resets the center of rotation to the center of the protein structure. Only one sequence variant can be zoomed at one time.

11. *Ligand player* enables interactive visualization of the ligands of a selected binding site. It goes over the ligands in sequence and displays each ligand in the 3D viewer bound to the protein structure. This is enabled by the following buttons:
    (a) ◀◀ (*backward* button) shows the previous ligand.
    (b) ▶ (*play* button) displays all ligands of the selected binding site one by one with the speed of one ligand per second. Another click on this button stops the ligand playing at the currently displayed ligand.
    (c) ■ (*stop* button) stops the playing of ligands and reverts the view to the first ligand in the queue.
    (d) ▶▶ (*forward* button) shows the next ligand.
    (e) **X** (*exit* button) closes the ligand player.

The first cell of the ligand player shows the name of the ligand and its serial number in the currently selected binding site.

3.4 Use of the LiSiCA Stand-Alone Program

Here, we demonstrate how to use LiSICA stand-alone within the command line tool to perform a ligand-based virtual screening procedure, as is often used in the first steps of a drug discovery pipeline. We will use an example molecule and to safe time a database of only a few small molecules.

1. Download the compiled version of LiSiCA from http://insilab.org/lisica/. LiSiCA is available for three operating systems, i.e., Linux, Windows, and Mac OS; however, this example will be based on the version compiled for the Linux operating system. From the same web page download the example reference molecule and the target molecules, available as "reference.mol2" and "database.mol2" and "database3d.mol2" files, respectively.
2. Change to the directory "cd" in which the downloaded files were saved and mark the "lisica" file as executable:

```
$ chmod +x lisica
```

3. Then, from the same directory, run the 2D (default) virtual screening process using the command (*see* **Notes 20** and **21**):

```
$ ./lisica -R reference.mol2 -T database.mol2
```

This will compare every molecule within the "database.mol2" file to the reference molecule contained in the "reference.mol2" based on their topological (2D) similarities. The subsequent results will be expressed using Tanimoto coefficients, which can have a value between zero and one; where zero corresponds to a pair of molecules that have no substructure in common, while the value one means that the molecules are identical. When the screening is finished the results are outputted to the standard output. The 3D version of LiSiCA can simply be used by adding the "-d 3" option to the command: (*see* **Note 22**):

```
$ ./lisica -R reference.mol2 -T database3d.mol2 -d 3
```

Due to the effectiveness of LiSiCA's maximum clique algorithm, the screening process of this small database should be practically instantaneous on modern computers. As with the 2D case, the results expressed with Tanimoto coefficient will be written to the standard output.

3.5 Use of the LiSiCA Graphical User Interface

The graphical user interface for LiSiCA is available as a plugin for the widely used PyMOL molecular visualization software. Currently, we support PyMOL version 1.8 or lower. The LiSiCA plugin allows all the features of the stand-alone tool, but makes the use of the tool very user-friendly. This example assumes that PyMOL is already installed on the computer.

1. Download the "lisica.py" file from http://insilab.org/lisica-plugin/ web page. Also download to the same directory the example reference molecule and the target molecule databases in the ".mol2" format (*see* the first step in the previous subsection).
2. Open the PyMOL viewer and go to the *Plugin* → *Plugin Manager.*
3. Select *Install New Plugin* category and click on the *Choose file*... button.
4. Add the "lisica.py" file. In the Select Plugin Directory window, select the directory and click *OK.*
5. Close the Plugin Manager window. LiSiCA plugin should appear under *Plugin* → *LiSiCA.*
6. When LiSiCA plugin is started for the first time it downloads additional required files from our server, which may take a while (*see* **Notes 23** and **24**).
7. As default the *Inputs tab* window opens. Here, it is possible to specify all the required options for LiSiCA and change the default options. Upon clicking *Browse* near the *Reference Ligand* title, a new window that enables to open the previously downloaded *reference.mol2* file. Similarly, clicking the *Browse*

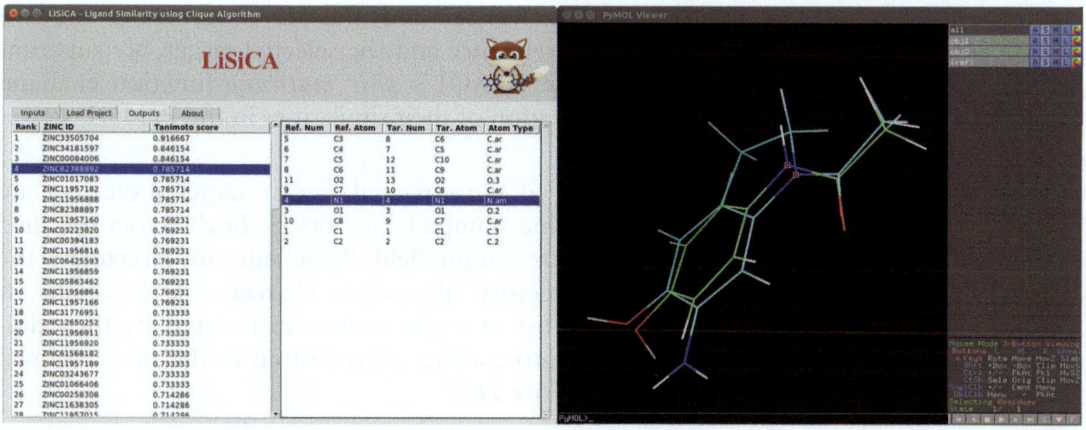

Fig. 4 LiSiCA plugin for PyMOL. Three-dimensional molecular structures (the reference and the currently selected target) are superimposed using the PyMOL's *pair matching* function enabling easy visualization of their similarities in the PyMOL viewer screen. The left table contains the names of molecules as they are found in the .mol2 database file, while the right table shows atom pairs that were found by LiSiCA to be similar in both the reference and the current target molecule. Clicking on these atom pairs will automatically select the corresponding atoms in the PyMOL viewer window

button near the *Target Ligand(s)* title, it is possible to select and open the *database3d.mol2* file (*see* **Note 20**).

8. Near the *Product Graph Dimensions* option we select the *3D Dimensional Screening* option. Using this option, the program will search for similarities in molecule structures based on spatial distances between atoms.
9. We leave other options as they are by default (*see* **Note 21**) and press the *GO* button. This will initiate screening, and may take a few moments, depending on each computer capability.
10. After the screening is complete the *Outputs* tab will automatically open (Fig. 4).
11. In the *Outputs* tab, there are two lists, side-by-side. The first contains the ZINC IDs and the Tanimoto Coefficients of the target molecules in the decreasing order of the Tanimoto coefficient values. Any single target molecule can be selected in this list by clicking on the corresponding row in the list or by using up and down arrow keys. The selected target molecule is displayed together with the reference molecule in the separate PyMOL viewer window. Depending on the target molecule chosen, the atoms that were found to form a similar substructure in the reference and in the target molecule are displayed in the second list. Any single pair of corresponding atoms can be selected on this list using a mouse click or using up and down arrow keys. The selected pair of atoms is highlighted in the compared molecular structures in the PyMOL viewer window.

12. If 3D screening was selected, the 3D structures of the two molecules, the reference and the selected target, are superimposed using the PyMOL's *pair matching* function enabling efficient visualization of their similarities in the PyMOL viewer (*see* **Note 25**).
13. The mol2 files of the screened similar target molecules are written into a time-stamped directory in the directory specified in the *Save results in* input field. By default, this directory is the user's home directory. Also, a text file named lisica_results.txt containing the list of target molecules ordered by their descending Tanimoto coefficients is written to the time-stamped directory (*see* **Note 26**).

3.6 Use of the BoBER Web Server

Here, we demonstrate the use of BoBER web server to optimize an existing inhibitor of the MurF enzyme by finding bioisosteric replacement fragments.

1. Load the web server http://bober.insilab.org/.
2. In the *Input drug structure*, it is possible to draw an input molecule using a molecular editor, on which we wish to perform bioisosteric replacements. Click the *Example molecule* button to automatically local an example MurF inhibitor structure into the molecular editor. The *Submit query* button then executes the fragmenting of the molecule and the new *Fragment selection* window will be displayed.
3. In the *Fragment selection* window, we click on the fragment (tetrahydro-benzothiophene) that we wish to replace within the input molecule structure. We select the *Intra-Family* radio button within the *Custom options.* We also select *Use specific structure as query* and as the structure is quite unique, we also select the *Interchangeable join atoms* radio button to expand the chemical space from which the bioisosteric replacements are sought (*see* **Note 27**). All other options are left as they are (*see* **Note 28**). Clicking the *Submit query* button will open the *Results* window.
4. BoBER found, in this case, one bioisosteric replacement (tetrahydrohydro-thienopyridine) for the selected fragment in the input molecule. The equally colored join atoms (red) *N.pl* and *N.sp2* were found to overlap in the BoBER database. Clicking the *glyphicon icon* shows a dropdown menu listing the overlapping join atoms (in this case only one pair) at which the replacement should take place (Fig. 5).
5. Selecting the join pair connects the new bioisosteric fragment with the rest of the molecule and opens the molecular editor where the new molecular structure containing the bioisosteric replacement is shown.

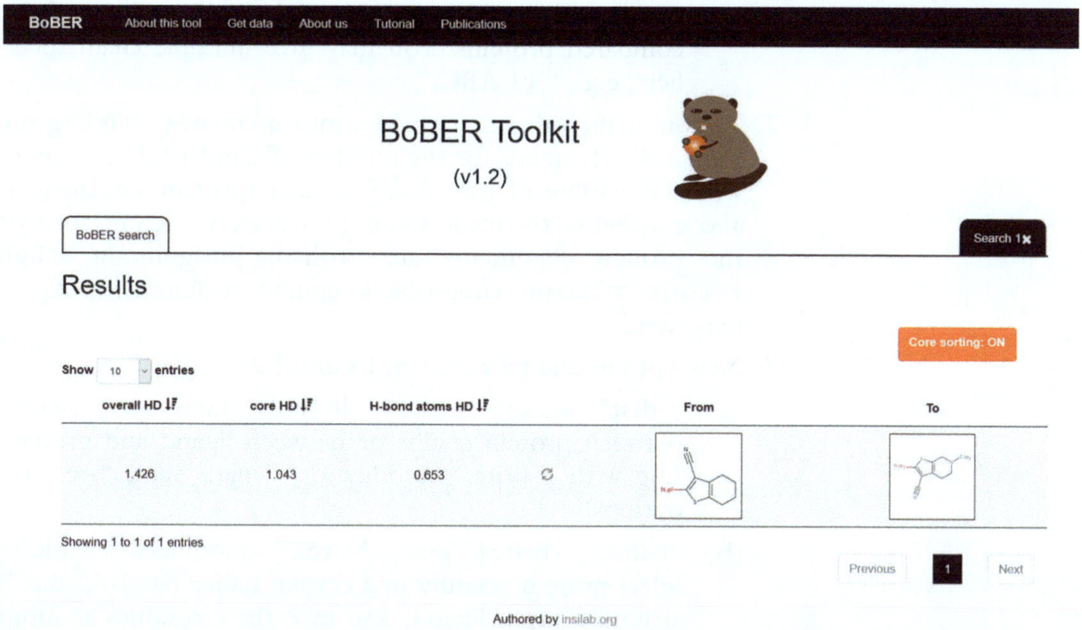

Fig. 5 BoBER web server. The bioisosteric fragment (column "*To*") that BoBER found with the query tetrahydro-benzothiophenic fragment (column "*From*"). In the Core HD column, the Hausdorff distance, which is a measure of overlap between the two fragments in superimposed binding sites, is specified. The smaller this distance, the better the overlap of fragments within the database. Clicking on the glyphicon in the second column will open a new window, where the drug structure will be drawn with the bioisosteric replacement performed

4 Notes

1. New options and modifiers in Example 1:
   (a) "**-compare**"—option—compare two protein surfaces (.pdb or .srf files). If you use .pdb files, surfaces will be computed first. Output their local structural alignments in a .nosql file. Each alignment consists of a rotational matrix, alignment scores, and aligned residues of the compared proteins.
   (b) "**-super**"—modifier—find local structural alignments between two proteins (use with "-compare" option) and superimpose the two proteins according to all found alignments. For each alignment output the .rota.pdb file with the proteins superimposed according to this alignment.
   (c) "**-f1**" and "**-f2**"—modifiers—protein files to compare (.pdb or .srf format).

(d) "**-c1**" and "**-c2**"—modifiers—chain identifiers of the compared proteins. You may give multiple chain specifiers, e.g., "-c1 ABC."

2. Z-score is the value that enables measurement of binding site similarities. It stands for the number of standard deviations by which the value of the observed data (protein similarity) is above or below the mean value. Low z-score (<2) means that the protein alignments are probably insignificant. High z-scores (>2) represent probable significant alignments of protein pairs.
3. New options and modifiers in Example 2:

   (a) "**-dist**"—modifier—the allowed minimal distance between protein chains or between ligand and protein. Use with "-bsite" modifier or "-mark" and "-results" options.

   (b) "**bsite**," "**-bsite1**" and "**-bsite2**"—modifiers—enable to select protein residues in a certain radius (set by "-dist") around a given ligand, and take these residues as input (use with "-extract" or "-compare" options). If used with compare, it only works with .pdb files (not .srf). For example: "-bsite ATP.305.A"—ATP (residue name), 305 (residue number) and A (chain id).

   (c) "**-motif**," "**-motif1**," and "**-motif2**"—modifiers—select residues to be used as a query instead of the whole protein structure (use with "-extract" or "-compare" options). This will generate a .srf file with only the selected residues. To select some residues on chains A and B, for example, of the input protein use "-motif [:A and (14,57,69-71) or :B and (33,34,50)]". Note that chain IDs are case sensitive. Square brackets are mandatory.

4. New options and modifiers used in Example 3:

   (a) "**-extract**"—option—calculate the surface of a protein. Redirect the output, which is the protein surface, to a surface (.srf) file. Surface files can be used instead of . pdb files together with "-compare" and "-surfdb" options, which improves performance when doing repetitive comparisons, since protein surfaces do not need to be recalculated for each comparison. Option "-surfdb" works with .srf files exclusively.

   (b) "**-surfdb**"—option—compare the query protein surface (.srf) with other protein surfaces listed in the SURF_FILE (see "-sfile" modifier). This does the same calculation as the "-compare" option, but faster, due to protein surfaces being already pre-calculated (see "-extract") option. This option also supports parallel computation on multiple CPUs. Output is the same as in "-compare."

(c) "**-sfile**"—modifier—specify file that contains names of .srf files to be compared with the query protein (see "-surfdb") option. Each line must contain one .srf file name. Example:
 - protein1.srf A
 - protein2.srf B
 - protein3.srf A

(d) "**-ncpu**"—modifier—specify the number of concurrent processor threads.

(e) "**-results**"—option—read alignments from an .nosql file, filter them according to their scores, and calculate fingerprint residues, which can also be used as a filter. Output results in the JSON format. Additionally, replace the B-factors in the query protein's PDB files with degrees of structural conservation. If used with "-ligdir" modifier, output ligands in JSON format as well.

(f) "**-align**"—option—read a rotational matrix of an alignment from an .nosql file and superimpose the two given proteins accordingly (first run "-compare" or "-surfdb"). Output the superimposed proteins' coordinates in a .pdb file. Both .pdb files that are to be superimposed (see "-f1," "-c1," "f2," "-c2" modifiers) and an alignment number (see "-alno" modifier) must be provided.

(g) "**.alno**"—modifier—specify the alignment number. Each comparison of a pair of proteins may result in many different structural alignments.

5. The expression "${i:0:4}" is replaced with the first four letters of each line from the proteins.txt file. This is part of the standard Bash notation.
6. "**mpiexec**" will run the MPI program. MPI is a protocol for passing parallel computations to host computers. The arguments used here have the following meaning:

 (a) "**-v**"—verbose—include additional information in the output. Can be useful for debugging purposes.

 (b) "**-hostfile**"—specify the file containing the list of nodes to be used in parallel computation, one node per line.

 (c) "**-np**"—specify the number of processes to start.

7. JSON is a language independent data format using human-readable text to communicate objects consisting of key-value pairs or arrays. It is supported by all major programming languages. More on its format and use is available on http://www.json.org/.

8. The ∗.cons.pdb contains data identical to the normal .pdb files, except that the beta-factor column is replaced by the degrees of structural conservation, which can take values between zero and one. A higher value represents a more conserved residue.

9. This is a homodimeric structure where the ligand binding site is completely within one chain, therefore it is enough to enter only a single chain (e.g., "A") to perform database query.

10. New options and modifiers used in Example 4:

    (a) "**-local**"—modifier—use this to perform local alignments search (with "-compare" and "surfdb" options). By default, after the local alignment is found with the maximum clique algorithm, an attempt is made to extend this alignment along the backbones of the compared proteins. In this way, parts of proteins that adopt different conformations can be aligned. These residues are marked with "fix" in the "alignments.json" file.

11. Literature shows that π–π interactions are of great importance in butyrylcholinesterase enzymes. These interactions can form across considerable distances; therefore, we increase the Distance parameter to a higher value.

12. In the Advanced options, the *Local Alignments Only* modifier is used by default. ProBiS will therefore find only near perfect local structural alignments with about <2 Å root mean square deviation (RMSD) between the aligned residues. Otherwise, residues with up to 7 Å RMSD will be considered in the alignments, which allows finding similar, but flexible residues in compared proteins.

13. The nonredundant PDB currently contains approximately 40,000 unique protein chains. The database is obtained by identifying protein chains having >40% sequence similarity to each other and only retaining one representative chain from each such cluster.

14. On the right side, the server shows two main tabs: *Predicted Ligands* and *Similar Proteins.* The Similar Proteins tab displays protein structures found within the PDB that show high structural similarity to our query protein which we entered in the search box. Clicking on the *View 3D* button will show a superimposition between the query protein and the selected similar protein. The *Predicted Ligands* tab shows ligands obtained from the binding sites of PDB proteins, where these binding sites show structural similarity to our query protein binding site. Based on the binding site superimpositions, the ligands are transferred (translated and rotated) from the query binding site to the predicted one. In addition to small molecule ligands, the web server can also predict other ligands, such as nucleic acids,

ions, and also peptides and proteins. More than one binding sites can also be detected, as we can observe by clicking on the *Proteins* or the *Ions* subtabs.

15. CHARMMing uses the standard CHARMM force-field parameters [26] to minimize the complex, as well as to calculate the predicted energy of protein-ligand binding. The ligand parameters are obtained from the ParamChem web server [27]. We should be careful at interpreting these results, as small-molecule ligand structures often require additional manual parameterization efforts. Nevertheless, this automated procedure was shown to be useful in predicting holo version of the proteins, when only apo structures are available [24].
16. Label and zoom buttons are synchronized with the label and zoom buttons in the Table of Binding sites and draggable Table and Sequence Variants.
17. Binding site tabs are synchronized with binding site tabs in Table of Ligands.
18. Binding site tabs are synchronized with binding site tabs in Table of Binding Sites.
19. Label and zoom buttons are synchronized with the label and zoom buttons in the Table of Sequence Variants and in the Table of Binding Sites.
20. Both the reference ("-R" specifier) and the target compound ("-T" specifier) files must be in the Tripos ".mol2" format. Alternatively, the target database file can be a ".mol2" file compressed with gzip [mol2.gz]. The LiSiCA similarity search depends on comparing SYBYL atom types (e.g., C.2, C.3, and O.3); therefore, these must be present in the fifth column. The reference file should contain only one (reference) compound; the target file may contain many compounds. If the reference file contains more than one compound, the first molecule is used as a reference. Molecules in the target file having the same name (the line after the "@MOLECULE" line) are considered as different conformers of the same molecule. Only the best-scoring (by Tanimoto coefficient) conformer will be shown by default in the final output for the 3D screening option. To save the final output (ranked list by Tanimoto coefficient) to a file in Linux, use the bash standard output redirection function.
21. Besides the "-R" and "-T" specifiers, LiSiCA has the following options available:
    (a) **-n**—the number of CPUs to use. The default is to try to detect the number of CPUs and use all of them or, failing that, use 1.
    (b) **-d**—product graph dimensions. Possible values are 2 or 3. The default value is 2. The 2D option is based on finding similarities between ligands which have the same

(-s option = 0) or similar (-s option > 0) number of covalent bonds between the atoms of the same type. Accordingly, the 3D option is based on finding similarities between ligands where spatial distance between the same type of atoms is similar (or the same if the -m option = 0.0). See the subsequent explanations of the -s and -m options.

(c) **-s**—maximum allowed shortest path difference for the 2D product graph measured in the number of covalent bonds between atoms. The default value is 1. This option corresponds to the maximum allowed difference in shortest-path lengths between atoms of two compared product graph vertices. Lesser values correspond to a more rigorous screening criterion. This option can only be used in combination with the -d 2 (2D screening) option.

(d) **-m**—maximum allowed atom spatial distance difference for the 3D product graph measured in Å. The default value is 1.0. This option corresponds to the maximum allowed difference in spatial distances between atoms of the two compared product graph vertices. Smaller values correspond to a more rigorous screening. This option can only be used in combination with the -d 3 (3D screening option) option.

(e) **-h**—consider hydrogen atoms. This option lets the user choose if hydrogen atoms are to be considered for the calculation of the similarity using the maximum clique algorithm. By default, hydrogen atoms are not considered in finding the largest substructure common to the reference and the target molecules to speed-up the calculation.

(f) **-w**—the number of highest ranked molecules to output. The default value is 0. If the option is set to a value >0, LiSiCA will create .mol2 files of the highest scoring target molecules with a comment section at the end of each file listing the matching atom pairs found.

(g) **-c**—maximum allowed number of highest scoring conformations to be outputted. The default value is 1.

22. The database containing different conformers of the same molecule has to be prepared in advance. This can easily be done with a number of third-party software, such as Open Babel [32].

23. For complete installation of the plugin, a stable internet connection is required. If a "ssh timeout" error occurs, it could mean that our server is slow to respond. If this happens, retry installation a couple of times.

24. If installation fails with the "pmg_tk" error or similar, one or both of the tkinter and ttk libraries are not installed. To install

them, open the terminal and type as root "yum install tkinter python-pip" and "pip install pyttk" (in CentOS) or "apt-get install tkinter python-pip" and then "pip install pyttk" (in Ubuntu).

25. For 2D screening, the two molecules, the reference and the selected target, are visualized side-by-side in the PyMOL viewer screen.
26. The plugin also has a feature to load saved results. In the *Load Project* tab, the user can choose the directory with the saved results (containing .mol2 files of each target and the reference) and the "lisica_results.txt" file. When the load button is clicked, the results will be loaded onto the output tab and the PyMOL Viewer window.
27. We define *join atoms* as atoms at which the rotatable covalent bonds were broken during the fragmenting process at the creation of the BoBER database of bioisosteres. *Core atoms* are all atoms that are not join atoms.
28. BoBER has two sets of general options: *Basic options* and *Custom options.* Under *Custom options*, the user can select options concerning:
    (a) Protein SCOP families
    - The *Intra-family* option limits the displayed fragment pairs to the part of the database obtained from superimposing protein structures belonging to the same SCOP family.
    - The *Inter-family* option outputs fragments that originate from proteins that are of different SCOP families or when one or both of the protein families are unspecified in the SCOP database.
    - If *Both* is selected fragment pairs independently of the SCOP family criteria are displayed.

    (b) Interchangeability of join atom types
    - The *interchangeable join atom types* option enables overlapping join atoms of the bioisosteric replacement to be exchangeable with atoms of the same or similar hybridization type (e.g., C.sp3 with N.sp3). This broadens the chemical space in which replacements are sought for, as it permits fragments with at least one similar overlapping join atom to be considered replaceable. The *non-interchangeable join atom types* option requires at least one overlapping join atom type of the bioisosteric structure to be exactly the same as in the query one.

(c) If common cores should be used as queries?

- The *Use structures with common core as queries* option enables BoBER to use every fragment that has the same core structure as the inputted fragment as query. Join atoms on the bioisosteric fragment will be ignored. This significantly broadens the chemical space in which bioisosteric fragments are sought for.
- The *Use specific structure as query* options only allow to use the exact same structure as was inputted (including join atoms) as a query for searching.

(d) The cutoff value of the Hausdorff distance. Hausdorff distance is a measure of fragment overlap obtained during the generation of the BoBER database. Fragment pairs with the Hausdorff distance value higher than specified here will not be displayed.

By choosing *Basic options*, we have two options available, which represent two useful settings that can also be manually selected within the Custom options:

(a) *Loose filtering*—using this filter, fragment replacements are obtained from co-crystallized ligand based on superimposed proteins independent of their SCOP family classification. This corresponds to the *Both* option in the *Custom options.* Join atoms of the bioisosteric fragments are ignored. The bioisosteric fragments will be searched for only based on their core structure overlap. The reconnection is based on mapping the join atoms from the query fragment to the bioisosteric one. Corresponds to the *Use structures with common core as queries* option in *Custom options.* As common cores are used here as queries, *the Interchangeability of join atoms* option has no meaning here. The Hausdorff distance cutoff value is 1.5 Å.

(b) *Rigorous filtering*—using this filter, fragment replacements are obtained from co-crystallized ligands based only on superimposed proteins that belong to the same SCOP family classification. This corresponds to the *Intra-family* option in the *Custom options.* Join atoms of the fragments are taken into account. For fragments to be replaceable, they have to have at least one overlapping join atom of the same type. This corresponds to the *Non-interchangeable join atoms* option in the *Custom options.* The search for bioisosteric replacements is based on the specific query defined at the beginning (including join atoms). The Hausdorff distance cutoff value is 1.5 Å.

Acknowledgments

Financial support through Slovenian Research Agency grant L7-8269 is gratefully acknowledged.

References

1. Sliwoski G, Kothiwale S, Meiler J, Lowe EW (2014) Computational methods in drug discovery. Pharmacol Rev 66:334–395. https://doi.org/10.1124/pr.112.007336
2. Macalino SJY, Gosu V, Hong S, Choi S (2015) Role of computer-aided drug design in modern drug discovery. Arch Pharm Res 38:1686–1701. https://doi.org/10.1007/s12272-015-0640-5
3. Konc J, Janežič D (2007) An improved branch and bound algorithm for the maximum clique problem. MATCH Commun Math Comput Chem 58:569–590
4. Burley SK, Berman HM, Kleywegt GJ et al (2017) Protein Data Bank (PDB): the single global macromolecular structure archive. In: Wlodawer A, Dauter Z, Jaskolski M (eds) Protein crystallography: methods and protocols. Springer, New York, NY, pp 627–641
5. Konc J, Hodoščck M, Ogrizek M et al (2013) Structure-based function prediction of uncharacterized protein using binding sites comparison. PLoS Comput Biol 9:e1003341. https://doi.org/10.1371/journal.pcbi.1003341
6. Konc J, Skrlj B, Erzen N et al (2017) GenProBiS: web server for mapping of sequence variants to protein binding sites. Nucleic Acids Res 45:W253–W259. https://doi.org/10.1093/nar/gkx420
7. Štular T, Lešnik S, Rožman K et al (2016) Discovery of Mycobacterium tuberculosis InhA inhibitors by binding sites comparison and ligands prediction. J Med Chem 59:11069–11078. https://doi.org/10.1021/acs.jmedchem.6b01277
8. Konc J, Janežič D (2010) ProBiS algorithm for detection of structurally similar protein binding sites by local structural alignment. Bioinformatics 26:1160–1168. https://doi.org/10.1093/bioinformatics/btq100
9. Konc J, Janežič D (2014) ProBiS-ligands: a web server for prediction of ligands by examination of protein binding sites. Nucleic Acids Res 42:W215–W220. https://doi.org/10.1093/nar/gku460
10. Konc J, Janežič D (2010) ProBiS: a web server for detection of structurally similar protein binding sites. Nucleic Acids Res 38: W436–W440. https://doi.org/10.1093/nar/gkq479
11. Konc J, Janežič D (2012) ProBiS-2012: web server and web services for detection of structurally similar binding sites in proteins. Nucleic Acids Res 40:W214–W221. https://doi.org/10.1093/nar/gks435
12. Konc J, Janežič D (2017) ProBiS tools (algorithm, database, and web servers) for predicting and modeling of biologically interesting proteins. Prog Biophys Mol Biol 128:24–32. https://doi.org/10.1016/j.pbiomolbio.2017.02.005
13. Konc J, Česnik T, Konc JT et al (2012) ProBiS-database: precalculated binding site similarities and local pairwise alignments of PDB structures. J Chem Inf Model 52:604–612. https://doi.org/10.1021/ci2005687
14. Konc J, Depolli M, Trobec R et al (2012) Parallel-ProBiS: fast parallel algorithm for local structural comparison of protein structures and binding sites. J Comput Chem 33:2199–2203. https://doi.org/10.1002/jcc.23048
15. Konc J, Lešnik S, Janežič D (2015) Modeling enzyme-ligand binding in drug discovery. J Chem 7:48. https://doi.org/10.1186/s13321-015-0096-0
16. Miller BT, Singh RP, Klauda JB et al (2008) CHARMMing: a new, flexible web portal for CHARMM. J Chem Inf Model 48:1920–1929. https://doi.org/10.1021/ci800133b
17. Konc J, Miller BT, Štular T et al (2015) ProBiS-CHARMMing: web interface for prediction and optimization of ligands in protein binding sites. J Chem Inf Model 55:2308–2314. https://doi.org/10.1021/acs.jcim.5b00534
18. Lešnik S, Štular T, Brus B et al (2015) LiSiCA: a software for ligand-based virtual screening and its application for the discovery of butyrylcholinesterase inhibitors. J Chem Inf Model 55:1521–1528. https://doi.org/10.1021/acs.jcim.5b00136
19. Sterling T, Irwin JJ (2015) ZINC 15—ligand discovery for everyone. J Chem Inf Model 55:2324–2337. https://doi.org/10.1021/acs.jcim.5b00559

20. DeLano WL (2002) The PyMOL molecular graphics system. http://www.Pymol.Org
21. Dilip A, Lešnik S, Štular T et al (2016) Ligand-based virtual screening interface between PyMOL and LiSiCA. J Chem 8:46. https://doi.org/10.1186/s13321-016-0157-z
22. Lešnik S, Škrlj B, Eržen N et al (2017) BoBER: web interface to the base of bioisosterically exchangeable replacements. J Chem 9:62. https://doi.org/10.1186/s13321-017-0251-x
23. Lešnik S, Konc J, Janežič D (2016) Scaffold hopping and bioisosteric replacements based on binding site alignments. Croat Chem Acta 89:431–437. https://doi.org/10.5562/cca2993
24. Rožman K, Lešnik S, Brus B et al (2017) Discovery of new MurA inhibitors using induced-fit simulation and docking. Bioorg Med Chem Lett 27:944–949. https://doi.org/10.1016/j.bmcl.2016.12.082
25. Reigada C, Valera-Vera EA, Sayé M et al (2017) Trypanocidal effect of isotretinoin through the inhibition of polyamine and amino acid transporters in *Trypanosoma cruzi*. PLoS Negl Trop Dis 11:e0005472. https://doi.org/10.1371/journal.pntd.0005472
26. Huang J, MacKerell AD (2013) CHARMM36 all-atom additive protein force field: validation based on comparison to NMR data. J Comput Chem 34:2135–2145. https://doi.org/10.1002/jcc.23354
27. Vanommeslaeghe K, Hatcher E, Acharya C et al (2010) CHARMM general force field: a force field for drug-like molecules compatible with the CHARMM all-atom additive biological force fields. J Comput Chem 31:671–690. https://doi.org/10.1002/jcc.21367
28. The UniProt Consortium (2015) UniProt: a hub for protein information. Nucleic Acids Res 43:D204–D212. https://doi.org/10.1093/nar/gku989
29. Forbes SA, Beare D, Gunasekaran P et al (2015) COSMIC: exploring the world's knowledge of somatic mutations in human cancer. Nucleic Acids Res 43:D805–D811. https://doi.org/10.1093/nar/gku1075
30. Landrum MJ, Lee JM, Benson M et al (2016) ClinVar: public archive of interpretations of clinically relevant variants. Nucleic Acids Res 44:D862–D868. https://doi.org/10.1093/nar/gkv1222
31. Thorn CF, Klein TE, Altman RB (2013) PharmGKB: the pharmacogenomics knowledge base. In: Innocenti F, van Schaik RHN (eds) Pharmacogenomics: methods and protocols. Humana Press, Totowa, NJ, pp 311–320
32. O'Boyle NM, Banck M, James CA et al (2011) Open Babel: an open chemical toolbox. J Chem 3(33). https://doi.org/10.1186/1758-2946-3-33

Chapter 2

The In Silico Fischer Lock-and-Key Model: The Combined Use of Molecular Descriptors and Docking Poses for the Repurposing of Old Drugs

Marco Tutone and Anna Maria Almerico

Abstract

Not always lead compound and/or derivatives are suitable for the specific biological target for which they are designed but, in some cases, discarded compounds proved to be good binders for other biological targets; therefore, drug repurposing constitute a valid alternative to avoid waste of human and financial resources. Our virtual lock-and-key methods, VLKA and Conf-VLKA, furnish a strong support to predict the efficacy of a designed drug a priori its biological evaluation, or the correct biological target for a set of the selected compounds, allowing thus the repurposing of known and unknown, active and inactive compounds.

Key words Lock-and-key model, Molecular docking, Descriptors, Drug repurposing, Statistics

1 Introduction

Modern medicinal chemistry takes advantage of computational methodologies to save time and, above all, money during the lead identification and/or optimization [1, 2]. However not always the designed lead, once screened, results suitable for the chosen biological target, and the alternative choice is either to change lead or to change biological target.

Moreover, also the discarded compounds could be good inhibitors for other biological targets. These considerations are also supported by several lines of evidence suggesting that drugs may have many physiological targets [3, 4].

For these reasons, in the last years, computational chemistry has been intensively used for a new drug design approach switching this process from the concept "one drug one target" to "one drug multiple target" known as polypharmacology [5–12].

Several computational methodologies are available to medicinal chemist researchers: i.e., molecular docking, induced fit docking,

Nikolaos E. Labrou (ed.), *Targeting Enzymes for Pharmaceutical Development: Methods and Protocols*, Methods in Molecular Biology, vol. 2089, https://doi.org/10.1007/978-1-0716-0163-1_2, © Springer Science+Business Media, LLC 2020

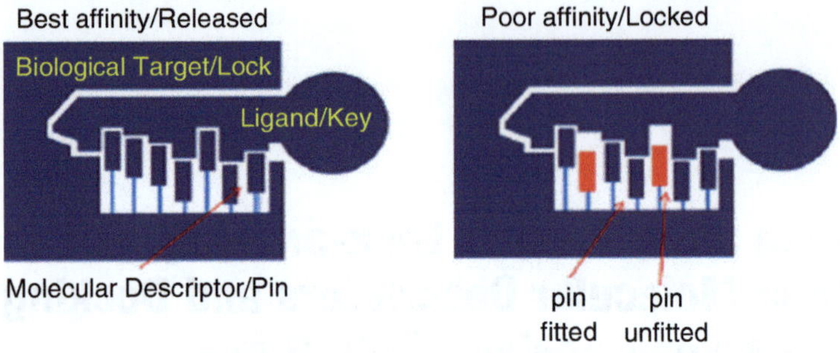

Fig. 1 Lock release mechanism

molecular dynamics, pharmacophore modeling, QSAR, and others, and all of them can be applied on biological fields. Some of them could be considered as derived from the old Emil Fischer lock-and-key model [13–15].

Taking into account all these considerations, we have proposed and developed an in silico methodology that can be for good reasons considered the heritage of original Fischer theory and that we have called "virtual lock-and-key approach" (VLKA) [16, 17].

The protocol allows to set up a "lock model" for a biological target, starting from the respectively known inhibitors. In order to release a real lock it is necessary that the pins of the lock fit the key (Fig. 1). We can use the molecular descriptors as pins, and a tested compound can be considered an inhibitor of a biological target if the values of its molecular descriptors fall in the calculated range values for the set of known inhibitors (*see* **Notes 1** and **2**).

Thus, the proposed protocol can transform a biological target into a "lock model" starting from its known inhibitors as Fisher suggested in his famous Lock-and-Key model.

In our works, we proved the real versatility of the VLKA protocol which is free user-defined. Compounds, biological targets, and molecular descriptors can be chosen by every scientist, which is interested in discovering new biological targets for old inhibitors or new inhibitors for old biological targets.

The application of statistics to biological data, testified also by recent results [18], revealed useful to provide clues to the classification of drugs whose target is unknown or controversial. In this kind of studies, all the property data are represented in the form of a matrix where each compound in each line is represented as an array characterized by a sequence of molecular descriptors values, in each of the matrix columns. In this fashion we developed the so-called BIOTA (BIOlogical Target Assignment) protocol with the aim to assign a correct biological target of designed molecular structures by using the multivariate analysis applied on the abovementioned type of molecular descriptors matrix [19]. The protocol resulted useful to hypothesize the biological target of a candidate drug prior to its biological evaluation or to repurpose old drugs.

Either the BIOTA or the VLKA approach has been used by us to successfully assess the biological activities of classes of inhibitors studied by us, such as molecules targeting heat shock protein 90 (Hsp90) [18] or topoisomerase II [20, 21].

The latest version of the protocol named "Conf-VLKA" introduced the use of other techniques such as docking to measure the capability of docking scoring function in correctly ranking compounds toward their own target, first. Secondly, the docked conformation was exploited for 3D molecular descriptors calculation.

This more sophisticated approach, based on the calculation of 3D molecular descriptors on the docked conformation of ligands, helps to predict the possible biological target for new molecules starting from the structural information contained in molecular descriptors calculated on a set of known inhibitors (*see* **Note 3**).

2 Materials

1. A great amount of information has been collected by the Binding DB [22] by using a continuous upload of biological data. The first step of the proposed protocol, virtual lock-and-key approach, is the random choice of a suitable set of heterogeneous biological targets with known inhibitors available in Binding DB (Table 1).

   Other databases of public access or developed in-house, of course, could be used. For example, drug screening data are available in the National Cancer Institute (NCI) Anti-Cancer Agent Mechanism (ACAM). In this case, for each designed ligands (our keys) the lock models can be prepared by using the available data included in the DB as measurement of their growth inhibition ability over a panel of about 60 human tumor cell lines. In particular, the database is constituted by 114 antitumor drugs ranked according to their MA (Mechanism of Action) belonging to each class of drugs (alkylating agents, antimitotic agents, topoisomerase I inhibitors, topoisomerase II inhibitors, RNA/DNA antimetabolites, and DNA antimetabolites).

2. A set of molecular descriptors for the inhibitors structure was calculated by CODESSA PRO software [23]. This package is able to calculate more than 900 molecular descriptors, but for the protocol aims only molecular descriptors without blanks, common for all the compounds and with a high variance, should be selected.

3. The structures of the drugs to be screened can be prepared through Ligprep software [24] for the 3D optimization.

Table 1
Selected biological targets for VLKA protocol

Biological target T_n (BindingDB acronym)	#Total[a]	# Lock[b]	Cut-off activity[c]
11-beta-hydroxysteroid dehydrogenase (11betaHSD1)	40	35	100
ABL kinase (ABL)	70	14	500
Adenosine A1 receptor (ARA1)	110	16	100
Aldose reductase (ALR2)	126	46	100
Aldosterone synthase (CYP11B2)	129	76	100
Androgen receptor (AR)	244	82	10
Angiotensin converting enzyme (ACE)	51	19	100
Angiotensin converting enzyme 2 (ACE2)	73	22	1000
Anthrax lethal factor (ALF)	130	36	1000
Aromatase (AROM)	440	66	100
Asparaginyl endopeptidase (AE)	27	15	100
Aurora kinase A (AurKA)	179	47	100
BCL-2 (BCL-2)	31	17	1000
BCL-xl (BCL-xl)	50	7	200
Ca-Moduline kinase 2 (CaMK2)	20	5	200
Cannabinoid receptor 2 (CB2)	104	58	100
Carbonic anhydrase 1 (CA-1)	305	12	100
Carbonic anhydrase 2 (CA-2)	402	183	100
Carbonic anhydrase 4 (CA-4)	203	64	100
Caspase-1 (CASP1)	83	12	10,000
Caspase-3 (CASP3)	226	42	100
Checkpoint kinase (CHEK1)	57	35	100
α-Chymotrypsin (CT)	33	10	500
Collagenase (CLG)	309	83	100
Corticotropin-releasing hormone receptor 1 (CRHR1)	62	46	100
Cyclin-dependent kinase (Cdk4)	631	52	100
Delta opioid receptor (DOR)	25	9	100
Diacylglycerol acyltransferase (DGAT-1)	14	13	100
Dihydrofolate reductase (DHFR)	144	25	100
Dopamine transporter (DAT)	58	11	100
EGFR tyrosine kinase (EGFR TK)	979	209	10

(continued)

Table 1
(continued)

Biological target T_n (BindingDB acronym)	#Total[a]	# Lock[b]	Cut-off activity[c]
ERK-2 kinase (ERK-2)	66	35	500
Estrogen receptor (ER-alpha)	199	45	100
Factor Xa (Fxa)	109	66	10
Ghrelin receptor(GHSR)	90	44	100
Glucocorticoid receptor (GR)	109	54	100
Glutaminyl cyclase (GC)	183	36	1000
Glycogen synthase kinase 3α (GSK3α)	229	48	100
Histone deacetylase 1 (HD1)	143	64	100
Matrix metalloproteinase 13 (MMP-13)	142	32	100
Matrix metalloproteinase 3 (MMP-3)	80	25	100
Neutrophil endopeptidase (NEP)	26	15	10
Phosphoinositide-dependent kinase (PDK1)	97	48	100
Phosphodiesterase 10A (PDE10A)	41	16	100
Plasmepsin 1 (PSP1)	51	17	10
Protein-tyrosine phosphatase (PTP1B)	336	48	100
Tyrosine kinase C-kit (TKC-kit)	96	40	100
Total inhibitors	7352	2000	

[a] Total number of inhibitors used for each biological target.
[b] Number of inhibitors used for each lock set selected by cut-off activity values espressed in c.
[c] cut-off activity are expressed in nM.

Different force field protocols, such as OPLS_2005, could be used and all possible states at the selected pH range were generated using Ionizer. The structures were desalted, all possible tautomers were generated, and specified chiralities were retained. Molecular descriptors selected are 1D and 2D, which are not affected by conformation variability. But for the calculation of 3D molecular descriptors, in spite of its approximation, global minimum conformations were selected. This approximation allows not to constrain the molecular structure geometry to the single biological target.

4. The matrix reporting the number of compounds (S_{iTn}) versus the calculated descriptors (D_j) is created. The compounds selection to define the "lock model" for each biological target (T_n) was performed by means biological activity sampling by applying the cut-off of biological data (K_i, IC_{50}, EC_{50})

(Table 1). About one-fourth of the compounds were selected for building the different lock models (training sets). Mean (μ) and standard deviation (σ) of the molecular descriptors values ($X_{i,j}$) for each biological target (T_n) were calculated.

5. In the case of Conf-VLKA:
   (a) Ligand structure similarity evaluation: To check the structural diversity of ligands set, preventing the enrichment of redundant molecular analogs, we set up a topological evaluation of the chosen database. For each target, ligand structures were submitted for calculation of radial fingerprint [25], molprint2D fingerprint [25], and MACCS keys [26] and then analyzed in terms of Tanimoto distance [27] using similarity matrix on CANVAS [28].
   (b) The 3D structures of the biological targets included in the VLKA have been downloaded from the RCSB Protein Databank (PDB) [29], complexed with co-crystalized ligands. The selected structures were submitted to the optimization and refinement process using Protein Preparation Wizard utility of Maestro Schrödinger suite [30]. During this process, bond orders were assigned, the missing hydrogens were added, the disulfide bonds were assigned, the water molecules were deleted, and the protonation of amino acids were determined. At the end, the hydrogen bonds of the proteins were optimized, and restrained minimization was carried out on heavy atoms converging to RMSD equal to 0.30 Å, and on the hydro gen atoms.
   (c) Glide software [31] was used to perform the molecular docking and experiments were carried out using the default parameters and the two different protocols: standard precision (SP) level of accuracy for the generation and scoring of 10 poses for each ligand, top-scored conformation are further re-docked by using the extra-precision (XP) algorithm. Further the compounds were submitted to the docking and scoring procedure versus the own target, and then versus the entire biological targets dataset. The best pose for each compound is selected according to Glide score and on the best pose retrieved 3D molecular descriptors were recalculated.

3 Methods

The individual steps necessary to carry out the technique are reported in the virtual lock-and-key approach flow chart [17].

The first step of the VLKA protocol consists in the conversion of the biological target in a "lock model" in which the keys (the structures) could be "fitted."

1. Calculate mean (μ) and standard deviation (σ) of the molecular descriptors values ($X_{i,j}$) for each biological target (T_n): the hypothesis is that the value of each molecular descriptor of a suitable inhibitor should be close to the molecular descriptors mean (μ) calculated for all the inhibitors of the same biological target.
2. Convert each molecular descriptor value [$X_{i,j}(T_n)$] in α coefficient in relation to closeness to μ according to Eq. 1:

$$\begin{aligned}&\text{if } X_{i,j}(T_n) > m \pm \sigma, \alpha = 0;\\&\text{if } (\mu - \tfrac{1}{2}\sigma) < X_{i,j}(T_n) < (\mu + \tfrac{1}{2}\sigma), \alpha = 1;\\&\text{if } -\sigma < X_{i,j}(T_n) < -\tfrac{1}{2}\sigma, \alpha = 0.5;\\&\text{if } +\tfrac{1}{2}\sigma < X_{i,j}(T_n) < +\sigma, \alpha = 0.5.\end{aligned} \tag{1}$$

   where X is the molecular descriptor value; i is related to the compound; j is related to the molecular descriptor; T_n is the biological target.
3. Molecular descriptors weighing by a coefficient for each biological target (T_n): this was carried out on the basis of the α coefficients determined for the lock set, by considering the sum of the α value for each descriptor (D_j) for all compounds, belonging to the specific biological target $\Sigma\alpha_{i,j}(T_n)$.
4. Normalization step by defining the ωD_j coefficients: The following step was to normalize these values by defining the ωD_j coefficients as reported in Eq. 2:

$$\omega_{Dj} = \frac{\sum \alpha_{i,j(Tn)}}{\max\left[\sum \alpha_{i,j(Tn)}\right]} \tag{2}$$

   where i, j, and T_n are defined as above and max represents the higher α sum of all molecular descriptors belonging to the specific biological target.
5. Partial scores φ calculation: The $\alpha_{i,j}(T_n)$ and ωD_j coefficients were used to calculate the affinity of all the compounds under investigation for each biological target. Thus, according to Eq. 3 the partial score φ was calculated:

$$\varphi_{i,j} = \alpha_{i,j(Tn)}\ \omega_{D_j} \tag{3}$$

6. Total score Φ calculation. The total score Φ was defined as sum of the partial score φ (Eq. 4):

$$\Phi_{i(Tn)} = \sum_{j=1}^{173} \varphi_{i,j(Tn)} \tag{4}$$

where $\varphi_{i,j}$ represents the partial score; Φ_i represents the total score; i, j, and T_n are defined in Eq. 1.

All the calculated scores Φ_i for all the structures for each biological target were converted into ranking positions. At the end, the Φ scores rank all the database compounds with respect to the biological targets. The final hypothesis is that inhibitors related to each biological target should occupy the higher rankings. To verify this hypothesis, the enrichment score ($E\%$), considered as the percentage of correct classification, was calculated according to Eq. 5:

$$E\% = \left(\frac{\sum W - \sum P}{\sum W - \sum B}\right) 100 \tag{5}$$

where ΣW represents the sum of hypothetical lowest rankings; ΣB represents the sum of hypothetical highest rankings; ΣP represents the sum of obtained rankings.

Because each biological target needs specific chemico-physical requests, it is wise to assume that some molecular descriptors could express better than the other structural requirements of the specific biological target. This is the crucial point in the design of a suitable inhibitor (*see* **Note 4**). Drug polypharmacology is tightly linked to the concept of the re-purposing of old drugs or inactive derivatives for new biological targets and drug re-purposing is one of the goals of VLKA computational approach. The more sophisticated procedure Conf-VLKA evaluated also the influence of 3D conformation of ligands on the accuracy of the prediction. The same algorithm of scoring and ranking was employed but, this time, combining it with a structure-based approach as docking. The docking protocol was used to retrieve docking scores, then, from the docked poses of each molecule, 3D descriptors were calculated (Conf-VLKA). While the use of the simple docking scores proved to be inadequate to improve compounds classification, the Conf-VLKA showed some interesting variations compared to the original VLKA. This was particularly true especially for targets whose ligands present a high number of rotamers. This study can be further completed using other techniques such as induced fit docking or molecular dynamics structure clustering to take into account the protein side chains adaptation to ligands structures.

4 Notes

1. The developed in-house virtual lock-and-key approach (VLKA) allowed evaluating target assignment starting from molecular descriptors calculated on known inhibitors used as an information source.

2. The use of molecular descriptors as the starting point to build lock models for biological targets was necessary because a simple analysis of structural similarity does not always imply similarity in the biological activity [32] and does not involve descriptors similarity [33].
3. For the correct development of the models, whereas by using 1D and 2D molecular descriptors it is not important to consider the conformation variability, in the calculation of 3D molecular descriptors, global minimum conformations were selected. Of course, this constitutes an approximation but it has the advantage not to constrain the molecular structure geometry to the single biological target.
4. The VLKA protocol predicts the correct biological target for the whole dataset with a good degree of reliability (80%), and proved experimentally, which was useful for the target fishing of unknown compounds.

   To be noted that drugs may have many physiological targets [3, 4, 34, 35], aspect called "polypharmacology," which is recognized to be therapeutically essential in the treatment of several types of diseases such as schizophrenia [36].

The importance of drug polypharmacology has pushed the efforts to predict and characterize drug-biological target associations [37–40]. The use of chemical similarities among molecules has allowed to identify drugs with multiple biological targets [41, 42], and early drug candidates are screened against biological target panels [43].

Acknowledgements

The work reported in this chapter is based on the reference [17] (Tutone M, Perricone U, Almerico AM (2017) Conf-VLKA: A structure-based revisitation of the virtual lock-and-key approach. J Mol Graph Model 71:50–57. doi: 10.1016/j.jmgm.2016.11.006) and was adapted with permission.

References

1. Dimasi JA, Feldman L, Seckler A, Wilson A (2010) Trends in risks associated with new drug development: success rates for investigational drugs. Clin Pharmacol Ther 87:272–277. https://doi.org/10.1038/clpt.2009.295
2. Dickson M, Gagnon JP (2004) Key factors in the rising cost of new drug discovery and development. Nat Rev Drug Discov 3:417–429. https://doi.org/10.1038/nrd1382
3. Peterson RT (2008) Chemical biology and the limits of reductionism. Nat Chem Biol 4:635–638. https://doi.org/10.1038/nchembio1108-635
4. Nobeli I, Favia AD, Thornton JM (2009) Protein promiscuity and its implications for biotechnology. Nat Biotechnol 27:157–167. https://doi.org/10.1038/nbt1519
5. Reddy AS, Zhang S (2013) Polypharmacology: drug discovery for the future. Expert Rev Clin

Pharmacol 6:41–77. https://doi.org/10.1586/ecp.12.74

6. Hopkins AL (2008) Network pharmacology: the next paradigm in drug discovery. Nat Chem Biol 4:682–690. https://doi.org/10.1038/nchembio.118
7. Peters JU (2013) Polypharmacology—foe or friend? J Med Chem 56:8955–8971. https://doi.org/10.1021/jm400856t
8. Overington JP, Al-Lazikani B, Hopkins AL (2006) How many drug targets are there? Nat Rev Drug Discov 5:993–996. https://doi.org/10.1038/nrd2199
9. Hopkins AL, Mason JS, Overington JP (2006) Can we rationally design promiscuous drugs? Curr Opin Struct Biol 15:104–111. https://doi.org/10.1016/j.sbi.2006.01.013
10. Aislyn DW, Boran RI (2010) Systems approaches to polypharmacology and drug discovery. Curr Opin Drug Discov Devel 13:297–309. https://doi.org/10.1126/scisignal.2001965.Introduction
11. Anighoro A, Bajorath J, Rastelli G (2014) Polypharmacology: challenges and opportunities in drug discovery. J Med Chem 57:7874–7887. https://doi.org/10.1021/jm5006463
12. Gujral TS, Peshkin L, Kirschner MW (2014) Exploiting polypharmacology for drug target deconvolution. Proc Natl Acad Sci 111:5048–5053. https://doi.org/10.1073/pnas.1403080111
13. Fischer E (1895) Ueber den Einfluss der Konfiguration auf die Wirkung der Enzyme III. Berichte der Dtsch Chem Gesellschaft 28:1429–1438. https://doi.org/10.1002/cber.18950280243
14. Forster MO (1920) Emil Fischer memorial lecture. J Chem Soc Trans 117:1157–1201. https://doi.org/10.1039/CT9201701157
15. Fischer E (1899) Bedeutung der Stereochemie für die Physiologie. Hoppe Seylers Z Physiol Chem. https://doi.org/10.1515/bchm2.1899.26.1-2.60
16. Lauria A, Tutone M, Almerico AM (2011) Virtual lock-and-key approach: the in silico revival of Fischer model by means of molecular descriptors. Eur J Med Chem 46:4274–4280. https://doi.org/10.1016/j.ejmech.2011.06.033
17. Tutone M, Perricone U, Almerico AM (2017) Conf-VLKA: a structure-based revisitation of the Virtual Lock-and-Key Approach. J Mol Graph Model 71:50–57. https://doi.org/10.1016/j.jmgm.2016.11.006
18. Lauria A, Ippolito M, Almerico AM (2009) Principal component analysis on molecular descriptors as an alternative point of view in the search of new Hsp90 inhibitors. Comput Biol Chem 33:386–390. https://doi.org/10.1016/j.compbiolchem.2009.07.010
19. Lauria A, Tutone M, Barone G, Almerico AM (2014) Multivariate analysis in the identification of biological targets for designed molecular structures: the BIOTA protocol. Eur J Med Chem 75:106–110. https://doi.org/10.1016/j.ejmech.2014.01.025
20. Lauria A, Patella C, Abbate I et al (2012) Lead optimization through VLAK protocol: new annelated pyrrolo-pyrimidine derivatives as antitumor agents. Eur J Med Chem 55:375–383. https://doi.org/10.1016/j.ejmech.2012.07.046
21. Lauria A, Abbate I, Patella C et al (2013) New annelated thieno[2,3-e][1,2,3]triazolo[1,5-a]pyrimidines, with potent anticancer activity, designed through VLAK protocol. Eur J Med Chem 62:416–424. https://doi.org/10.1016/j.ejmech.2013.01.019
22. Liu T, Lin Y, Wen X et al (2007) BindingDB: a web-accessible database of experimentally determined protein-ligand binding affinities. Nucleic Acids Res 35:D198–D201. https://doi.org/10.1093/nar/gkl999
23. Karelson M, Lobanov VS, Katritzky AR (1996) Quantum-chemical descriptors in QSAR/QSPR studies. Chem Rev 96:1027–1044. https://doi.org/10.1021/cr950202r
24. LigPrep, version 2.5. In: Suite. Schrödinger, LLC, New York NY; 2012
25. Rogers D, Brown RD, Hahn M (2005) Using extended-connectivity fingerprints with Laplacian-modified Bayesian analysis in high-throughput screening follow-up. J Biomol Screen 10:682–686. https://doi.org/10.1177/1087057105281365
26. Duan J, Dixon SL, Lowrie JF, Sherman W (2010) Analysis and comparison of 2D fingerprints: insights into database screening performance using eight fingerprint methods. J Mol Graph Model 29:157–170. https://doi.org/10.1016/j.jmgm.2010.05.008
27. Gilbert G (1972) Distance between sets. Nature 239:174. https://doi.org/10.1038/239174c0
28. Sastry M, Lowrie JF, Dixon SL, Sherman W (2010) Large-scale systematic analysis of 2D fingerprint methods and parameters to improve virtual screening enrichments. J Chem Inf Model 50:771–784. https://doi.org/10.1021/ci100062n
29. Dutta S, Berman MH, Bluhm FW (2005) RCSB Protein Data Bank. Curr Prot

Bioinformatics chapter 1: Unit 1.9. doi: https://doi.org/10.1002/0471250953.bi0109s20

30. Maestro, version 9.4, Schrödinger, LLC, New York, NY; 2013
31. Halgren TA, Murphy RB, Friesner RA et al (2004) Glide: a new approach for rapid, accurate docking and scoring. 2. Enrichment factors in database screening. J Med Chem 47:1750–1759. https://doi.org/10.1021/jm030644s
32. Martin YC, Kofron JL, Traphagen LM (2002) Do structurally similar molecules have similar biological activity? J Med Chem 45:4350–4358. https://doi.org/10.1021/jm020155c
33. Kubinyi H (2002) Chemical similarity and biological activities. J Braz Chem Soc 13:717–726. https://doi.org/10.1590/S0103-50532002000600002
34. Marona-Lewicka D, Nichols DE (2007) Further evidence that the delayed temporal dopaminergic effects of LSD are mediated by a mechanism different than the first temporal phase of action. Pharmacol Biochem Behav 87:453–461. https://doi.org/10.1016/j.pbb.2007.06.001
35. Marona-Lewicka D, Nichols DE (2009) WAY 100635 produces discriminative stimulus effects in rats mediated by dopamine D4 receptor activation. Behav Pharmacol 20:114–118. https://doi.org/10.1097/FBP.0b013e3283242f1a
36. Roth BL, Sheffer DJ, Kroeze WK (2004) Magic shotguns versus magic bullets: selectively non-selective drugs for mood disorders and schizophrenia. Nat Rev Drug Discov 3:353–359. https://doi.org/10.1038/nrd1346
37. Bajorath J (2008) Computational analysis of ligand relationships within target families. Curr Opin Chem Biol 12:352–358. https://doi.org/10.1016/j.cbpa.2008.01.044
38. Oprea TI, Tropsha A, Faulon JL, Rintoul MD (2007) Systems chemical biology. Nat Chem Biol 3:447–450. https://doi.org/10.1038/nchembio0807-447
39. Newman DJ (2008) Natural products as leads to potential drugs: an old process or the new hope for drug discovery? J Med Chem 51:2589–2599. https://doi.org/10.1021/jm0704090
40. Siegel MG, Vieth M (2007) Drugs in other drugs: a new look at drugs as fragments. Drug Discov Today 12:71–79. https://doi.org/10.1016/j.drudis.2006.11.011
41. Young DW, Bender A, Hoyt J et al (2008) Integrating high-content screening and ligand-target prediction to identify mechanism of action. Nat Chem Biol 4:59–68. https://doi.org/10.1038/nchembio.2007.53
42. Wagner BK, Kitami T, Gilbert TJ et al (2008) Large-scale chemical dissection of mitochondrial function. Nat Biotechnol 26:343–351. https://doi.org/10.1038/nbt1387
43. Krejsa CM, Horvath D, Rogalski SL et al (2003) Predicting ADME properties and side effects: the BioPrint approach. Curr Opin Drug Discov Devel 6:470–480

Chapter 3

Determination of Half-Maximal Inhibitory Concentration of an Enzyme Inhibitor

Nikolaos Georgakis, Elisavet Ioannou, Christina Varotsou, Georgios Premetis, Evangelia G. Chronopoulou, and Nikolaos E. Labrou

Abstract

Drug development is the process of bringing a new pharmaceutical drug to the market once a lead compound has been identified through the process of drug discovery. Enzymes are one of the most important groups of drug targets; thus, enzyme inhibition is widely used for the treatment of certain disorders. The assessment of an inhibitor against an enzyme is predominantly based on two different parameters: the half-maximal inhibitory concentration (IC_{50}) and the inhibition constant (K_i). This chapter describes an experimental procedure for the determination of the IC_{50} value of an enzyme inhibitor. The relationship between IC_{50} and K_i is also discussed.

Key words Half-maximal inhibitory concentration (IC_{50}), Enzyme inhibitor, Enzyme kinetics, Inhibition potency

1 Introduction

The potency of a candidate enzyme inhibitor can be estimated by the decrease of the rate of an enzyme-catalyzed reaction. The concentration of an inhibitor capable of causing the decrease of enzyme activity by 50%, under specified assay conditions, is called the half-maximal inhibitory concentration (IC_{50}). In other words, the IC_{50} value indicates how much of an inhibitor is needed to inhibit the activity of a given enzyme by half. The IC_{50} value is typically expressed as molar concentration.

Pharmacological and drug discovery studies are often in need of quantitative calculations in order to compare the inhibitory potency of a large number of compounds against a specific target enzyme [1]. In such cases, the experimentally determined IC_{50} values are widely used to rank the inhibition potency of different inhibitors for a given enzyme or different enzymes for a common inhibitor. However, the IC_{50}-based ranking can be misleading since the IC_{50} values are dependent on conditions under which they are

Nikolaos E. Labrou (ed.), *Targeting Enzymes for Pharmaceutical Development: Methods and Protocols*, Methods in Molecular Biology, vol. 2089, https://doi.org/10.1007/978-1-0716-0163-1_3, © Springer Science+Business Media, LLC 2020

Table 1
Relationship between type of inhibition, IC_{50}, and K_i

Inhibitor type	IC_{50}	Inhibition (%)
Competitive	$\frac{K_i}{[I]}\left(1+\frac{[S]}{K_m}\right)$	$\frac{100}{\frac{K_i}{[I]}\left(1+\frac{[S]}{K_m}\right)}$
Uncompetitive	$K_i\left(1+\frac{K_m}{[S]}\right)$	$\frac{100}{\frac{K_i}{[I]}\left(1+\frac{K_m}{[S]}\right)}$
Noncompetitive	K_i	$\frac{100}{\left(1+\frac{K_i}{[I]}\right)}$
Mixed	$\frac{[S]+K_m}{\frac{[S]}{K_i'}+\frac{K_m}{K_i}}$	$\frac{100}{\frac{(K_m+[S])+\frac{K_m}{K_i}+\frac{[S]}{K_i'}}{\frac{K_m}{K_i}+\frac{[S]}{K_i'}}}$

measured. For example, they are dependent on the concentrations of the enzyme, the inhibitor, and the ligand. In addition, the IC_{50} values are dependent on the type of inhibition in the case of reversible inhibitors [2–5]. Alternatively, and for more accurate comparison of a range of inhibitors, the inhibition constant (K_i) should be preferred, as it is independent of the concentrations but depends on the enzyme (target) and inhibitor. However, for the determination of K_i values, more time-consuming assays are required. Therefore, the determination of K_i is usually done only with the most promising candidates. The relationship between IC_{50}/K_i and the type of inhibition is summarized in Table 1 [5] (*see* **Note 1**).

In the present protocol, we describe a simple experimental procedure for the determination of the IC_{50} value for a given inhibitor. It is assumed that all of the substrate- and inhibitor-binding reactions are reversible. Comparison of IC_{50} values for a set of inhibitors toward a given enzyme is only assumed to be valid when they are evaluated under identical experimental conditions.

2 Materials

All solutions should be prepared using ultrapure water and analytical grade reagents.

1. Reaction buffer: Prepare a buffer according to enzyme-specific standard assay conditions. Store at 4 °C.
2. Substrates: Prepare stock solutions of the standard enzyme substrates for measuring its catalytic activity (*see* **Notes 2** and **3**). In this protocol, we assume that the enzyme has two substrates (A and B) and catalyze the reaction: $A + B \rightarrow C + D$.

3. Test compound (inhibitor): Prepare an initial stock solution of relatively high concentration for the inhibitor (10×) that is about to be tested. Prepare diluted solutions for the different concentrations which will be implemented in the assays (*see* **Note 4**).
4. Enzyme solution: Prepare a stock solution of the purified enzyme, preferably diluted in the reaction buffer. Keep at 4 °C.
5. Use appropriate instrumentation and consumables for the experiment (*see* **Note 5**).

3 Methods

Carry out the assay at 25 °C unless otherwise specified. The final volume of the reaction is 1000 μL (*see* **Note 6**). In most experimental studies of enzyme kinetics, the total concentrations of substrates and inhibitor used are in excess of the enzyme concentration to make their free and total concentrations essentially the same.

1. Add the appropriate volume of the reaction buffer (Table 2).
2. Add Substrate A at the required concentration of the implemented assay.
3. Add the test compound in the sample reaction.
4. Add a proper amount of enzyme solution.
5. Initiate the reaction by adding the required amount of Substrate *B*.
6. Monitor the initial reaction rate ($\Delta\varsigma/\Delta t$) at the desirable wavelength for a preset amount of time (1–5 min).
7. Calculate the percentage of remaining enzyme activity (or the percentage of enzyme inhibition) using as 100%, the rate obtained in the absence of the inhibitor that is tested.
8. The IC_{50} can be determined graphically using a dose-response curve and plotting the concentration of the inhibitor on the *X*-axis, against the remaining enzyme activity (%) on the *Y*-axis. The IC_{50} can be calculated using an appropriate computer program (e.g., GraphPad Prism and GraFit) (*see* **Notes** 7 and **8**) (Fig. 1).

4 Notes

1. A web server tool is available for the calculation of K_i values from IC_{50} values (http://botdb.abcc.ncifcrf.gov/toxin/kiConverter.jsp) [4]. User-defined required input values

Table 2
Components for a routine enzyme assay

Components	Control (μL)	Sample test (μL)
Reaction buffer	$1000 - (A + y + B)$	$1000 - (A + x + y + B)$
Substrate *A*	*A*	*A*
Test compound	–	*x*
Enzyme	*y*	*y*
Substrate *B*	*B*	*B*

All but one of the components are added in the reaction vessel (cuvette) and the reaction is initiated by adding the missing component (one of the substrates, e.g., substrate *B*)

include the total enzyme and substrate concentrations, the K_m value of the enzyme-substrate and the IC_{50} value.

2. The IC_{50} value may be defined as a relative measure for defining an inhibitor [1]. The substrate concentration affects the IC_{50} value. In a screening experiment, where a library of compounds are assessed for inhibitory potency against an enzyme, it is recommended the use of substrates at concentrations equal to K_m [6].

3. Make sure to follow the manufacturer's instructions for the proper use and handling of the substrates required for the enzyme reaction. Bear in mind the solvent, the storage temperature, and the protection of light that might be important for the stability of the solutions.

4. Prepare at least ten different concentrations of the inhibitor under study. The concentrations which will be used should be properly allocated covering the whole range of inhibition (e.g., 0–100%) for an accurate estimation of the IC_{50} value. Furthermore, it is desirable that the estimated IC_{50} value (derived from the plot) should be surrounded by at least two different inhibitor concentrations.

5. An IC_{50} experiment requires the use of a suitable spectrophotometer or appropriate UV/Vis microplate reader or disposable plastic cuvettes. Concerning plastic cuvettes, it is important to take into consideration the wavelength of the enzymatic reaction that is to be used.

6. Each concentration of the tested inhibitor should be examined at least in triplicate and include the standard error of the calculated value in the graphic depiction of the IC_{50}.

7. The graphic depiction of the derived curve may be presented using the inhibitor concentrations or the logarithmic values of the corresponding inhibitor concentrations. The shape of the

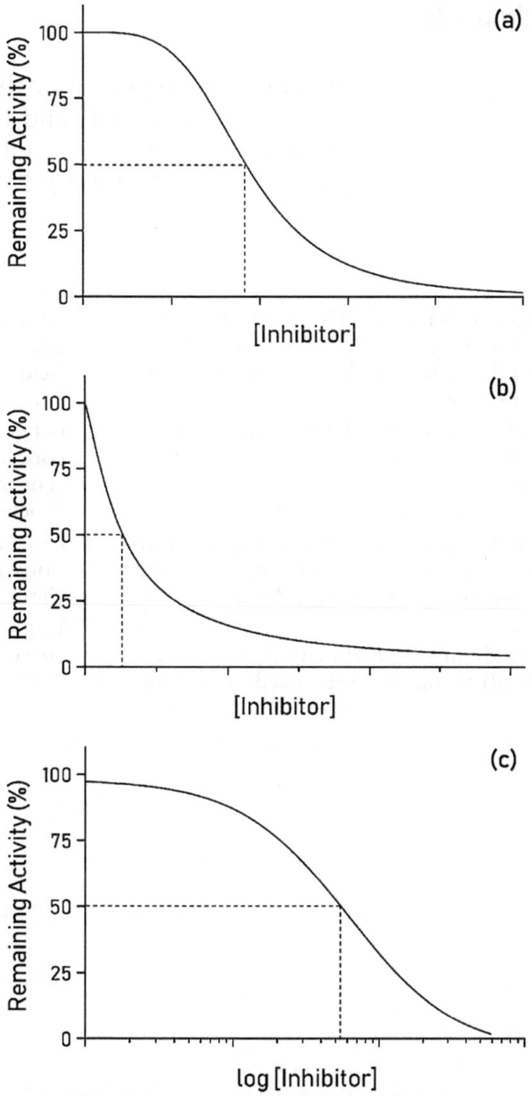

Fig. 1 Three typical dose-response curves produced from an IC_{50} experiment. (**a**) A common depiction of the enzyme activity against different concentrations of an inhibitor. (**b**) A more complex diagram produced under the same conditions as (**a**), indicating a Hill coefficient less than 1. (**c**) An alternative version of (**b**) having transformed the concentrations of the inhibitor to its corresponding logarithmic values

generated curve and, consequently, the Hill coefficient depend on the inhibitor employed [7] (*see* Fig. 1).

8. An IC_{50} experiment comprises a dose-response curve. The IC_{50} value is usually calculated by fitting the obtained data on standard four-parameter logistic models. While the curve is generated through an automated data processing computer program, the researcher has to warrant the quality of the provided data and ensure the reliability of the results.

Acknowledgments

This work was supported by the Hellenic Foundation for Research and Innovation (HFRI) and the General Secretariat for Research and Technology (GSRT), under the HFRI PhD Fellowship grant (GA. No. 8904/22-09-2017).

References

1. Burlingham BT, Widlanski TS (2003) An intuitive look at the relationship of K_i and IC_{50}: a more general use for the Dixon plot. J Chem Educ 80:214–218
2. Sebaugh JL (2011) Guidelines for accurate EC50/IC50 estimation. Pharm Stat 10:128–134
3. Ramsay RR, Tipton KF (2017) Assessment of enzyme inhibition: a review with examples from the development of monoamine oxidase and cholinesterase inhibitory drugs. Molecules 22 (7):E1192
4. Cer RZ, Mudunuri U, Stephens R, Lebeda FJ (2009) IC50-to-K_i: a web-based tool for converting IC50 to K_i values for inhibitors of enzyme activity and ligand binding. Nucleic Acids Res 37:W441–W445. (Web Server issue).
5. Cheng Y, Prusoff WH (1973) Relationship between the inhibition constant (K_i) and the concentration of inhibitor which causes 50 percent inhibition (IC_{50}) of an enzymatic reaction. Biochem Pharmacol 22:3099–3108
6. Copeland RA (2003) Mechanistic considerations in high-throughput screening. Anal Biochem 320:1–12
7. Prinz H (2010) Hill coefficients, dose–response curves and allosteric mechanisms. J Chem Biol 3:37–44

Chapter 4

Applications of Differential Scanning Fluorometry and Related Technologies in Characterization of Protein–Ligand Interactions

Bolormaa Baljinnyam, Michael Ronzetti, Adam Yasgar, and Anton Simeonov

Abstract

Differential scanning fluorometry (DSF) is an efficient and high-throughput method to analyze protein stability, as well as detect ligand interactions through perturbations of the protein's melting temperature. The method monitors protein unfolding by observing the fluorescence changes of a sample, whether through an environmentally sensitive fluorophore or by intrinsic protein fluorescence, while a temperature gradient is applied. Here, we describe in detail how to develop and optimize DSF assays to identify protein–ligand interactions while exploring different buffer and additive conditions. Analysis of the data and further applications of the method are also discussed.

Key words Differential scanning fluorometry, DSF, Thermal shift assay, Thermal denaturation, Protein stability, Label-free DSF, NanoDSF, Protein–ligand interaction

1 Introduction

Differential scanning fluorometry (DSF), also known as the thermal shift assay (TSA), is a versatile method to monitor protein unfolding in response to temperature increase and characterize protein stability under various conditions, including upon interaction with a ligand [1, 2]. The protein conformational changes during thermal denaturation can be examined by monitoring the fluorescence emission of an environmentally sensitive fluorescent dye or intrinsic protein fluorescence (label-free DSF). The most commonly used fluorescent dye is SYPRO Orange, which undergoes a significant increase in quantum yield upon binding to hydrophobic sites that have become exposed during protein unfolding. Label-free DSF exploits the high sensitivity of tryptophan fluorescence, a dominant source of intrinsic protein fluorescence, to its local environment [3]. Tryptophan fluorescence maximum and

Nikolaos E. Labrou (ed.), *Targeting Enzymes for Pharmaceutical Development: Methods and Protocols*, Methods in Molecular Biology, vol. 2089, https://doi.org/10.1007/978-1-0716-0163-1_4, © Springer Science+Business Media, LLC 2020

intensity are highly influenced by the polarity of its microenvironment, hydrogen bonding, and other non-covalent interactions, and display a red shift in increasing polarity.

The Gibb's free energy of unfolding, ΔG_u, provides a value related to the stability of a protein, which is temperature dependent [4]. As the temperature of a system increases, ΔG_u will decrease until the value is zero at equilibrium, at which point the concentrations of folded and unfolded protein are equal [5]. The point at which there are equal populations of folded and unfolded protein is also known as the melting temperature, T_m. In the case of ligand binding-induced stabilization, the free energy contributions from direct ligand–protein interaction will increase the ΔG_u value, and in turn may cause an increase in T_m.

Analysis of DSF thermograms is straightforward; when the fluorescence intensity is plotted as a function of temperature, a sigmoidal curve is produced whose midpoint signifies the T_m of the protein. This inflection point at the T_m is often calculated using the Boltzmann equation:

$$y = LL + \frac{(UL - LL)}{1 + \exp\left(\frac{T_m - x}{a}\right)}$$

The "*LL*" and "*UL*" denote the minimum and maximum intensities of the sigmoid, and "*a*" is the slope of the curve at the T_m. A simpler way to calculate the T_m is to determine the maximum or minimum of the first derivative of the sigmoid, or the point where the second derivative crosses the *x*-axis. It is worth noting that binding between ligand and protein may cause an increase in ΔG_u with no discernible increase in T_m, and the magnitude of the T_m shift for a ligand does not necessarily correlate with its affinity for the protein. Similarly, a range of affinities can account for the same shift in melting temperature, as each ligand may have different enthalpic and entropic contributions.

The distinct measurement strategies employed by DSF and label-free DSF may sometimes reveal different thermal denaturation profiles of the same protein and/or changes in the melting curve upon ligand binding. For example, aldehyde dehydrogenase 1A1 (ALDH1A1) has a typical single transition curve in a DSF experiment using SYPRO Orange. When comparing the first derivates of the ALDH1A1 melting profile alone or in presence of the cofactor NAD+ and/or a known inhibitor NCT-501, there is no apparent Tm shift (Fig. 1a) [6]. In contrast, data collected by a label-free method shows a bimodal thermogram, whose first peak was right-shifted by the cofactor NAD^+ and completely eliminated when NCT-501 was added (Fig. 1b). Therefore, testing an enzyme with both methods may reveal a greater breadth of information than one technique alone.

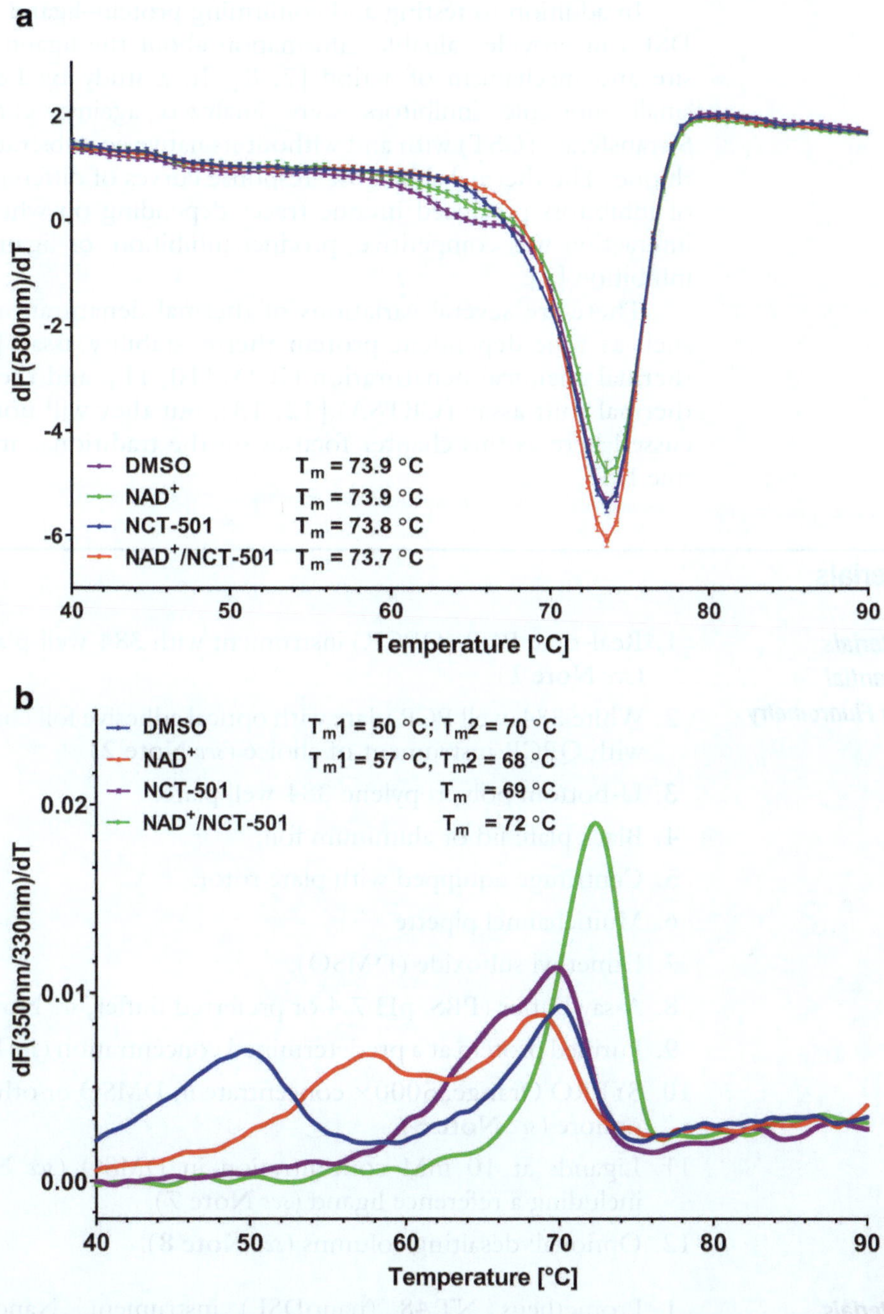

Fig. 1 Comparison of the melting profiles of aldehyde dehydrogenase 1A1 (ALDH1A1) in the absence and presence of its cofactor, NAD$^+$, and/or a known inhibitor, NCT-501, acquired by DSF or label-free DSF. (**a**) DSF experiment was performed with 4 μM ALDH1A1 and 4× SYPRO Orange in 100 mM HEPES, pH 7.5 with or without 1 mM NAD$^+$ and/or 100 μM NCT-501 were indicated. Data was acquired at a temperature ramp of 0.11 °C/s with five acquisitions per °C and the first derivative of the fluorescence emission at 580 nm is plotted as a function of temperature. (**b**) Label-free DSF experiment was performed at a temperature ramp of 1.0 °C/min, with 1.75 μM (0.1 mg/mL) ALDH1A1 in 100 mM HEPES, pH 7.5 with or without 1 mM NAD$^+$ and/or 100 μM NCT-501. The protein melting curves are shown as the first derivative of the 350 nm/330 nm emission ratio as a function of temperature

In addition to testing and confirming protein–ligand binding, DSF can provide valuable information about the ligand binding site and mechanism of action [7, 8]. In a study by Lea et al., small molecule inhibitors were analyzed against glutathione S-transferase (GST) with and without its native co-substrate, glutathione. The thermal shift dose–response curves of different classes of inhibitors produced unique traces depending on whether the interaction was competitive, product inhibition, or ligandin-type inhibition [7].

There are several variations of thermal denaturation assays, such as time-dependent protein thermostability assay [9], isothermal chemical denaturation (ICD) [10, 11], and the cellular thermal shift assay (CETSA) [12, 13], but they will not be discussed here as this chapter focuses on the traditional and label-free DSF.

2 Materials

2.1 Materials for Differential Scanning Fluorometry

1. Real-time PCR (QPCR) instrument with 384-well-plate block (*see* **Note 1**).
2. White 384-well PCR plate with optical adhesive foil compatible with QPCR instrument of choice (*see* **Note 2**).
3. U-bottom polypropylene 384-well plate.
4. Black plate lid or aluminum foil.
5. Centrifuge equipped with plate rotor.
6. Multichannel pipette.
7. Dimethyl sulfoxide (DMSO).
8. Assay buffer (PBS, pH 7.4 or preferred buffer, *see* **Note 3**).
9. Purified protein at a predetermined concentration (*see* **Note 4**).
10. SYPRO Orange, 5000× concentrate in DMSO or other fluorophore (*see* **Note 5**).
11. Ligands at 10 mM concentration in DMSO (*see* **Note 6**), including a reference ligand (*see* **Note 7**).
12. Optional: desalting columns (*see* **Note 8**).

2.2 Materials for Label-Free Differential Scanning Fluorometry

1. Prometheus NT.48 (nanoDSF) instrument, NanoTemper Technologies (*see* **Note 9**).
2. Standard or high-sensitivity capillaries for Prometheus NT.48 (*see* **Note 10**).
3. PCR tubes, tube strips, or 96-well plate.
4. Centrifuge equipped with tube strip or plate rotor.

5. DMSO.
6. Assay buffer (PBS, pH 7.4 or preferred buffer, *see* **Note 3**).
7. Purified protein at a predetermined concentration (*see* **Note 4**).
8. Ligand at 10 mM concentration in DMSO (*see* **Note 6**), including a reference ligand (*see* **Note 7**).
9. Optional: desalting columns (*see* **Note 8**).

3 Methods

3.1 Methods for Differential Scanning Fluorometry

*3.1.1 Determination of Protein-to-Dye Ratio (See **Note 11**)*

1. Prepare a 200× stock of SYPRO Orange in assay buffer.
2. Prepare a dilution series of SYPRO Orange in assay buffer from 40× to 2× (twofold of final concentration) in row A of a U-bottom plate (plate 1). Prepare 100 μL of each dilution, which is sufficient for triplicate reactions at five different protein concentrations + "no protein"/"buffer only" control.
3. Cover plate 1 with a black lid or aluminum foil.
4. Prepare protein dilution series in assay buffer from 20 to 1 μM (twofold of final concentration) in column 1 of a U-bottom plate (plate 2). Include a "no protein"/"buffer only" control. Prepare 100 μL of each sample, which is sufficient for reactions at six different dye concentrations in triplicate.
5. If the protein is stable at room temperature, plate 2 can be kept at room temperature. If not, plate 2 should be kept on ice.
6. Using a multichannel pipette, transfer 5 μL from row A of plate 1 into the first six rows of the PCR plate (*see* Table 2).
7. Transfer 5 μL from column 1 of plate 2 into the first six columns of the PCR plate using multichannel pipette (*see* Table 2).
8. Choose two different areas of the PCR plate and repeat **steps 7** and **8** for a triplicate setup of the dye–protein cross-titration reaction.
9. Seal the PCR plate with an optical adhesive foil.
10. Centrifuge the plate for 1 min at 200 × *g*.
11. Place the PCR plate into the PCR instrument and run the temperature scan from 25 to 95 °C at a chosen ramp speed (*see* **Note 12**).
12. Analyze the data and choose the best protein-to-dye ratio (*see* **Note 13**).

3.1.2 Characterization of Protein–Ligand Interaction

1. Prepare 1 mM ligand solutions (*see* **Note 14**) + 10% (v/v) DMSO control sample (*see* **Note 15**) in assay buffer in a U-bottom plate.

2. Prepare a mixture of protein and SYPRO Orange in assay buffer at 10/9th of the final concentrations (*see* **Notes 13** and **16**). Prepare a sufficient amount for triplicate measurements of each ligand + control (i.e., 480 μL is sufficient to test 15 different ligands + DMSO control).
3. Dispense 9 μL of the protein-dye mixture into a white PCR plate.
4. Transfer 1 μL of compound to the protein-dye mixture.
5. Seal the PCR plate with an optical adhesive foil.
6. Centrifuge the plate for 1 min at 200 × *g*.
7. Incubate for 15 min at room temperature.
8. Run a temperature scan for the predetermined temperature range, at chosen ramp speed (*see* **Note 12**).
9. Analyze data (*see* **Note 17**).

3.1.3 Validation of Interacting Ligands

1. Prepare 8-point serial dilution of a ligand in DMSO starting at 10 mM in first column of U-bottom plate. Preparing the serial dilution in DMSO ensures constant concentration of DMSO in all final samples. Include 100% (v/v) DMSO control sample.
2. Dispense 9 μL of assay buffer into the second column of a U-bottom plate.
3. Transfer 1 μL of sample from column 1 into column 2 using a multichannel pipette and mix well.
4. Centrifuge the plate for 1 min at 200 × *g* and cover.
5. Prepare a mixture of protein and SYPRO Orange in assay buffer at 10/9th of final concentrations and of sufficient amount for triplicate reactions at each compound concentration.
6. Dispense 9 μL of the protein-dye mixture column wise into a white PCR plate.
7. Transfer 1 μL of sample from column 2 of the U-bottom plate to the protein-dye mixture in a PCR plate.
8. Seal the PCR plate with an optical adhesive foil.
9. Centrifuge the plate for 1 min at 200 × *g*.
10. Incubate for 15 min at room temperature.
11. Run a temperature scan for the predetermined temperature range, at chosen ramp speed (*see* **Note 12**).
12. Analyze data (*see* **Note 18**).

3.2 Methods for Label-Free Differential Scanning Fluorometry

3.2.1 Determination of Optimal Conditions

1. Prepare a 4-point protein dilution series in assay buffer in PCR tubes or a tube strip. Prepare 20 μL for each concentration, which is sufficient for duplicate measurements for each concentration.
2. Fill capillaries and place them into the instrument (*see* **Note 19**).
3. Take initial fluorescence scan and set up the LED power (*see* **Note 20**).
4. Run a temperature scan from 25 to 95 °C (*see* **Note 12**).
5. Analyze data using instrument analysis software and choose the optimal enzyme concentration (*see* **Note 21**).

3.2.2 Characterization of Ligands

1. Prepare 1 mM ligand solution (*see* **Note 14**) + 10% (v/v) DMSO control sample (*see* **Note 15**) in assay buffer in a tube strip.
2. Prepare protein solution at 10/9th of final concentration in assay buffer sufficient for triplicate measurements at 10 μL for each sample (i.e., 480 μL is sufficient to test 15 different ligands + DMSO control).
3. Dispense 9 μL of protein into a 96-well PCR plate.
4. Transfer to each well containing protein 1 μL of ligand and DMSO control, respectively.
5. Centrifuge and cover the plate.
6. Incubate for 15 min at room temperature.
7. Fill capillaries and place into the instrument.
8. Run a temperature scan for the predetermined temperature range, at chosen ramp speed (*see* **Note 12**).
9. Analyze data using instrument analysis software (*see* **Note 21**).

   For further characterization, a dose–response analysis of selected ligands can be set up similarly as described in Subheading 3.1.3.

4 Notes

1. DSF experiments can be run with any real-time PCR machine with the appropriate spectral settings necessary for reporter dye signal detection (Table 1). The protocols in this chapter can be scaled to 96-well format with a total reaction volume of 100 μL.
2. White PCR plates are preferred over translucent plates to minimize signal leakage from neighboring wells. Some batches of polypropylene PCR plates are reported to induce a SYPRO Orange fluorescence signal. Always include "no protein"/

Table 1
Selected fluorescence reporter dyes used in DSF with their physical properties and tolerance for detergents in solution

Fluorophore	Excitation/ emission (nm)	Fluorescence increase due to	Detergent compatibility
bis-ANS	350/492	Hydrophobic milieu	Below CMC
Dapoxyl sulfonic acid	395/601	Hydrophobic milieu	Below CMC
CPM	384/470	Thiol-binding	Yes
DCVJ	450/505	Decreased viscosity	Yes
SYPRO Orange	470/570	Hydrophobic milieu	No
BODIPY FL L-cysteine	494/518	Thiol-binding	Yes
Nile red	530/635	Hydrophobic milieu	No

Bis-ANS (4,4′-bis(1-anilinonaphthalene 8-sulfonate)), CPM (7-diethylamino-3-(4′-maleimidylphenyl)-4-methylcoumarin), DCVJ (9-(2,2-Dicyanovinyl)julolidine)

"buffer only" control in each run to monitor the signal of the fluorophore in the absence of protein.

3. The selection of assay buffer is highly dependent on the target protein. Previously established buffer conditions for the target can serve as a starting point. If buffer conditions are not well established, we strongly recommend running a buffer and additive screen using DSF. Label-free DSF is the method of choice for buffer screening for several reasons: it is compatible with a wide range of buffer and additives and has low sample consumption, easy experimental setup, and short run times. Figure 2 shows an example of a buffer and detergent screen to improve protein stability. A single-pass transmembrane protein was purified in HEPES buffer, pH 8.0, and exhibited a tendency to aggregate. The switch to PBS, pH 7.4 greatly enhanced protein stability and solubility. Screening of commonly used detergents showed no improvement over PBS, and some of the detergents, i.e., CHAPS, caused destabilization of protein. Based on the result of this buffer screen, DSF and other in vitro experiments with this protein were performed in PBS. 0.05% Tween-20 was added for instances where detergent was needed, because it had the smallest effect on protein stability in PBS. Buffer additive screens are also useful for proteins with high melting points, where the stabilizing effect of a ligand could be hard to detect. Choosing a suboptimal condition for DSF can increase the window between the T_m of protein in the absence and presence of ligand.

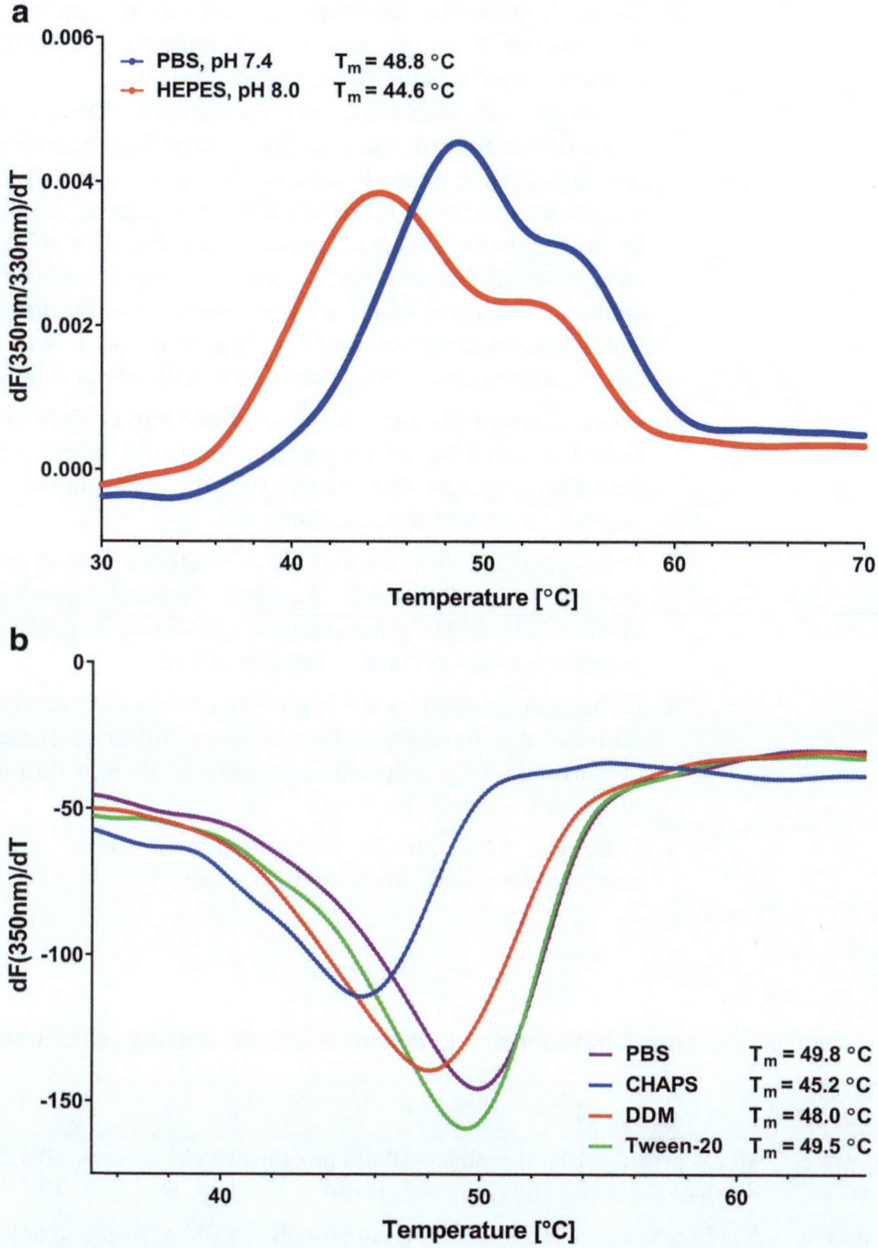

Fig. 2 The thermal stability of a single-pass transmembrane protein in different buffer conditions tested by label-free DSF. (**a**) Protein melting profile in PBS, pH 7.4 or in HEPES buffer, pH 8.0. (**b**) Screening of protein stability against common detergents at their critical micelle concentration in PBS, pH 7.4. Samples were run in triplicate at a concentration of 0.5 mg/mL with a temperature ramp speed of 1.0 °C/min

4. It is critical to use high-purity recombinant protein without carriers, such as bovine serum albumin (BSA), or contaminants, to avoid fluorescence signal interference.

5. Depending on the target protein and assay conditions, other dyes might be more suitable. A collection of fluorescence reporters used in DSF are listed in Table 1.

   In general, dyes which exhibit different fluorescence properties depending on the polarity of their local environment are less compatible with hydrophobic ligands, lipid-binding or membrane proteins. High initial fluorescence of the dye caused by hydrophobic ligands, proteins, or micelles of detergent, used to solubilize membrane proteins, can interfere with data analysis. The thiol-reactive dyes have their limitations, such as the pH dependence of thiol binding, interaction with reducing agents, and interference of the probe with a ligand binding site.

6. Most compounds are soluble in DMSO at 10 mM concentration. However, before preparing compound stocks, refer to the data sheet, or test the solubility of the compound and adjust the stock concentration accordingly.

7. It is recommended to include a reference ligand as positive control for a detectable T_m shift. Natural ligands, such as GDP/GTP, NAD^+/NADH, or previously reported ligands or peptides, can serve as a good reference.

8. If the protein stock is at a high concentration, it can be directly diluted into the assay buffer. If not, a buffer exchange can be performed using a desalting column or dialysis cup following the manufacturer's protocol.

9. Label-free DSF can be performed on "Uncle" and "Unit" instruments from Unchained Labs as well.

Table 2
Template protein and dye cross-titration matrix to determine optimal working concentrations

	1	2	3	4	5	6
A	10 μM protein 1× dye	10 μM protein 2× dye	10 μM protein 5× dye	10 μM protein 7.5× dye	10 μM protein 10× dye	10 μM protein 20× dye
B	5 μM protein 1× dye	5 μM protein 2× dye	5 μM protein 5× dye	5 μM protein 7.5× dye	5 μM protein 10× dye	5 μM protein 20× dye
C	2.5 μM protein 1× dye	2.5 μM protein 2× dye	2.5 μM protein 5× dye	2.5 μM protein 7.5× dye	2.5 μM protein 10× dye	2.5 μM protein 20× dye
D	1 μM protein 1× dye	1.0 μM protein 2× dye	1.0 μM protein 5× dye	1.0 μM protein 7.5× dye	1.0 μM protein 10× dye	1.0 μM protein 20× dye
E	0.5 μM protein 1× dye	0.5 μM protein 2× dye	0.5 μM protein 5× dye	0.5 μM protein 7.5× dye	0.5 μM protein 10× dye	0.5 μM protein 20× dye
F	Buffer 1× dye	Buffer 2× dye	Buffer 5× dye	Buffer 7.5× dye	Buffer 10× dye	Buffer 20× dye

10. Nanotemper Technologies supply two different capillaries with their instrument: standard and high sensitivity. Standard capillaries will work with the majority of targets, while it is recommended to use the high-sensitivity capillaries when working with protein concentrations ≤200 g/mL.
11. Optimizing the protein and dye concentrations is a critical first step in producing reliable and reproducible data. The first derivative of a thermogram in a DSF experiment using a dye should ideally produce a single, sharp peak. An example matrix to screen for the optimal protein-to-dye ratio is provided in Table 2.
12. For the initial protein melting experiment, it is recommended to choose a wide temperature range (i.e., 25–95 °C) to determine the T_m of the target protein. Subsequent experiments can have a narrowed temperature range based on the predetermined T_m. The choice of temperature ramp speed of the thermal shift assay is a compromise between the number of data points read at each temperature and the length of the assay. Once a ramp rate is chosen, it should be held constant, as the

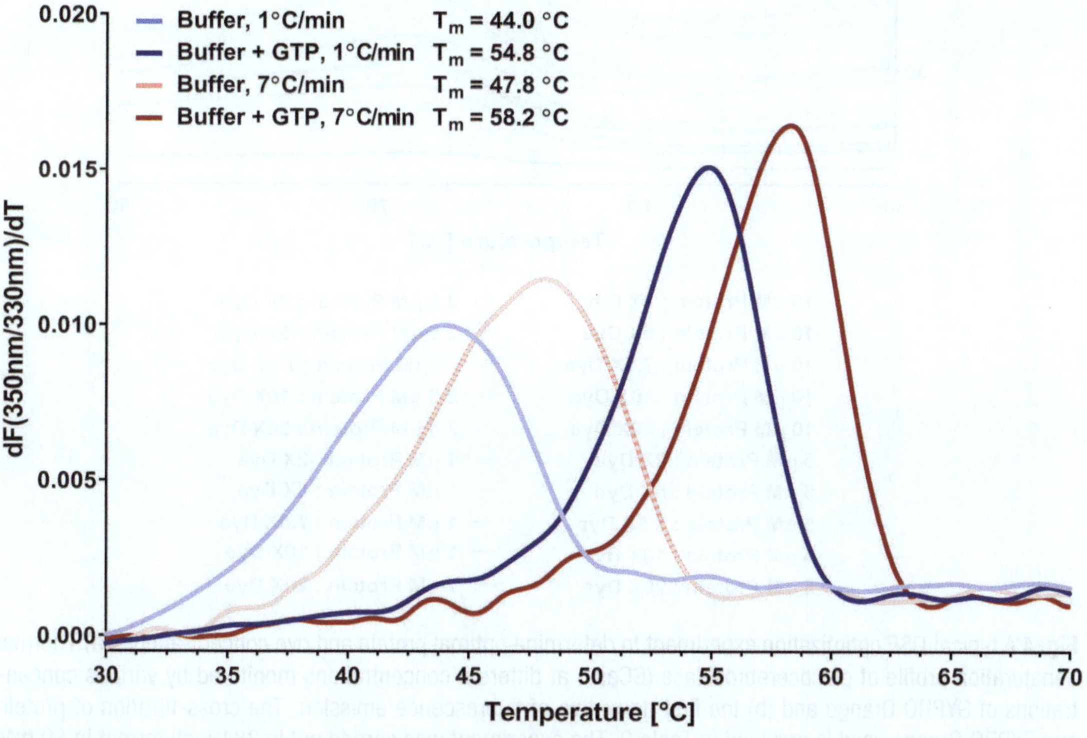

Fig. 3 Effect of different temperature ramp speed on T_m and ΔT_m. Thermal denaturation of a protein was monitored by label-free DSF in the absence or presence of its natural ligand GTP at (**a**) 1 °C/min or (**b**) 7 °C/min temperature ramp speeds. Samples were run in triplicate at a concentration of 0.25 mg/mL in assay buffer (25 mM HEPES, pH 7.5, 10 mM $MgCl_2$, 2 mM TCEP) with or without 100 μM GTP

T_m can change considerably depending on the speed of the temperature ramp. Figure 3 illustrates this phenomenon where the T_m of a protein is shifted with a faster ramp speed. However, the ΔT_m induced by the ligand binding remains the same.

13. The fastest way to analyze DSF data is to display the first derivative of the fluorescence signal over temperature and evaluate these curves. The ideal protein-to-dye ratio should produce a curve with a single, sharp peak. As an example, the raw

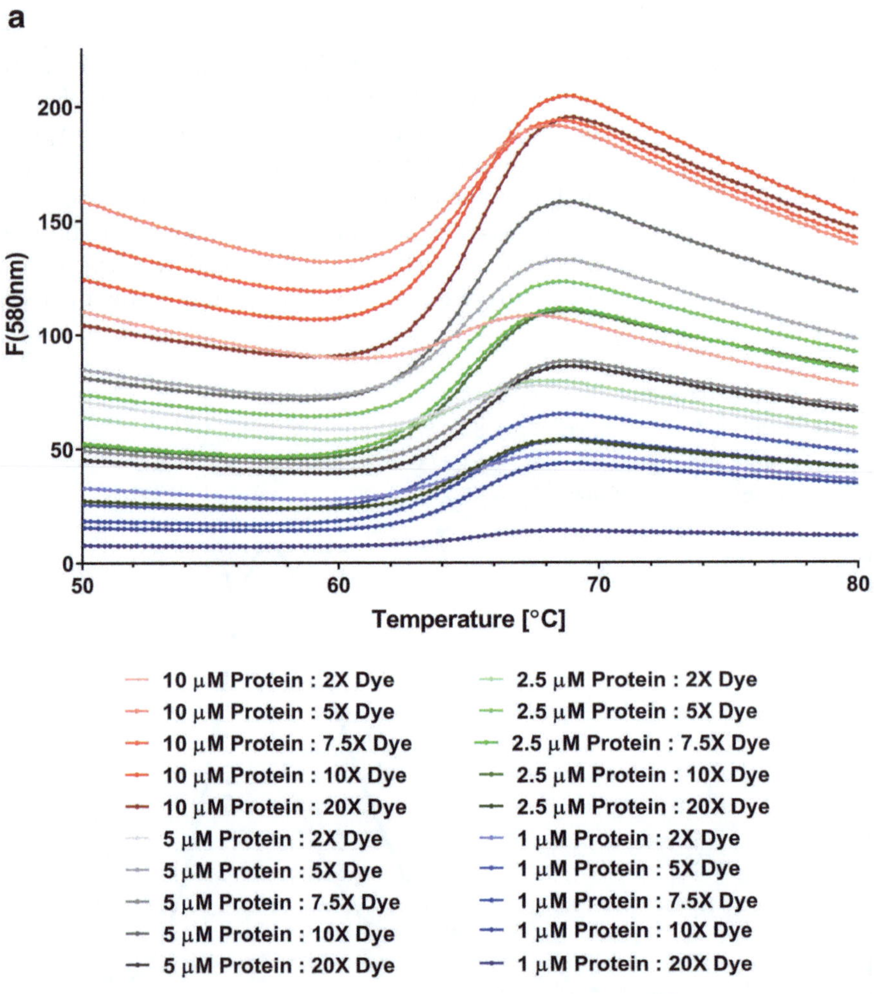

Fig. 4 A typical DSF optimization experiment to determine optimal protein and dye concentrations. (**a**) Thermal denaturation profile of glucocerebrosidase (GCase) at different concentrations monitored by various concentrations of SYPRO Orange and (**b**) the first derivative of fluorescence emission. The cross-titration of protein and SYPRO Orange used is provided in Table 2. The experiment was carried out in 384-well format in 50 mM citric acid, pH 5.9, with a temperature ramp speed of 0.11 °C/s and five acquisitions per °C. (**c**) The first derivative curves of GCase at 2.5 μM in the presence of different concentrations of SYPRO Orange are selected for better visualization and analysis to choose the best experimental conditions

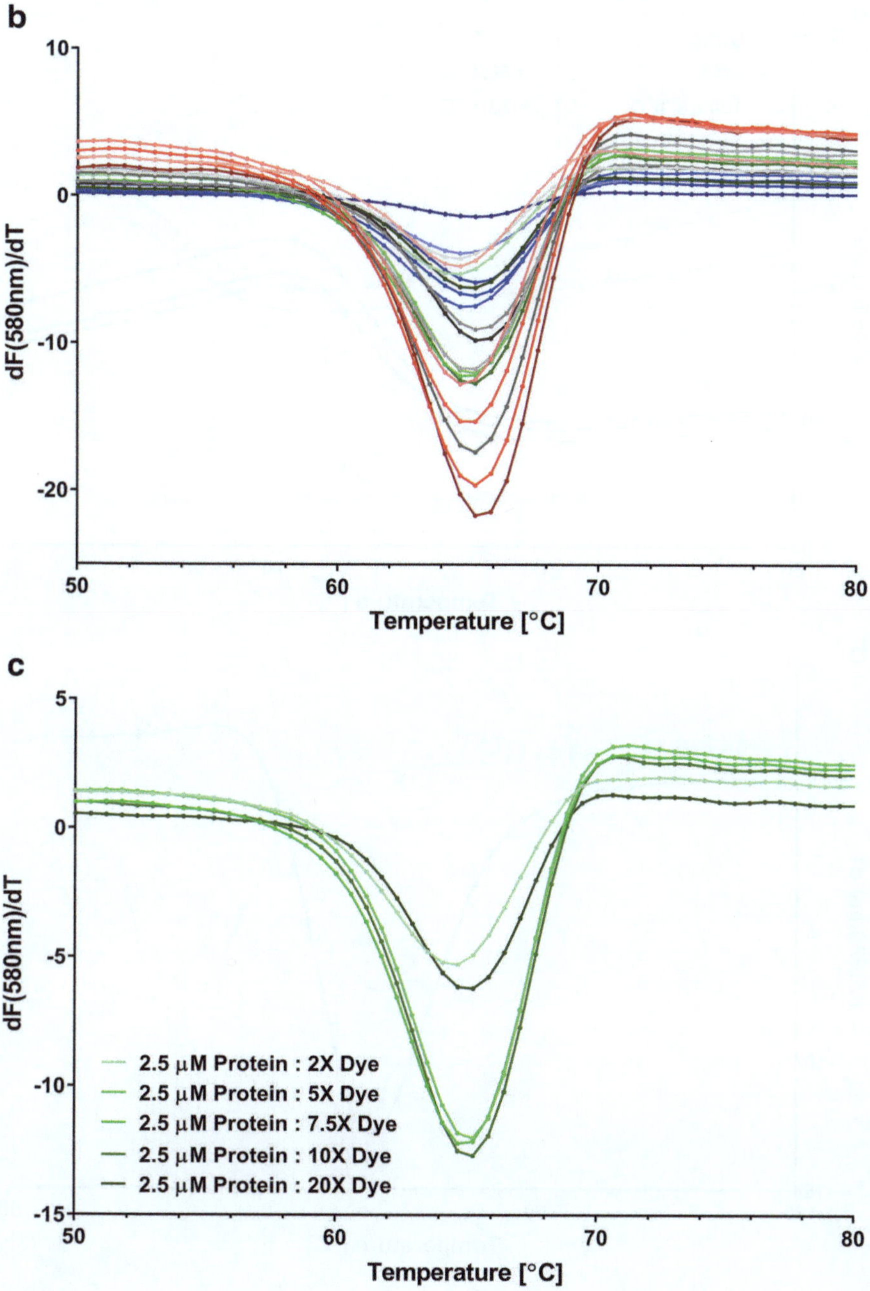

Fig. 4 (continued)

data from a cross-titration of glucocerebrosidase and SYPRO Orange, and the subsequent analysis to determine optimal protein-to-dye ratio, is shown in Fig. 4. First derivative of the fluorescence signal at all protein concentration, except at 1 μM, produce curves with a single, sharp peak (Fig. 4b). Further

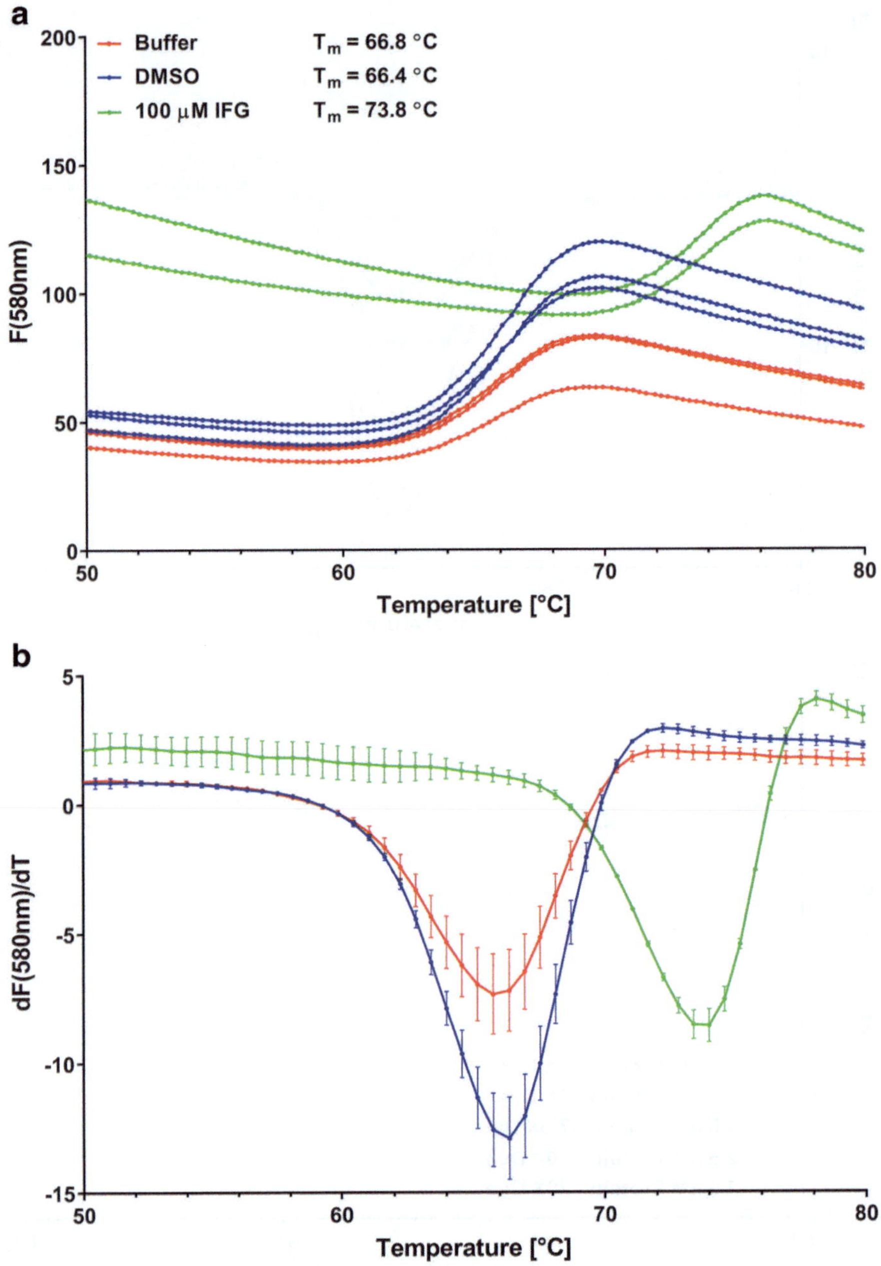

Fig. 5 A single-concentration DSF experiment to characterize protein–ligand interaction. (**a**) Melting profile of GCase in the presence of 100 μM isofagomine (IFG), diluent control (DMSO), or in assay buffer (50 mM citric acid, pH 5.9) and (**b**) the first derivative of fluorescence emission plotted as a function of temperature to determine T_m. For each condition, triplicate samples of 2.5 μM GCase and 5× SYPRO Orange were used

evaluation of the melting curves of glucocerebrosidase at 2.5 μM shows that 5× dilution of SYPRO Orange is sufficient for a reliable fluorescence signal (Fig. 4c).

14. The T_m shift caused by ligand binding has a simple thermodynamic meaning only when the receptor is saturated. At less than saturating ligand concentration, melting can appear to be multiphasic due to ligand redistribution. Current protocol for initial ligand testing at a single-concentration point proposes 100 μM of ligand (final concentration), but other concentrations can be chosen based on previously known information of the specific protein–ligand interaction from other functional assays and/or ligand solubility. Our recommended option is to set up the initial ligand testing experiment at two ligand concentrations, e.g., 20 and 100 μM to address the above concern. Figure 5 provides an example of a single-concentration DSF experiment monitoring the stabilization of glucocerebrosidase by a known inhibitor isofagomine (IFG). Here, 100 μM isofagomine stabilizes glucocerebrosidase by 7.4 °C. Though this significant ΔT_m suggests that there is an enzyme–ligand interaction, an experiment using a range of concentrations will yield an appropriate verification (*see* **Note 16**).
15. It is critical to include a DMSO control in the experiments to rule out T_m shifts that may arise from DMSO interactions, as DMSO may perturb the structure of a protein and influence its thermal and chemical stability. As a general rule, the final DMSO concentration should not exceed 2% (v/v).
16. SYPRO Orange is a proprietary dye, supplied as a concentrated stock solution in DMSO, most commonly at 5000×. The combined total concentration of DMSO in the reaction contributed by the SYPRO Orange stock, as well as by the compound, has to be taken into account.
17. Real-time PCR instruments from certain vendors, such as Applied Biosystems and Roche, are accompanied by software for thermal shift data analysis. These software packages can automate the rank ordering of compounds and perform mode of action analysis of the DSF data. If DSF-specific software is not available, raw thermograms and the first derivatives of those curves can be analyzed in Microsoft Excel or any data analysis software, such as GraphPad Prism using the Boltzmann equation, or calculating the first derivative. Independent of how the data will be analyzed, we recommend to examine the raw thermal graphs manually for secondary transitions, bumps and other noises.
18. To determine and validate a "true" binder which induced a T_m shift in the single-point experiment, it is crucial to rescreen the ligand in a dose–response experiment. An example of one such experiment is shown in Fig. 6, where the effect of different

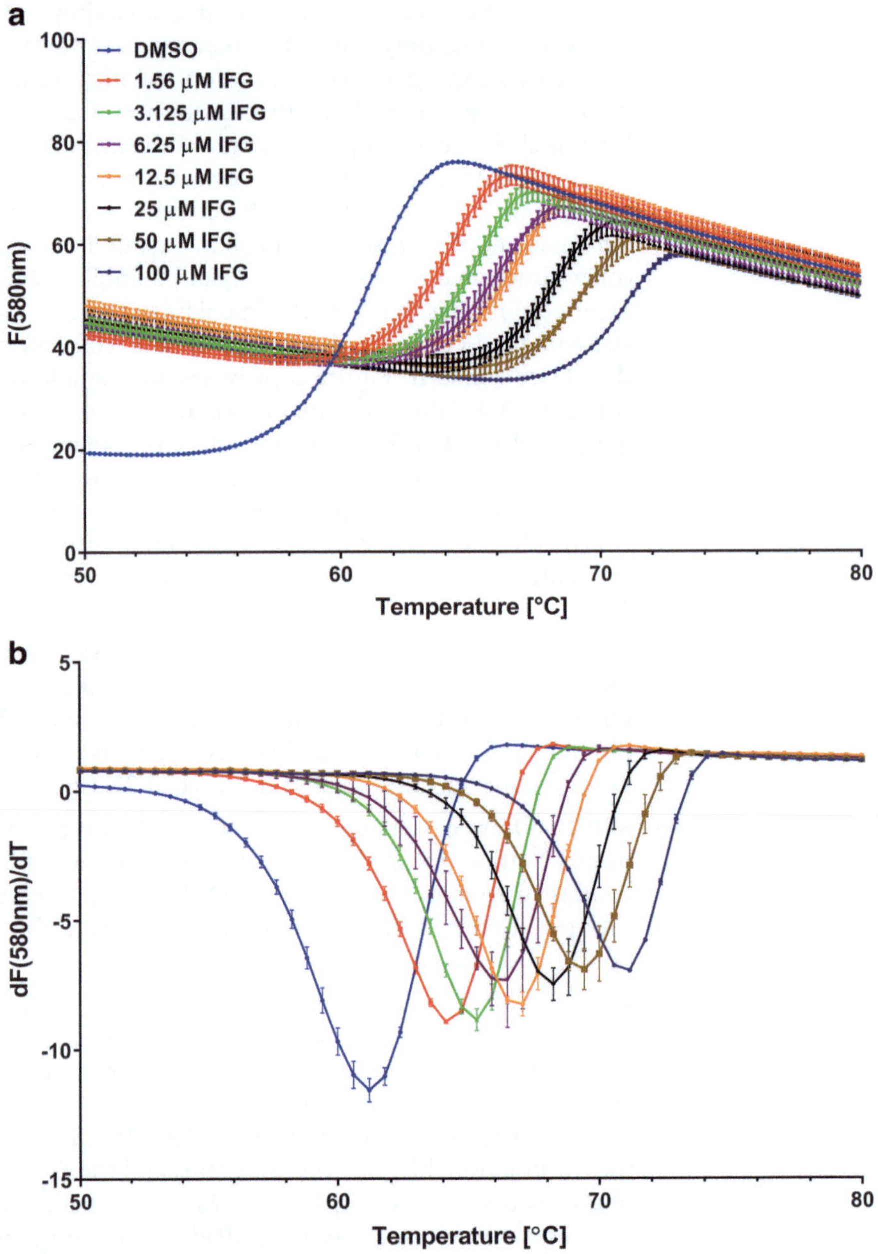

Fig. 6 Concentration-dependent stabilization of GCase by IFG: (**a**) melting profile of GCase in the presence of increasing concentrations of IFG, (**b**) first derivative of fluorescence emission, (**c**) T_m values of GCase in the presence of given IFG concentrations, and (**d**) exponential curve fit to ΔT_m values plotted against the ligand concentration

concentrations of isofagomine on the stability of glucocerebrosidase has been investigated. Protein stability (Fig. 6b, c) and ΔT_m (Fig. 6d) are increased with the increasing ligand

c

IFG (μM)	T_m (C)
0	60.9
1.56	63.9
3.125	64.9
6.25	65.7
12.5	66.6
25	67.9
50	69.2
100	70.6

d

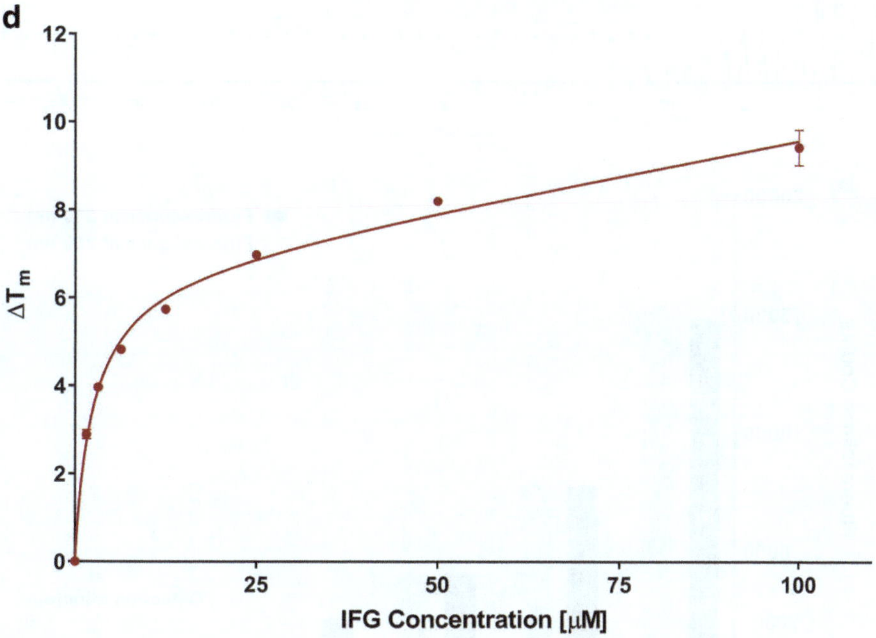

Fig. 6 (continued)

concentration, confirming the ligand-induced stabilization that was indicated in a single-dose experiment.

19. Ten microliters of sample are required for each Prometheus NT.48 capillary (instruments from Unchained Labs have the same sample size requirements), and capillaries are filled by dipping into sample and automatically filling by capillary forces. It is important to remove any bubbles present in the solution, and ensure that no solution is present on the outside of the capillary.

20. Examining initial fluorescence readings of the samples and choosing the correct protein concentration is necessary to obtain high-quality, reproducible data. Each instrument will

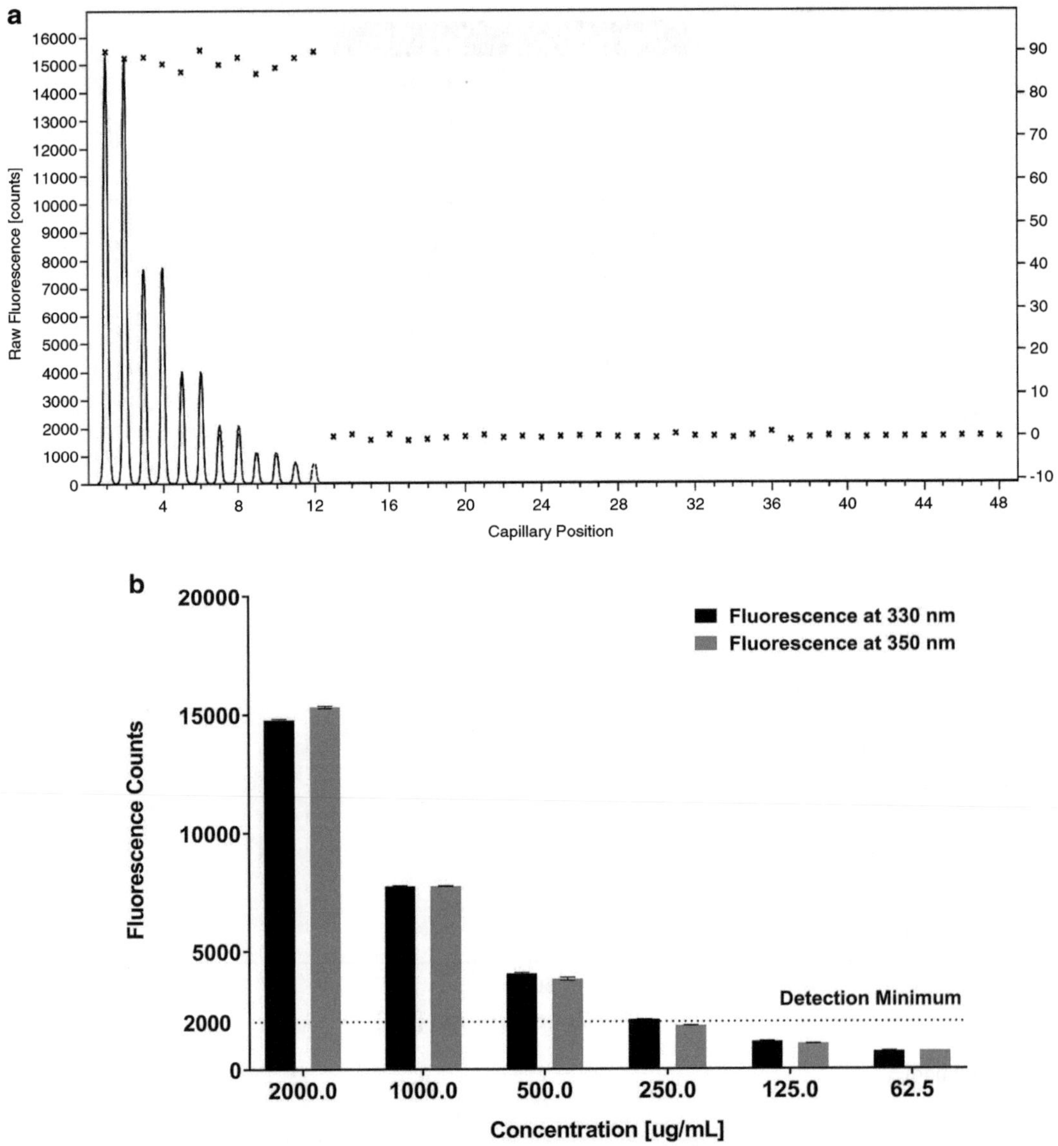

Fig. 7 A typical label-free DSF optimization experiment using the Prometheus NT.48 instrument. (**a**) Screenshot of Prometheus control software showing the initial fluorescence read of a protein titration at 25% LED power. (**b**) Plot of the initial fluorescence signal at 330 and 350 nm. The recommended detection minimum for the Prometheus (~2000 counts) is given by the dotted line

have unique detection limits; for the Prometheus NT.48 instrument, the recommended minimum fluorescence signal for reliable data collection is 2000 fluorescence counts. The LED power of the instrument can be varied to increase fluorescence counts at a given protein concentration, and it is recommended to use as low a LED power as your material requirements allow. Prometheus NT.48 detects fluorescence

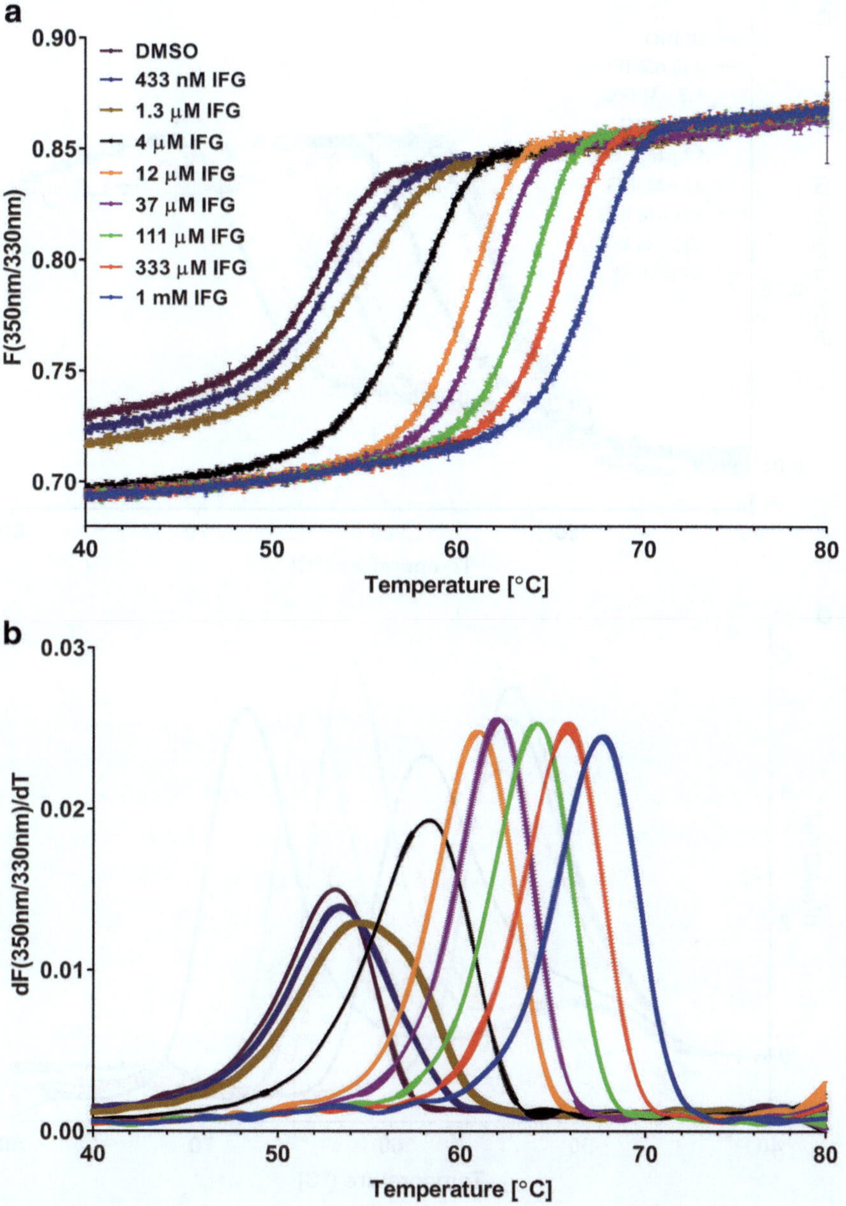

Fig. 8 Label-free DSF experiment examining the stabilization of GCase by IFG in dose–response: (**a**) melting profile of GCase derived from the fluorescence emission ratio of 350 to 330 nm in the presence of given IFG concentration, and (**b**) the first derivative of these traces. Simultaneous with the fluorescence emission, the Prometheus NT.48 collects protein aggregation data through a backscattering sensor: (**c**) thermal aggregation profile of GCase in the presence of IFG at given concentration and (**d**) the first derivative of these traces. (**e**) A summary of T_m and T_{agg} values determined from above data. (**f**) An exponential fit curve to the ΔT_m values plotted against the ligand concentration. Samples were run in triplicate at a protein concentration of 4 μM (0.25 mg/mL) in 50 mM citric acid, pH 5.9, with a temperature ramp of 1.0 °C/min

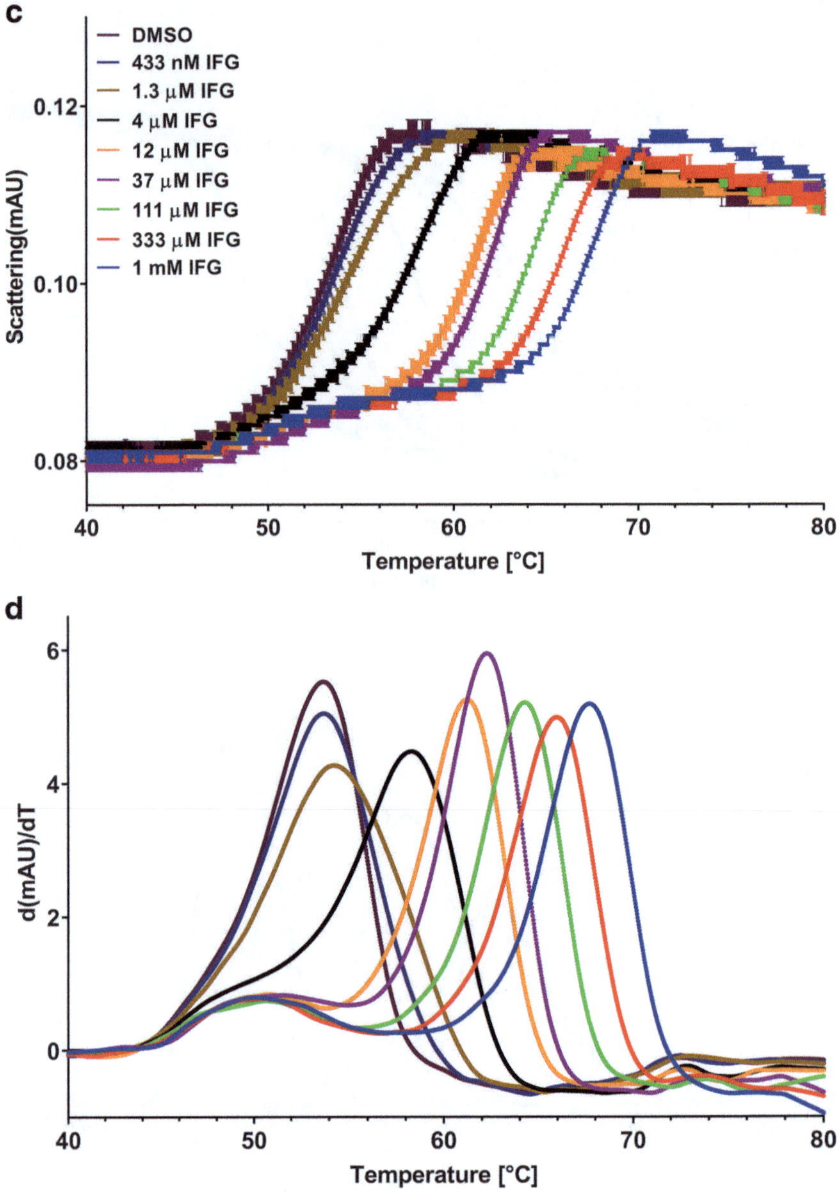

Fig. 8 (continued)

intensity at two wavelengths, 330 and 350 nm, and, when equipped, can measure back-scattering to assess protein aggregation. A screenshot of initial fluorescence reading of a protein at different concentrations reported by Prometheus software is shown in Fig. 7a. All measurements taken during this initial reading at 25% LED power are summarized in Fig. 7b. The protein sample at 0.5 mg/mL concentration provides adequate signal for data collection with low material consumption.

e

IFG (μM)	T_m (°C)	T_{agg} (°C)
0	53.0	53.1
0.433	53.4	53.5
1.3	54.7	54.9
4	58.1	58.3
12	60.9	61.2
37	62.0	62.2
111	64.0	64.3
333	65.7	65.8
1000	67.6	67.7

f

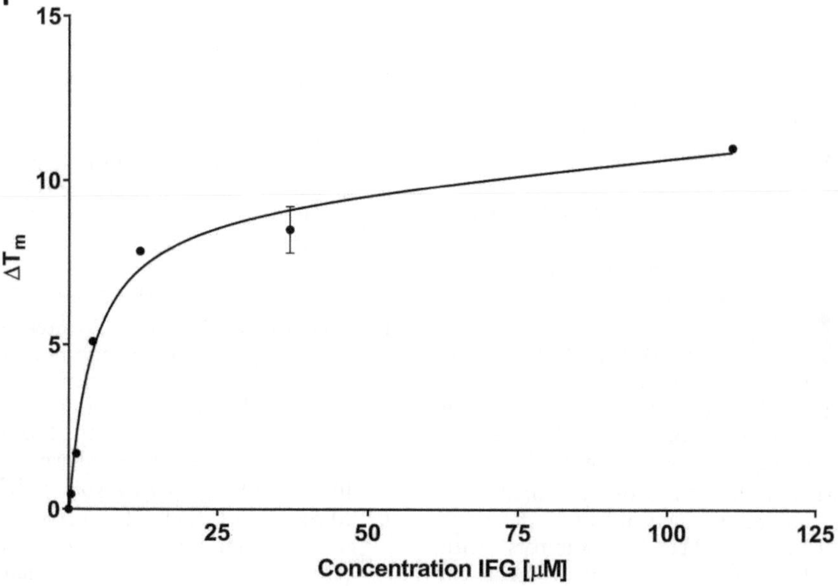

Fig. 8 (continued)

21. As mentioned above, the intrinsic fluorescence of a protein is measured by Prometheus NT.48 at two wavelengths: 330 and 350 nm. The instrument software reports the tryptophan fluorescence emission shift during thermal unfolding at individual wavelengths and at a ratio between 350 and 330 nm. The simplest method to analyze the data is to compare T_m values obtained from the first derivative of the 350 nm/330 nm ratio. Typically, the 350 nm/330 nm ratio gives a well-defined transition even if the single wavelengths do not exhibit clear unfolding curve. However, depending on the target and experimental conditions, a single-wavelength curve can be used for data analysis if it provides cleaner signal than the ratio. Prometheus software calculates and reports protein unfolding onset

temperature (T_{onset}) and aggregation temperature (T_{agg}), as well. Representative thermal unfolding and aggregation curves of glucocerebrosidase in the presence of different concentrations of its inhibitor isofagomine monitored by label-free DSF, data analysis, and summary are presented in Fig. 8.

References

1. Scott AD (2018) Chapter 8 Fluorescent thermal shift assays for identifying small molecule ligands. In: Canales A (ed) Biophysical techniques in drug discovery. The Royal Society of Chemistry, Cambridge, pp 208–238. https://doi.org/10.1039/9781788010016-00208
2. Simeonov A (2013) Recent developments in the use of differential scanning fluorometry in protein and small molecule discovery and characterization. Expert Opin Drug Discov 8:1071–1082. https://doi.org/10.1517/17460441.2013.806479
3. Teale FW, Weber G (1957) Ultraviolet fluorescence of the aromatic amino acids. Biochem J 65:476–482
4. Privalov PL (1979) Stability of proteins: small globular proteins. Adv Protein Chem 33:167–241
5. Niesen FH, Berglund H, Vedadi M (2007) The use of differential scanning fluorimetry to detect ligand interactions that promote protein stability. Nature Protocols 2:2212–2221. https://doi.org/10.1038/nprot.2007.321
6. Yang S-M et al (2018) Discovery of orally bioavailable, quinoline-based aldehyde dehydrogenase 1A1 (ALDH1A1) inhibitors with potent cellular activity. Journal of medicinal chemistry 61:4883–4903. https://doi.org/10.1021/acs.jmedchem.8b00270
7. Lea WA, Simeonov A (2012) Differential scanning fluorometry signatures as indicators of enzyme inhibitor mode of action: case study of glutathione S-transferase. PLoS One 7: e36219. https://doi.org/10.1371/journal.pone.0036219
8. Senisterra G, Chau I, Vedadi M (2012) Thermal denaturation assays in chemical biology. Assay Drug Dev Technol 10:128–136. https://doi.org/10.1089/adt.2011.0390
9. Vandecaetsbeek I, Vangheluwe P (2016) Time-dependent protein thermostability assay. Methods Mol Biol 1377:79–85. https://doi.org/10.1007/978-1-4939-3179-8_9
10. Moggridge J, Biggar K, Dawson N, Storey KB (2017) Sensitive detection of immunoglobulin G stability using in real-time isothermal differential scanning fluorimetry: determinants of protein stability for antibody-based therapeutics. Technol Cancer Res Treat. doi: https://doi.org/10.1177/1533034617714149.
11. Svilenov H, Markoja U, Winter G (2018) Isothermal chemical denaturation as a complementary tool to overcome limitations of thermal differential scanning fluorimetry in predicting physical stability of protein formulations. Eur J Pharm Biopharm 125:106–113. https://doi.org/10.1016/j.ejpb.2018.01.004
12. Martinez NJ et al (2018) A widely-applicable high-throughput cellular thermal shift assay (CETSA) using split Nano Luciferase. Scientific reports 8:9472–9472. https://doi.org/10.1038/s41598-018-27834-y
13. Molina DM et al (2013) Monitoring drug target engagement in cells and tissues using the cellular thermal shift assay. Science 341:84–87. https://doi.org/10.1126/science.1233606

Chapter 5

High-Throughput Differential Scanning Fluorimetry of GFP-Tagged Proteins

Alanna E. Sorenson and Patrick M. Schaeffer

Abstract

Differential scanning fluorimetry is useful for a wide variety of applications including characterization of protein function, structure–activity relationships, drug screening, and optimization of buffer conditions for protein purification, enzyme activity, and crystallization. A limitation of classic differential scanning fluorimetry is its reliance on highly purified protein samples. This limitation is overcome through differential scanning fluorimetry of GFP-tagged proteins (DSF-GTP). DSF-GTP specifically measures the unfolding and aggregation of a target protein fused to GFP through its proximal perturbation effects on GFP fluorescence. As a result of this unique principle, DSF-GTP can specifically measure the thermal stability of a target protein in the presence of other proteins. Additionally, the GFP provides a unique in-assay quality control measure. Here, we describe the workflow, steps, and important considerations for executing a DSF-GTP experiment in a 96-well plate format.

Key words Fluorimetry, High-throughput screening, Thermal shift assay, Green fluorescent protein, Ligand binding, Enzyme inhibitors, Selective protein unfolding, Drug discovery

1 Introduction

Standard differential scanning fluorimetry techniques have been used to screen compounds for activity against target enzymes in drug discovery programs [1]. However, these techniques require solvatochromic dyes, such as 1,8-ANS and SYPRO Orange, which can interfere with additives, ligands, and the target protein. Differential scanning fluorimetry of GFP-tagged proteins (DSF-GTP) does not require any dye, making it ideal for studying protein-ligand interactions [2]. DSF-GTP is simple, fast, robust, and insensitive to variations in reaction volumes, with a useful temperature and pH range of 25–75 °C and 5–11, respectively. The thermal stability of a target protein is obtained within 1–2 h by simply subjecting a GFP-tagged protein in a given condition to a standard melt curve protocol in a real-time thermal cycler (Fig. 1) [2–6]. DSF-GTP has been applied for the high-throughput screening

Nikolaos E. Labrou (ed.), *Targeting Enzymes for Pharmaceutical Development: Methods and Protocols*, Methods in Molecular Biology, vol. 2089, https://doi.org/10.1007/978-1-0716-0163-1_5, © Springer Science+Business Media, LLC 2020

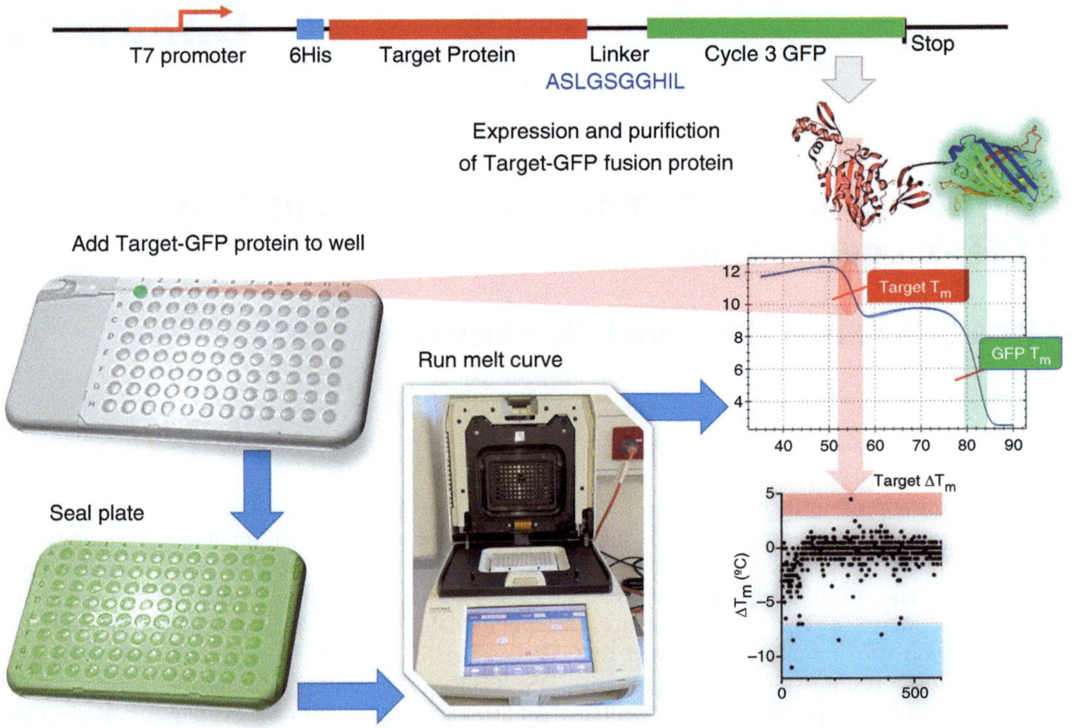

Fig. 1 Standard HT-DSF-GTP workflow. The target protein sequence is cloned into a suitable GFP vector with a flexible 10 aa linker (example linker shown). Target-GFP protein is expressed and purified. Target-GFP protein is added to a well with or without a test compound. The plate is sealed and a melt curve program is run in a real-time thermal cycler. The transition midpoints (T_m) for the target protein and GFP are determined. The T_m of GFP provides an internal quality control. The change in T_m (ΔT_m) induced by the binding of a test compound is determined relative to the T_m of free target protein. Compounds can stabilize or destabilize the target protein, resulting in a positive (highlighted in red) or negative (highlighted in blue) ΔT_m

of optimal buffer conditions such as pH and salt, the identification of protein-specific substrates, cofactors, and other complex ligands (e.g., DNA and RNA), the screening of small molecule libraries as well as structure–activity relationships and off-target effects, and the functional characterization of proteins [2–6]. DSF-GTP is the only DSF method that can selectively measure the transition midpoint (T_m) of a target protein in the presence of other proteins, enabling assessment of target engagement and off-target serum interaction studies [7]. Additionally, the C-terminal GFP tag provides a unique intra- and inter-assay quality control reference GFP fluorescence T_m peak that is unique to this method and allows detection of buffer effects and inconsistencies as well as nonspecific ligand effects.

2 Materials

2.1 Cloning of Target Protein in a GFP Vector

1. pET-based GFP vector for cloning and expression of a target protein with N-terminal hexahistidine (6-His) tag, inter-protein linker, and C-terminal GFP (*see* **Note 1**, Fig. 1).
2. Single Step (KRX) Competent Cells (*see* **Note 2**). Alternatively, competent *E. coli* cloning strains such as DH12s and DH5α for vector amplification.
3. Appropriate antibiotics for selective growth media (e.g., ampicillin and chloramphenicol).
4. Luria-Bertani (LB) or SOC broth [8].
5. LB agar plates supplemented with (0.4% w/v) glucose and antibiotic.
6. Water bath (for transformation) and incubators.

2.2 Cell Culture for Protein Expression

Protein expression requires the fresh transformation of a pET-based GFP vector containing the target coding sequence in KRX or an alternative T7 expression strain such as BL21(DE3) or derivative thereof [9, 10].

1. Appropriate antibiotics for selective growth media (e.g., ampicillin and chloramphenicol).
2. Appropriate inducer (e.g., rhamnose, IPTG, or lactose, *see* **Note 3**).
3. LB broth and LB agar plates supplemented with (0.4% w/v) glucose and antibiotic [8].
4. Terrific broth (TB) or auto-induction medium [11] supplemented with antibiotic.
5. A refrigerated centrifuge capable of processing 50 mL polypropylene tubes at >2000 × *g*.
6. A laboratory refrigerated incubator shaker suitable for 100 mL and 1 L conical flasks.

2.3 Purification

1. Refrigerated high-speed centrifuge and rotors (40,000 × *g*) for polypropylene/polycarbonate tubes up to 50 mL (depending on culture lysate volume).
2. Cell lysis system (e.g., French pressure cell press, *see* **Note 4**).
3. Gravity flow chromatography columns depending on cell lysate volume.
4. SDS-PAGE system and reagents to assess protein expression, solubility, and purity.
5. Lysis buffer: 50 mM sodium phosphate (pH 7.8), 300 mM NaCl, 10 mM imidazole (optional), 2 mM β-mercaptoethanol (optional), and 10% glycerol v/v (optional).

6. Elution buffer: 50 mM sodium phosphate (pH 7.8), 300 mM NaCl, 200 mM imidazole, 2 mM β-mercaptoethanol (optional), and 10% glycerol v/v (optional).
7. Immobilized metal affinity chromatography (IMAC) nickel resin (e.g., BioRad Profinity IMAC nickel-charged resin).
8. Ammonium sulfate for protein precipitation (optional).
9. Appropriate resuspension buffer (dependent on target protein requirements for solubility and stability if known, or lysis buffer if requirements are unknown).
10. Spectrophotometer and protein quantification assay (e.g., Bradford and BCA).

2.4 Transition Midpoint (T_m) Peak Determination

1. Real-time thermal cycler capable of performing melt curve analyses (e.g., Bio-Rad CFX96 or equivalent).
2. Clear 96-well PCR plates, tubes, or strips.
3. PCR plate sealing films, caps, or cap strips.
4. Soluble GFP-tagged target protein ($\geq$1 μM in resuspension buffer).
5. Assay buffer at 2× concentration (dependent on target protein requirements for activity if known, or lysis buffer if unknown, *see* **Note 5**).
6. Test reagents (*see* **Note 5**).
7. Compound screening: hard-shell clear 96-well skirted PCR plates and seals for high-throughput screening (*see* **Note 6**).
8. Compounds (5 mM in DMSO).

3 Methods

3.1 Cloning of Target Protein Coding Sequence in a GFP Fusion Vector

The pET-uvGFP vector (pIM013) [12] enables the directional cloning of a target protein coding sequence without a stop codon flanked with 5′ EcoRV and 3′ NheI sites and its expression as a C-terminal GFP fusion protein. The pIM013 vector is available on request.

1. Obtain target protein coding sequence that is codon optimized for *E. coli* (optional, *see* **Note 7**).
2. Clone target protein coding sequence into a GFP expression vector such as pIM013 [12].
3. Transform the recombinant vector in KRX cells and select transformants on LB agar supplemented with glucose and ampicillin (*see* **Note 8**).
4. It is good practice to confirm presence and integrity of target sequence before storage and use (optional, *see* **Note 9**).

3.2 Small-Scale Cultures for Evaluation of Protein Expression in KRX Cells

We advise to first assess a few clones for GFP fusion protein expression in soluble and fluorescent form using a small-scale auto-induction system culture to identify the highest yielding colonies.

1. Always freshly transform competent KRX cells with recombinant expression vector.
2. Select transformants (i.e., single colonies) on LB glucose + ampicillin plates (*see* **Note 10**). Important: re-plate the selected colonies on a fresh master LB glucose + ampicillin plate for large-scale protein expression and temporary storage.
3. Inoculate a small-scale test culture with a single colony (10 mL TB media + ampicillin in a 100 mL flask) and incubate at 37 °C with shaking at 200 RPM until OD_{600} reaches ~1.
4. Take a preinduction sample, centrifuge, and combine with SDS-PAGE sample buffer for subsequent target protein production assessment.
5. To the remaining culture, add rhamnose to 0.1% (w/v) final concentration and reduce temperature of shaker to 16 °C (*see* **Note 11**) with shaking at 200 RPM for 2 days.
6. Take a post-induction sample as per **step 4** (*see* **Note 12**).
7. Harvest cells by centrifugation (*see* **Note 13**) and discard supernatant (optional: freeze the pellet for small-scale purification if desired).

3.3 Large-Scale Protein Expression

Identification of the best producing clone is described in Subheading 3.2. The colony is selected from a fresh LB glucose + ampicillin replica plate to inoculate the culture. A 100 mL culture yields sufficient proteins for at least 1000 DSF-GTP reactions.

1. Inoculate 100 mL TB media + ampicillin in a 1 L flask with a full inoculation loop of bacteria that was scrapped off an overnight master LB glucose + ampicillin plate and incubate at 37 °C with shaking at 200 RPM until OD_{600} reaches 1–1.5.
2. Add rhamnose to 0.1% final concentration (or perform autoinduction) and reduce temperature of shaker to 16 °C with shaking at 200 RPM for 2 days (*see* **Note 11**).
3. Harvest cells by centrifugation and discard supernatant (*see* **Note 14**).

3.4 Protein Purification

All purification steps are performed on ice or at 4 °C.

1. Resuspend cell pellet with 7 mL of ice-cold lysis buffer per gram of wet cell paste.
2. Lyse cells by passing the cell suspension twice through an ice-cold French pressure cell with the French press set at 10,000–12,000 psi. For optimal cell lysis, we slowly open the valve of the pressure cell to allow the cell suspension to drip out

at a rate of one to two drops per second (*see* **Note 4**). Set aside a small aliquot.

3. Clarify lysate by centrifugation at 40,000 × *g* for 20–40 min at 4 °C. Keep the supernatant and set aside a small aliquot.
4. Load the clarified lysate into a gravity column packed with 1–2 mL of IMAC nickel resin or a similar resin. Follow the manufacturer's guidelines for purification (*see* **Note 15**). Using this setup, we generally recover and combine about 4–8 mL of GFP fusion protein containing fractions and keep a small aliquot for each fraction to assess protein purity and integrity by SDS-PAGE (*see* **Note 16**).
5. Fractions containing GFP fusion proteins can be precipitated with 0.5 g/mL ammonium sulfate on ice with gentle mixing for 1 h (optional).
6. Protein precipitate is recovered by centrifugation at 40,000 × *g* for 20–40 min at 4 °C and supernatant is checked for GFP fluorescence before discarding it (optional: only necessary if **step 5** is performed).
7. Resuspend protein pellet in lysis buffer or alternative buffer suitable for the stability of the target protein at a final concentration of 50–100 μM (optional: only necessary if **steps 5** and **6** are performed) (*see* **Note 17**).
8. Further purification and dialysis can be performed if desired (*see* **Note 18**).
9. Determine protein concentration (*see* **Note 19**).

3.5 Validation of Target-GFP Fusion Protein by DSF-GTP

Preliminary screens can be run with any fraction containing the Target-GFP protein, i.e., straight after elution from the column. Initial T_m determination can be performed with as little as 20–50 μL of a 1–3 μM Target-GFP protein solution corresponding to an initial RFU value of 4000–8000 (*see* **Note 19**). Further buffer optimization can then be performed to identify ideal buffer conditions. If no T_m is apparent, the protein may be in an unstable or nonhomogenous oligomeric form and may require the addition of cofactors (*see* **Note 20**) to produce a homogenous complex capable of producing a T_m peak [2]. Thus, for proteins with known or suspected ligands and/or cofactors, reactions should be supplemented with specific ligands and/or cofactors at concentrations ranging from 1 to 10 mM. Alternatively, the protein may have a T_m greater than that of GFP which might require the addition of destabilizing molecules to lower the T_m of the target protein.

1. Reactions containing 1–3 μM Target-GFP protein equivalent to initial RFUs ~4000–8000 in elution buffer (20–50 μL final volume) are set up in 96-well plates and sealed with PCR plate sealing film (*see* **Note 21**). At this stage, the effect of a cofactor or ligand can be tested.

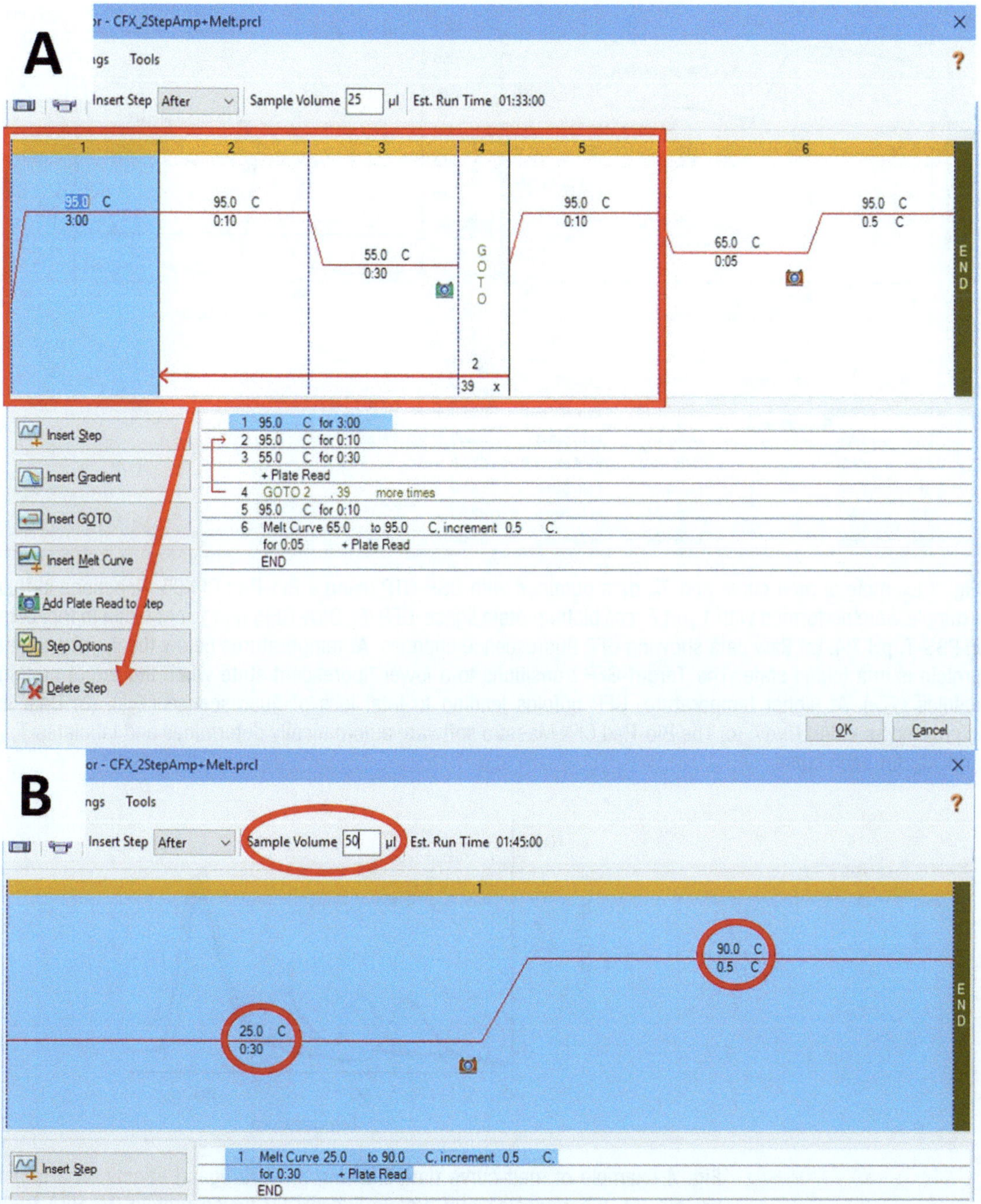

Fig. 2 Example of melt curve setup for DSF-GTP using Bio-Rad CFX-96. (**a**) After selecting a generic qPCR protocol with melt curve, the preceding PCR steps are deleted (red box and arrow). (**b**) The melt curve can then be edited to suit the DSF-GTP run by changing the reaction volume, start and end temperature, and dwell time (red circles). The FAM channel is selected with DSF-GTP and other channels can be deselected. The plate template can be exported and then imported with the sample identification if required (most useful for drug screening where 94 compounds are tested per run)

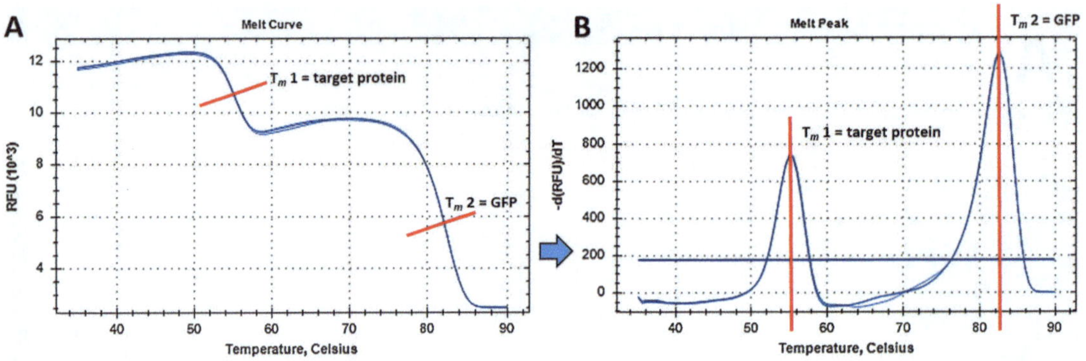

C

Well	Fluor	Target	Content	Sample	Melt Temp	
D04	FAM (Reaction 1)		Unkn-04	1mM biotin	82.50	→ GFP T_m 2
D04	FAM (Reaction 1)		Unkn-04	1mM biotin	55.00	→ Target T_m 1
D05	FAM		Unkn-04	1mM biotin	82.50	
D05	FAM		Unkn-04	1mM biotin	55.00	
D06	FAM		Unkn-04	1mM biotin	82.50	
D06	FAM		Unkn-04	1mM biotin	55.00	

Fig. 3 Example of melt curve and T_m data obtained with DSF-GTP using a Bio-Rad CFX96. Reactions in this example were performed with 1 μM *E. coli* biotin protein ligase-GFP (Ec BirA-GFP) and 1 mM biotin in triplicate in PBS-T, pH 7.4. (**a**) Raw data showing GFP fluorescence changes. At temperatures below the T_m, the target protein is in a folded state. The Target-GFP transitions to a lower fluorescent state when the target protein unfolds (T_{m1}). At higher temperature, GFP unfolds leading to total loss of fluorescence (T_{m2}). (**b**) Data is expressed as −d(RFU)/dT. (**c**) The Bio-Rad CFX Maestro software automatically determines and tabulates T_{m1} and T_{m2} for each curve

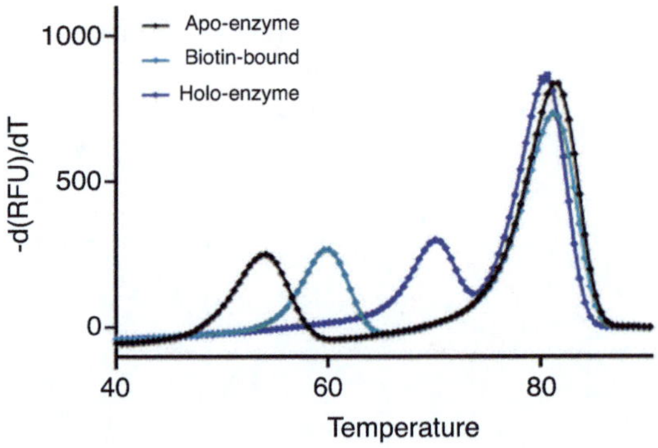

Fig. 4 Example of melt-curve data showing positive ligand effects on a target protein. *M. tuberculosis* biotin protein ligase-GFP fusion protein (Mt BirA-GFP) with and without important ligands and cofactors. The melt curves show an increase in T_{m1} of Mt BirA-GFP when combined with its ligand and cofactor. Reactions in this example were run in PBS-T with 1 μM Mt BirA-GFP. Data are presented using GraphPad Prism 7

2. Set up a melt curve program ranging from 25 to 90 °C with 0.5 °C increments for 30 s in a suitable real-time PCR thermal cycler (*see* **Note 22** and Fig. 2).
3. Run the melt curve program with the FAM channel selected (*see* **Note 23**).
4. T_m curves are obtained and T_m values are recorded for each reaction (*see* **Note 24** and Fig. 3). All data and graphs can be exported and processed further (*see* Fig. 4).

3.6 High-Throughput Screening by DSF-GTP

DSF-GTP can be used to screen for compounds that interact directly with a target protein in 96-well plate format (*see* **Note 6**). The T_m value of GFP provides a unique in-assay quality control measure which is useful to detect interferences or nonspecific interactions that can occur with a given compound or test condition. Applications such as hit or ligand identification and structure–activity relationship investigation are within the scope of DSF-GTP. Of special note, DSF-GTP has been shown to be the only DSF method suited for the detection of off-target effects such as nonspecific binding or neutralization of a compound in a complex biological sample such as serum or cell lysate [7]. Below, we provide an example of high-throughput experimental setup that was applied to BirA-GFP [7].

1. Target-GFP fusion protein is diluted to 1 μM in a suitable diluent buffer. For a 96-well plate, 5 mL of protein solution at 1 μM will be required.
2. Compounds are distributed in the 96-well PCR plate in identical solvent and concentration (e.g., 1 μL of a 5 mM solution of compound in DMSO).
3. A volume of 49 μL diluted protein is aliquoted and mixed in each well containing 1 μL of a 5 mM solution of compound in DMSO. Two control wells contain 1 μL of pure DMSO (*see* **Note 25**).
4. Seal and equilibrate the plate at room temperature for 10 min.
5. Run a melt curve and export data as described in Subheading 3.5.
6. Determine ΔT_m by subtracting the T_m of the fusion protein alone from T_m of the fusion protein with compound in identical buffer conditions.
7. Determine an appropriate $\pm\Delta T_m$ cutoff value and select hits for further investigation (*see* **Note 26**).

3.7 Salt and pH Dependence

Ionic strength and pH of buffers affect the stability of proteins and protein:ligand complexes. DSF-GTP can be applied to evaluate the effect of ionic strength and pH on a particular protein or protein: ligand complex. Reactions can be run with the fusion protein in the presence or absence of a ligand or cofactor of interest at different pH or salt concentrations (*see* **Note 27**). Buffers with pH ranging from 5 to 11 have been successfully used in DSF-GTP (*see* **Note 28**). An example of experimental setup to determine the effect of NaCl is provided below:

1. Dilute protein sample to 2 μM in buffer at twice the intended buffer concentration.
2. Dilute NaCl in water at twice the intended final concentration (*see* **Note 29**). Pure water is combined with the protein solution in the control reaction.
3. Combine equal volumes of protein and NaCl solutions in a well (50 μL reaction volume).
4. Seal and equilibrate the plate at RT for 10 min.
5. Run a melt curve and export data as described in Subheading 3.5.

3.8 Cofactors and Coenzymes

The binding of a cofactor or coenzyme to a protein can significantly affect its thermal stability. Examples of cofactors that can easily be assessed with DSF-GTP include divalent metal ions, coenzymes, nucleotides, and vitamins. Care must be taken with transition metals as they can significantly affect GFP at low mM concentrations (*see* **Note 30**).

1. Dilute protein sample to 2 μM in assay buffer (*see* **Note 17**) at twice the intended final concentration.
2. Dilute cofactor in water at twice the intended final concentration.
3. Combine equal volumes of protein and cofactor solutions in a well (50 μL reaction volume). Pure water is combined with the protein solution in the control reaction (*see* **Note 31**).
4. Seal and equilibrate the plate at RT for 10 min.
5. Run a melt curve and export data as described in Subheading 3.5.

3.9 Effect of Ligand Binding Affinity and Concentration on Target Protein T_m

In this section, we illustrate how to set up a concentration dependence experiment to inform about ligand binding affinity with a target protein. The compound concentrations should range between 1 and 1000 μM. We recommend a twofold serial dilution.

1. Dilute protein sample to 2 μM in buffer at twice the intended buffer concentration.
2. Prepare a twofold serial dilution of the compound (*see* **Note 29**). For this, dilute the compound in water at twice the intended final concentration.

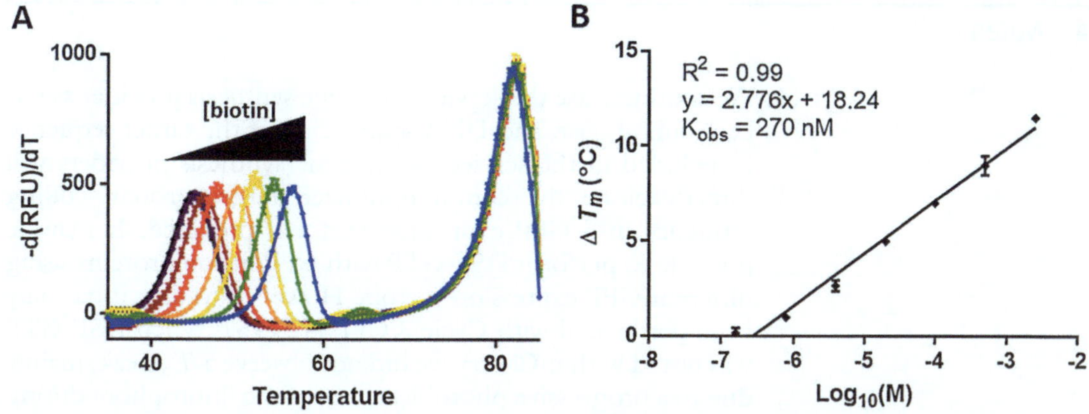

Fig. 5 Determination of K_{obs} using DSF-GTP. (**a**) In this example, the increase in thermostability of Ec BirA-GFP (1 μM in PBS-T) upon binding to increasing concentrations of biotin (160 nM to 2.5 mM) can be seen with the rightward shift of T_m peaks. (**b**) The K_{obs} for biotin binding to Ec BirA-GFP was derived from plotting ΔT_m against $\log_{10}$ of biotin concentrations and fitting to a linear regression model using GraphPad Prism 7 (*see* **Note 32**)

3. Combine equal volumes of protein and compound solutions in a well (50 μL reaction volume). Pure water is combined with the protein solution in the control reaction.
4. Seal and equilibrate the plate at RT for 10 min.
5. Run a melt curve and export data as described above.
6. Determine the net change in T_m (ΔT_m) between the T_m values obtained for a control reaction (with water) and a test condition (with compound) for each concentration (*see* **Note 32** and Fig. 5).
7. Plot ΔT_m values vs. $\log_{10}$ of compound concentrations (Fig. 5b) using a curve fitting software such as GraphPad.
8. Perform linear regression analysis of the linear portion of the curve and determine the compound concentration value at $\Delta T_m = 0$ (*see* **Note 33** and Fig. 5b).

3.10 Special Considerations for Nucleic Acid Interactions

Proteins binding to single- and/or double-stranded RNA or DNA can be assessed by DSF-GTP. Nucleic acids binding to a protein usually stabilize the protein [2–4, 13]. An important point to consider when examining these complex interactions by DSF-GTP is that in the case of RNA or DNA their secondary structures can melt during the melt curve run whether they are single or double stranded. It is nevertheless possible to perform such reactions as long as these specific limitations are considered [3]. It is always advisable to confirm RNA or DNA binding to the Target-GFP using EMSA or GFP-EMSA [5]. Ensure the buffer used does not interfere with binding [14].

4 Notes

1. We routinely use the services of a gene synthesis provider as it is economical, fast, and DNA sequencing of the target sequence is included in the service. Some gene synthesis providers will directly clone the codon optimized target protein coding sequence in a GFP expression vector if provided. It may be possible to perform DSF-GTP with GFP fusion proteins using different GFP expression vectors. However, to date it has only been performed with Cycle 3 GFP [12, 15]. When DSF-GTP was tested with mCherry we did not observe a T_m peak, mainly due to a progressive photobleaching of the fluorophore during the run.
2. KRX competent cells are useful as they can be used for both cloning and over-expression [16]. The tight expression control is also useful when the target protein is likely to be toxic (or has unknown toxicity).
3. Rhamnose and lactose are used for auto-induction of KRX and BL21(DE3) expression strains, respectively [11, 17]. Typical induction with IPTG can be achieved with BL21(DE3) strains, but requires monitoring of optical density for optimal induction of expression.
4. Alternative cell lysis methods can be used such as lysozyme digestion and/or osmotic shock. Sonication is not recommended.
5. Assay buffers (e.g., Tris and HEPES), and reagents such as salts (e.g., NaCl and KCl) divalent metals (e.g., $MgCl_2$ and $ZnCl_2$), co-factors, co-enzymes and ligands (e.g., nucleotides, NADH, and biotin), and small molecules obtained from compound libraries (stocks at 5 mM in DMSO) can be rapidly screened in high throughput using a 96-well plate format.
6. HT-DSF-GTP for screening compound interactions with target proteins has only been performed in hard-shell skirted PCR plates to date. This is due to the requirements of the dispensing equipment of the compound supplier we use. If the validation of compound application can be performed using other liquid handling systems such as the QIAsymphony, it would be possible to perform HT-DSF-GTP using a Rotor-Gene Q. While we have not trialed HT-DSF-GTP in 384-well PCR plates to date, DSF-GTP has been performed successfully in 25 μL aliquots indicating upscaling of the high-throughput capacity is possible.
7. Having the coding sequence for the target protein synthesized enables codon optimization, which maximizes protein expression in the host expression system. An additional benefit is the

ability to eliminate potentially problematic restriction sites if further cloning is desired while ensuring the protein sequence is not affected.

8. Glucose repression of the rhamnose promoter prevents leaky expression. Alternatively, standard cloning strains can be used for initial transformation. Transformants positive for the target protein sequence can be identified by colony PCR. This can be performed via conventional PCR using primers for the T7 promoter and GFP coding sequence.
9. Confirm target protein coding sequence insertion by restriction digest and sequencing if the target protein sequence has been amplified by PCR.
10. We typically select four transformants for assessment, with re-plating on a master plate divided into quadrants. This provides sufficient bacterial biomass for large-scale expression and short-term storage of desired clones.
11. Auto-induction of protein expression can be performed if desired [17]. In this case, glucose and rhamnose are added to 0.1% (w/v) final concentrations and preinduction samples are taken prior to temperature reduction. We find that performing protein expression at lower temperature reduces the rate of expression, enabling improved protein production (correct folding, increased solubility, and overall yield).
12. Pre- and post-induction samples can be assessed for relative yield and correct fusion protein size by SDS-PAGE.
13. Pellets can be checked for fluorescence by exciting with long-wave UV or blue light using a transilluminator or flashlight to provide a rapid assessment of fusion protein production in soluble form [18].
14. We recommend centrifuging 50 mL polypropylene tubes at 4000 × g for 5–10 min. Weigh the tube before and after the centrifugation step to determine the wet cell pellet biomass. Optional step: freeze the cell pellet in liquid nitrogen and store at −80 °C for later purification if desired.
15. Fractions containing the fusion protein are easily identifiable due to the GFP tag. Most fusion proteins we have produced have been eluted at a sufficient concentration to be visible to the naked eye due to GFP fluorescence.
16. For SDS-PAGE analysis we typically keep small aliquots of pre- and post- clarification steps (protein solubility), column flow-through (protein integrity and retention), initial and final wash steps, as well as each eluted fractions (protein purity and integrity).
17. It is important to remove as much supernatant as possible from the ammonium sulfate precipitation. Re-centrifugation is

recommended to ensure maximum buffer removal. Buffers can interfere with downstream testing. For example, assessment of divalent metal specificity can be impaired in phosphate buffer due to the formation of insoluble metal phosphate salts [19]. Protein pellets should be resuspended in a buffer that is compatible with subsequent assay conditions [14]. The addition of glycerol generally increases protein stability [20]. Resuspending the precipitated protein from 4–8 mL of eluted fractions in 1–2 mL of buffer will generally result in 50–100 μM target protein concentration (can be more or less depending on expression yield).

18. If needed, proteins can be further purified, desalted, and dialyzed. However, this is not essential for DSF-GTP, which can be performed with crude samples (e.g., If protein yields are low).
19. Protein concentration can be determined by Bradford assay or other standard colorimetric assays. Protein concentration can also be determined using a real-time PCR thermal cycler set on the FAM channel to record initial GFP fluorescence data (i.e., background corrected relative fluorescence value of a well). A short isothermal run at 25 °C for 2–5 min is sufficient to produce a standard curve of a serial dilution of a known Target-GFP.
20. For example, an enzyme known to bind ATP and Mg^{2+} ions can be tested in the presence or absence of ATP and Mg^{2+} individually or in combination.
21. Non-skirted 96-well PCR plates and plate sealing film can be cut to size, reducing wastage. Hard-shell skirted PCR plates can be reused, by cleaning with a reagent capable of removing protein aggregates, rinsing with RO water, and drying thoroughly.
22. This protocol is based on the Bio-Rad CFX-96, but DSF-GTP has been performed with a Rotor-Gene Q, and can be performed in other real-time PCR thermal cyclers with melt curve capability.
23. Initial screens are performed from 25 to 90 °C in order to record the T_m values of the target protein and GFP in the chosen buffer system. Once the target protein T_m has been determined the melt curve window can be reduced, so long as the GFP T_m can still be determined. For example, the window can be reduced to 35–80 °C or 35–75 °C depending on the buffer conditions (e.g., the GFP T_m may drop in lower pH buffers). Theoretically, if a protein is capable of refolding following thermal denaturation, it should show an identical T_m peak in the subsequent melt curve as long as the GFP fluorophore was not denatured in the first run (i.e., the first run is

stopped before GFP unfolding starts). Most proteins are incapable of refolding following thermal denaturation and in our hands this has never occurred, but this should be tested as it provides an additional control to confirm the independent unfolding of a target protein which can then be further evaluated using GFP-Basta or similar aggregation-based protein assays [12, 21].

24. The T_m obtained by DSF-GTP has been confirmed to correspond to the target protein unfolding and aggregation [2, 12]. Real-time thermal cyclers, such as the Bio-Rad CFX96, automatically determine, display, and provide an option to export T_m data (*see* Fig. 3.).
25. Take care not to introduce bubbles in the wells as they will interfere with the light path of both the excitation LED and GFP emission spectra. If bubbles have formed, they should be eliminated prior to plate sealing (easily broken with a needle or by centrifugation).
26. DSF-GTP can also be performed in diluted serum to determine nonspecific binding of a hit to serum proteins and its concentration effects [7].
27. If the pH dependence of ligand and/or cofactor interactions is being assessed, it is recommended to prepare these at twice the concentration of the final assay in order to combine the protein:ligand complex 1:1 with the buffer samples.
28. It is best to avoid buffering systems that have strong temperature dependence such as Tris. Phosphate is preferable as it has a lower temperature coefficient, and a gradient is easily prepared with different ratios of monobasic and dibasic sodium phosphate salts. However, this may not suit the desired pH range or divalent metal requirements of your target protein. Select a buffer with an appropriate pH range that has the lowest temperature coefficient possible. Use a buffer temperature coefficient table to determine the actual pH at the protein T_m [22].
29. When testing a particular compound, it can be useful to prepare serial dilutions of the test compound that can then be combined with the protein solution.
30. Low concentrations of divalent Mg^{2+} and Mn^{2+} ions (i.e., up to 10 mM) can be tested in phosphate buffer systems containing NaCl. However, divalent metal ions often form insoluble phosphate salts [19] and thus phosphate buffers should be avoided. Buffer systems should be assessed for suitability in terms of pH range and metal binding capacity [14]. Some transition metal ions (Cu^{2+} and Zn^{2+}) can significantly affect the GFP fluorophore at relatively low concentrations (50–100 μM).

31. Multiple compounds can be tested simultaneously, so long as the final buffer concentration is the same as in the control reaction.

32. We typically use a general spreadsheet software such as Excel for general calculations. Graphs can also be produced in Excel; however, we prefer using GraphPad due to the more extensive curve fitting options available.

33. To determine the equilibrium dissociation constant (K_D), values of T_m against $\log_{10}$ of ligand concentrations can be plotted and fitted to various mathematical models depending on the type of affinity and stoichiometry of complex formation [23, 24]. In most cases, the concentration of the target protein is not negligible and the reaction is not performed in standard temperature conditions; hence, the data cannot simply be fitted to a Langmuir binding isotherm equation. This becomes particularly obvious with high-affinity ligands. In most cases, we have been able to derive a value for the affinity of a protein: ligand complex (K_{obs}) by simply determining the *x*-intercept value of the tangent (i.e., linear part) of the curve obtained by plotting ΔT_m values against $\log_{10}$ of ligand concentration. While not directly comparable, we found that K_{obs} is proportional to K_D and provides a good estimate of the stability of a protein:ligand complex. In general and for a variety of target proteins, K_{obs} values are in good agreement with their reported K_D [2–6]. Determination of K_{obs} requires at least four consecutive T_m data over a concentration range of at least two orders of magnitude for adequate determination (*see* Fig. 5b).

References

1. Pantoliano MW, Petrella EC, Kwasnoski JD, Lobanov VS, Myslik J, Graf E, Carver T, Asel E, Springer BA, Lane P, Salemme FR (2001) High-density miniaturized thermal shift assays as a general strategy for drug discovery. J Biomol Screen 6(6):429–440. https://doi.org/10.1177/108705710100600609
2. Moreau MJ, Morin I, Askin S, Cooper A, Moreland NJ, Vasudevan SG, Schaeffer PM (2012) Rapid determination of protein stability and ligand binding by differential scanning fluorimetry of GFP-tagged proteins. RSC Adv 2:11892–11900
3. Moreau MJ, Schaeffer PM (2013) Dissecting the salt dependence of the Tus-Ter protein-DNA complexes by high-throughput differential scanning fluorimetry of a GFP-tagged Tus. Mol BioSyst 9(12):3146–3154. https://doi.org/10.1039/c3mb70426b
4. Askin SP, Bond TEH, Schaeffer PM (2016) Green fluorescent protein-based assays for high-throughput functional characterization and ligand-binding studies of biotin protein ligase. Anal Methods UK 8(2):418–424
5. Bond TEH, Sorenson AE, Schaeffer PM (2017) Functional characterisation of Burkholderia pseudomallei biotin protein ligase: a toolkit for anti-melioidosis drug development. Microbiol Res 199:40–48. https://doi.org/10.1016/j.micres.2017.03.007
6. Bond TEH, Sorenson AE, Schaeffer PM (2017) A green fluorescent protein-based assay for high-throughput ligand-binding studies of a mycobacterial biotin protein ligase. Microbiol Res 205:35–39. https://doi.org/10.1016/j.micres.2017.08.014
7. Askin S, Bond TEH, Sorenson AE, Moreau MJJ, Antony H, Davis RA, Schaeffer PM (2018) Selective protein unfolding: a universal

mechanism of action for the development of irreversible inhibitors. Chem Commun (Camb) 54(14):1738–1741. https://doi.org/10.1039/c8cc00090e

8. Green MR, Sambrook J (2012) Molecular cloning: a laboratory manual, vol 3, 4th edn. Cold Springs Hoarbour Laboratory Press, New York
9. Tabor S, Richardson CC (1985) A bacteriophage T7 RNA polymerase/promoter system for controlled exclusive expression of specific genes. Proc Natl Acad Sci U S A 82 (4):1074–1078
10. Dubendorff JW, Studier FW (1991) Controlling basal expression in an inducible T7 expression system by blocking the target T7 promoter with lac repressor. J Mol Biol 219(1):45–59
11. Studier FW (2005) Protein production by auto-induction in high density shaking cultures. Protein Expr Purif 41(1):207–234
12. Moreau MJ, Morin I, Schaeffer PM (2010) Quantitative determination of protein stability and ligand binding using a green fluorescent protein reporter system. Mol BioSyst 6 (7):1285–1292. https://doi.org/10.1039/c002001j
13. Morin I, Schaeffer PM (2012) Combining RNA-DNA swapping and quantitative polymerase chain reaction for the detection of influenza A nucleoprotein. Anal Biochem 420 (2):121–126. https://doi.org/10.1016/j.ab.2011.09.009
14. Ferreira CMH, Pinto ISS, Soares EV, Soares HMVM (2015) (Un)suitability of the use of pH buffers in biological, biochemical and environmental studies and their interaction with metal ions—a review. RSC Adv 5:30989–31003
15. Fukuda H, Arai M, Kuwajima K (2000) Folding of green fluorescent protein and the cycle3 mutant. Biochemistry 39(39):12025–12032
16. Hartnett J, Gracyalny J, Slater MR (2006) The single step (KRX) competent cells: efficient cloning and high protein yields. Promega Notes 94:27–30
17. Schagat T, Friedman R, Ohana OP, Hartnett J, Slater MR (2008) KRX autoinduction protocol: a convenient method for protein expression. Promega Notes 98:16–18
18. Waldo GS, Standish BM, Berendzen J, Terwilliger TC (1999) Rapid protein-folding assay using green fluorescent protein. Nat Biotechnol 17(7):691–695. https://doi.org/10.1038/10904
19. Hurst MO, Fortenberry RC (2015) Factors affecting the solubility of ionic compounds. Computat Theoret Chem 1069:132–137
20. Vagenende V, Yap MG, Trout BL (2009) Mechanisms of protein stabilization and prevention of protein aggregation by glycerol. Biochemistry 48(46):11084–11096. https://doi.org/10.1021/bi900649t
21. Antony H, Schaeffer PM (2013) A GFP-tagged nucleoprotein-based aggregation assay for anti-influenza drug discovery and antibody development. Analyst 138 (20):6073–6080. https://doi.org/10.1039/c3an01041d
22. Dawson RMC, Elliot DC, Elliot WH, Jones KM (1989) Data for biochemical research, 3rd edn. Oxford University Press, Oxford
23. Matulis D, Kranz JK, Salemme FR, Todd MJ (2005) Thermodynamic stability of carbonic anhydrase: measurements of binding affinity and stoichiometry using ThermoFluor. Biochemistry 44(13):5258–5266. https://doi.org/10.1021/bi048135v
24. Cimmperman P, Baranauskiene L, Jachimoviciute S, Jachno J, Torresan J, Michailoviene V, Matuliene J, Sereikaite J, Bumelis V, Matulis D (2008) A quantitative model of thermal stabilization and destabilization of proteins by ligands. Biophys J 95 (7):3222–3231. https://doi.org/10.1529/biophysj.108.134973

Chapter 6

Enzyme–Ligand Interaction Monitored by Synchrotron Radiation Circular Dichroism

Rohanah Hussain, Charlotte S. Hughes, and Giuliano Siligardi

Abstract

CD spectroscopy is the essential tool to quickly ascertain in the far-UV region the global conformational changes, the secondary structure content, and protein folding and in the near-UV region the local tertiary structure changes probed by the local environment of the aromatic side chains, prosthetic groups (hemes, flavones, carotenoids), the dihedral angle of disulfide bonds, and the ligand chromophore moieties, the latter occurring as a result of protein–ligand binding interaction. Qualitative and quantitative investigations into ligand-binding interactions in both the far- and near-UV regions using CD spectroscopy provide unique and direct information whether induced conformational changes upon ligand binding occur and of what nature that are unattainable with other techniques such as fluorescence, ITC, SPR, and AUC.

This chapter provides an overview of how to perform circular dichroism (CD) experiments, detailing methods, hints and tips for successful CD measurements. Descriptions of different experimental designs are discussed using CD to investigate ligand-binding interactions. This includes standard qualitative CD measurements conducted in both single-measurement mode and high-throughput 96-well plate mode, CD titrations, and UV protein denaturation assays with and without ligand.

The highly collimated micro-beam available at B23 beamline for synchrotron radiation circular dichroism (SRCD) at Diamond Light Source (DLS) offers many advantages to benchtop instruments. The synchrotron light source is ten times brighter than a standard xenon arc light source of benchtop instruments. The small diameter of the synchrotron beam can be up to 160 times smaller than that of benchtop light beams; this has enabled the use of small aperture cuvette cells and flat capillary tubes reducing substantially the amount of volume sample to be investigated. Methods, hints and tips, and golden rules to measure good quality, artifact-free SRCD and CD data will be described in this chapter in particular for the study of protein–ligand interactions and protein photostability.

Key words Circular dichroism, Ligand binding, Titration, Binding constant, UV denaturation, Protein stability, Data processing

1 Introduction

Circular dichroism (CD) is a powerful technique which enables the monitoring of local and global changes in the structure and

---

The original version of this chapter was revised. The correction to this chapter is available at https://doi.org/10.1007/978-1-0716-0163-1_19

Nikolaos E. Labrou (ed.), *Targeting Enzymes for Pharmaceutical Development: Methods and Protocols*, Methods in Molecular Biology, vol. 2089, https://doi.org/10.1007/978-1-0716-0163-1_6, © The Author(s) 2020

conformation of proteins. CD spectroscopy enables the selective monitoring of specific chromophores of the protein including peptide backbone in the far-UV region (180–250 nm), aromatic side chains of amino acid residues and dihedral angles of disulfide bonds in the near-UV region (250–350 nm), and prosthetic groups (hemes, flavones, carotenoids) in the visible region (400–800 nm). Bespoke benchtop CD instruments commonly use a xenon light source. Synchrotron radiation circular dichroism (SRCD) beamlines utilize the light produced at synchrotrons as the light source that is brighter than standard xenon light sources [1], and with much higher photon flux in the vacuum-UV region (130–200 nm). The highly collimated microbeam generated with the SRCD beamline B23 at Diamond Light Source (DLS) had enabled to reduce the amount of sample required for standard CD measurements by using lower volume cuvette cells with 1–2 mm diameter aperture [1, 2] and the high-throughput CD (HTCD) using 96 plates [3]. The higher photon flux of B23, however, can affect the photostability of the investigated protein by inducing structural denaturation [1]. Although this effect can be minimized or eliminated by reducing the slit width of the double-grating monochromator of B23 or rotating the cuvette cells around the axis parallel to the incident beamlight [4, 5], the UV protein denaturation has been used as an assay to determine the photostability and also qualitatively to determine ligand-binding interactions for UV-transparent ligands like fatty acids and carbohydrates [4, 5]. The rate of UV denaturation is significantly affected, usually decreased, if the ligand is interacting with the protein, otherwise, the rate with and without ligand is unaffected [4, 5]. In addition to qualitative investigations into ligand-binding interactions, recent advances saw the development of small-volume titration techniques that are most important when investigating the ligand-binding interactions of limited materials and has been well demonstrated with membrane proteins [6–15].

A large number of SRCD spectra can be collected during a single experiment, particularly when using the HTCD with 96 plates or UV denaturation assays at B23, and for Users who are allocated a specific number of 8-h shifts, it is important to be able to analyze on-the-fly to ensure that the experimental conditions are suitable for data collection. For this reason, software has been developed at B23 which allows for the analysis of CD data in .ols, .csv, and .txt format [16], making it compatible with data collected on other CD instruments off-site.

CD instruments measure the difference in absorbance of left- and right-circularly polarized light of chiral molecules as a function of wavelength expressed in nanometers (nm) otherwise known as CD spectroscopy [17]. Absorbance is therefore an important experimental parameter which must be considered when planning an experiment, and such considerations are made by application of the Beer-Lambert law $A = \varepsilon.c.l$, where A is absorbance, ε is specific molar extinction coefficient which is specific for the chromophore studied, c is concentration, and l is pathlength. In most cases an optimum A (also known as optical density OD) of 0.8 is recommended, with a maximum of 1.5. For proteins, the amide bond is the main chromophore that absorbs in the far-UV region (180–250 nm) [18], while the aromatic side chains of tryptophan, tyrosine, and phenylalanine amino acid residues and disulfide bonds are the chromophores that absorbs in the near-UV region. Prosthetic groups such as hemes, NAD and FAD cofactors, and carotenoid pigments are the chromophores that extend the absorption of light in the visible region (400–800 nm). The folding of proteins is promptly determined by SRCD/CD spectroscopy in the far-UV region for which the content of secondary structure can be estimated using a variety of algorithms [18–21] that are also available is several suite of programs [16, 22–24]. The CD/SRCD in the near-UV region is sensitive to the local environment of the aromatic side chains of Trp, Tyr, and Phe, and dihedral angle of disulfide bonds of cystine residues and has been used successfully to probe qualitatively and quantitatively ligand-binding interactions [25].

2 Materials

2.1 Fused Silica Cuvettes

Common cuvettes employed for measurements of SRCD/CD spectra in the far-UV region are cylindrical or rectangular cells with a 0.02 cm pathlength made of fused silica. Low-volume titrations are conducted with a cell with a small-diameter channel which leads down to a small-volume reservoir. The reservoir window is made of black fused silica to mask accordingly the area of the incident light passing through the sample.

For measurements conducted in the near-UV region using larger pathlengths, e.g., 1 cm, these cells have black walls and small aperture (2 mm × 2 mm) windows. This restricted surface for light entry permits the use of small sample volumes (for a 1 cm pathlength, 70 μL of solution is required for these specialized cuvettes compared to a standard cuvette requiring a sample volume >1 mL).

2.2 Buffer Systems Specificity

2.2.1 Buffering System of Choice

For measurements in the far-UV region, sulfonate- (HEPES, MES, MOPS, PIPES) [17] and carboxylate-rich (glycine, acetate, citrate) [26] buffering systems are not recommended as these buffers tend to absorb at 200 nm due to the n→π* electronic transitions of the S=O and C=O bond, respectively, and alternative buffers should be used instead. Regularly, phosphate buffer (concentrations of <25 mM) is suggested to be the most appropriate buffer system for CD measurements, and low concentrations of Tris or Tris-acetate-EDTA (TAE) (<25 mM) are also acceptable, provided that pH adjustment is not done with HCl as this would introduce or increase the existing content of chloride anions (Cl^-), which absorb in the far-UV region. pH adjustment of buffers should be achieved using more UV-transparent phosphoric acid, nitric acid, acetic acid, citric acid, and borate [17].

2.2.2 Salts

As discussed previously, the presence of high concentrations of Cl^- is undesirable for studies in the far-UV region below 200 nm. If monovalent salts are required for maintenance of the ionic stabilization in the system to be studied, UV-transparent fluoride anions (F^-) should be substituted for Cl^- [17, 25]. However, it is important to ascertain that the fluoride anions is not affecting the protein folding compared to that with chloride anions, which can be conducted by comparing the CD spectra in the overlapping transparent region that usually can be achieve down to 210–205 nm. Sulfate salts can also be used as an alternative [17, 25]. For a 0.02 cm and longer pathlengths, less salt concentration is better for CD measurements; otherwise the 200 nm or higher wavelength cutoff will prevent accurate secondary structure estimations. For pathlengths of 0.02 cm and longer, no more than 50 mM NaCl should be used, which could be increased to 100–150 mM NaCl with 0.01 cm. Higher concentrations of NaCl that need to be used (e.g., >500 mM) will require even narrower cell pathlengths that can be achieved only with demountable cells of 0.0050 or 0.0020 cm as it will be otherwise impossible to clean standard cuvette cells of these pathlengths. Of course, it will require higher protein concentration following the Beer's law. For example, as a rule of thumb, an average protein concentration of 0.4 mg/mL (for proteins highly helical the concentration could be as high as 0.5 mg/mL, while for disordered or highly beta-sheet proteins, the concentration would be 0.3 mg/mL as the UV absorption is higher) will enable a good signal-to-noise quality CD spectrum using a cuvette cell of 0.02 cm pathlength. Therefore, halving the pathlength to 0.01 cm will require double the protein concentration to obtain the same CD spectrum. With B23, the smallest pathlength available for standard measurements is 5 μm which requires a protein sample volume of 2 μL at a concentration of 16 mg/mL.

2.2.3 Buffer Additives

Addition of buffer additives should also be carefully assessed based on their UV absorption properties. For protein samples which have been purified using high concentrations of imidazole, such as from affinity purification columns, buffer exchange of the protein to remove all imidazole is required as it absorbs too strongly in the ~210 nm region due to its large extinction coefficient [17]. The addition of histidine in the protein formulation should be therefore avoided for the same reason. Similarly, urea and guanidine denaturant agents should be avoided unless the protein denaturation is the aim of the experiment. High concentrations of glycerol (<10% (v/v)) should also be avoided to enable the CD to be measured below 200 nm [22]. The four carboxylate groups of ethylenediaminetetraacetic acid (EDTA) absorb strongly below 200 nm and therefore should not be used in concentrations above 1 mM for measurements in the far-UV region using a 0.02 cm pathlength [17]. For higher EDTA concentration, a smaller pathlength should be used but a compromise has to be reached as the protein concentration should be increased as well that will decrease the EDTA-protein molar ratio.

If a reducing agent is required, beta-mercaptoethanol (BME) is preferred to dithiothreitol (DTT). DTT strongly absorbs across the UV region—DTT_{red} (the active for which maintains the reduced state of proteins) below 260 nm across the far-UV region and DTT_{ox} (the result of the eventual oxidization of DTT over time from the environment) below 340 nm and across the near-UV region [27, 28]. Tris (2-carboxymethyl) phosphine-HCl (TCEP-HCl) is also compatible with CD but is highly acidic; therefore, the suitability of a potentially acidic environment for the protein should be considered (no more than 1 mM) [29, 30]. DMSO should also be avoided for measurements in the far-UV region as it absorbs strongly in the UV region (λ_{max} 220 nm, cutoff is <260 nm) and therefore is unsuitable for far-UV measurements [31, 32]. If an organic solvent is required, most other types, e.g., hexane and trifluoroethanol, are transparent down to 185 nm; however, it is important to note that these solvents may induce structural changes to the protein being studied [22].

2.3 Protein Samples

Protein samples should be soluble, homogenous, and pure; therefore, purification techniques including size-exclusion chromatography (SEC) should be used in addition to standard immobilized metal affinity chromatography (IMAC) to ensure highly pure samples which can be checked using gel electrophoresis and mass spectroscopy [22]. Solid particles can lead to light scattering and distortion of measured CD spectra [17]; therefore, effort should be made to remove them from samples before data collection. Centrifugation of the sample and only using the supernatant for measurements or filtering samples through a 0.2 μm filter before loading into the cuvettes for data collection can be adopted to ensure samples are solid free [17]. Protein concentration determined spectroscopically absorbance should follow the Beer-Lambert law measuring either the absorbance at 280 nm for proteins containing aromatic amino acids or at 205 nm for nonaromatic amino acid containing peptides/proteins [18]. This should be performed with the final protein product (after purification and removal of solid particles) prior to any CD measurement.

2.4 CDApps Software

CDApps software has been developed to analyze SRCD and CD data (.ols, .csv, .txt format). Developed at Diamond B23 beamline, the software is available to use on-site for analysis, and can be copied on a USB stick along with video tutorial files for analysis and training off-site. Information regarding beamline software including manuals can be found on the beamlines software page http://www.diamond.ac.uk/Instruments/Soft-Condensed-Matter/B23/manual/Beamline-software.html.

The software has progressed from its original conception as detailed in [16] which focused on the planning and data processing of standard, thermal melt, UV denaturation and titration experiments to have a more intuitive interface as well as the processing capability for high-throughput CD (HTCD) data sets from a 96-well plate [3] and CD imaging applications [33].

3 Methods

Circular dichroism experiments conducted at B23, Diamond Light Source, can all be planned and processed using CDApps [16].

For all experiments, background measurements (background buffer equivalents) should be collected which can be subtracted from all experimental data files. The most recent upgrades for the data collection software include the option to collect UV absorption data which can be used as an internal control for each sample measured. Optimum absorbance readings are ~0.8, for titrations the range recommended is from 0.4 to a maximum of 1.5–1.6 [34]. Within these limits, the chromophore concentration will be sufficient for detection and without being too high to saturate the detector distorting the spectral profile. A recommendation before starting a series of measurements is to know the absorbance in terms of intensity magnitude and wavelength maxima for all components of the solutions such as solvent (buffer), solute (protein/peptide), additives, ligands, and any other chemical agent present therein (*see* Subheading 2.2.1 and **Notes 1–4**). CDApps can help to calculate the absorption of mixtures from the information input into the software (pathlength to be used, extinction coefficient at specific wavelengths, and molecular weights of materials) for the individual components during planning of the experiments [16].

Before commencing measurements, an air baseline should be recorded for the instrument once the experimental parameters (wavelength range, wavelength increment, integration time) have been set up. This will allow for the data collection software to calculate the absorbance simultaneously during CD signal collection.

3.1 Standard CD Measurements

Standard CD measurements can be performed at the B23 beamline at DLS. Compared to standard benchtop CD instruments, the synchrotron light source is comparatively tenfold brighter due to the higher photon flux in a smaller diameter beam [1]. These characteristics open up new opportunities for CD measurements which are unattainable using conventional CD instruments, some of which will be discussed throughout this chapter.

Standard CD measurements can be used to qualitatively assess the conformational property of proteins in terms of secondary structure and local tertiary structural changes occurring upon ligand-binding interactions.

Below will be described the basic experimental setup which can be changed for the sample available, using the principles of the Beer-Lambert law for guidance [35]. For protein solutions of a concentration higher than the standard described previously (about 0.4 mg/mL in 0.02 cm cell pathlength) smaller pathlengths are required, whereas for more dilute solutions larger pathlengths are used [35] (*see* **Notes 5** and **6**).

In the far-UV region, protein secondary structure conformational changes as a function of environment (solvent composition and polarity [8, 22], pH [8, 22], ionic strength [8, 22], surfactants [36], pressure (up to 200 MPa) [37], temperature (−170 °C to +350 °C) [38], and ligand interaction [8]) can be promptly monitored. Standard parameters are as follows:

1. A protein solution at concentration of 0.4 mg/mL, measured using a cuvette of 0.02 cm pathlength. These cuvettes (both cylindrical and rectangular designs) require to be filled at least with 40 μL of solution.
2. Default parameters for CD measurements in the far-UV region (180–260 nm) using B23 beamline:
   (a) 1 nm increments.
   (b) 1 s integration.
   (c) 1.2 nm bandwidth (bw) corresponding to the monochromator slit width of 0.5 mm (for benchtop CD instruments the recommended bw is 1 nm).
3. The number of scans can be tailored to the sample/experimental design (*see* **Note 6**).

In the near-UV region, protein conformation changes with respect to the local environment of aromatic side chains of Trp, Tyr, and Phe residues and disulfide bond of Cys residues as a function of solvent composition and polarity [8], pH [8], ionic strength [8], surfactants [39, 40], pressure (up to 200 MPa) [41], temperature (−170 °C to +350 °C) [42], and ligand [8] interaction can be promptly monitored. Standard parameters are as follows:

1. Protein solution with an absorbance at 280 nm of ~0.8 is desirable (for ligand titration, the recommended absorption range is from 0.4 to 1.6).
2. Data collection occurs in the near-UV region (260–350 nm) using the following parameters:
   (a) 1 nm increments.
   (b) 1 s integration.
   (c) 1.8 nm bandwidth (corresponding to slit width of 1 mm). For benchtop instruments, the recommended bandwidth is 2 nm.

3. The number of scans can be tailored to the sample/experimental design (*see* **Note 6**).

For all types of measurements concerning proteins, clean cuvettes by (*see* **Note** 7):

1. Flushing the cell with deionized water followed by 96% (v/v) ethanol.
2. Place the ethanol wet cell in conc. nitric acid in a fume hood, and leave for 10 min.
3. Remove the cell and flush with deionized water to remove the acid.
4. Wash with 96% (v/v) ethanol and dry under nitrogen or using a vacuum pump.

3.2 UV Denaturation of Proteins Monitored by Circular Dichroism in the Far-UV Region (180–260 nm)

In the far-UV and vacuum-UV regions, the intense photon flux of the highly collimated microbeam of B23 beamline has the potential to cause protein denaturation [5, 43–45]. UV photo denaturation experiments have been used successfully as a novel assay for the assessment of protein photostability that can be used to determine ligand-binding interactions, in particular for ligands with weak or devoid of any UV chromophores in the far-UV region [5].

For these measurements, samples are loaded into cuvettes following the guidelines mentioned in Subheading 2.1. Repeated continuous scans, usually 20, 30, or in case of stable materials up to 100 are collected from the sample (*see* **Note 8**).

The general protocol is as follows:

1. For a cuvette (either cylindrical or rectangular cell) of 0.02 cm pathlength, 40 μL of a protein solution at about 0.4 mg/mL is loaded into the cuvette.
2. Standard data collection parameters in the far-UV region (180–260 nm) include 1 nm increments, 1 s integration time, and 1.2 nm bandwidth. At this point a standard CD measurement can be collected.
3. For the UV denaturation method, repeated continuous measurements are scanned. The number of scans to be collected is dependent on the system under study as the protein sensitive to UV light (photostability) varies from protein to protein. A suggested range is 20–30 scans that correspond to a total time of 60–90 min in order to obtain a good denaturation rate trend within a reasonable time-scale which is often a key consideration especially for Users of the B23 beamline at Diamond Light Source where allocated experimental time is limited.
4. For data analyses *see* Subheading 3.4.

3.3 Low-Volume Titrations for Ligand Binding Monitored by Circular Dichroism

CD titration experiments enable the determination of the quantitative binding interactions of host-ligand systems. It has been successfully employed for the determination of the dissociation constant K_d for ligand binding to membrane proteins and soluble proteins [6–15]. This was done by monitoring the CD spectral changes of the protein upon the incremental addition of small but accurate aliquots of ligand corresponding to an increased molar ratio until no more changes were observed.

CD titration experiments conducted in the near-UV region (250–350 nm) have the potential to use a large volume of sample due to the larger pathlength which are often employed to monitor low-concentration solutions. A fused quartz cuvette (Starna) with a window aperture of 2 mm × 2 mm for a 1 cm pathlength requiring only 70 μL to be filled is routinely used at B23 due to the smaller footprint of the synchrotron light source beam that can be conveniently varied from 0.05 to 1 mm in diameter. The larger beamlight of benchtop CD instruments, from about 3 to 8 mm in diameter, makes the use of such a small aperture 1 cm cell very impractical. On the contrary, this cell is routinely used for CD titration with B23 in particular for membrane proteins of which yields are notoriously smaller than most soluble proteins under the same expression and purification conditions. Another advantage of the small aperture 1 cm cell is the fact that the addition of the ligand aliquots and their mixing with the protein can be conducted rather gently reducing the equilibration time when detergents are used to solubilize the membrane proteins. CDApps can also be used to calculate the absorption of the solution at each titration step point using the molar extinction coefficient for both the protein and the ligand to ensure the total absorption in the cuvette is within the acceptable 0.4–1.5 absorption limits [16] (*see* **Notes 9–11**).

For these experiments, the following protocol is used:

1. Measure the concentration of protein solution at A_{280}. Use this value to calculate the dilution factor for the stock solution for the working solution. The concentration of the protein to be used during the experiment is in part determined with respect to the UV absorbance contribution of the ligand at each titration point which should not be more than 1.5–1.6 (*see* **Notes 9–11**).
2. Aliquots of ligand are added to the host mixture in the cuvette, pipetting up and down to thoroughly combine the components (*see* **Notes 12** and **13**).
3. Mixture incubated for a user-defined period of time (e.g., 20 min) to allow for equilibration to occur (*see* **Note 14**) and ensure the appearance of the solution is homogenous before measuring (*see* **Notes 13**, **15** and **16**). This is particularly important in the presence of detergents as the addition of

aliquots affects the vesicle morphology and requires equilibration to obtain reproducible CD baselines. The way to determine the incubation time is to scan repeated consecutive CD spectra until the spectral shape is stable. The incubation time can be calculated as the product of the number of scans multiplied by the time to scan a single spectrum.

4. Measurements are collected for each titration point, after the determined incubation time.
5. CDApps is encouraged for the analysis, using the experimental plan as the template for the analysis as it contains information regarding volumes, concentration, and dilution factors of the mixture components in addition to pathlength(s) used to monitor the mixtures (*see* **Notes 10–12**).
6. CD spectrum of the Buffer under the same incubation time is measured to be used as baseline to be subtracted from each titration point spectrum, followed by the CD spectrum of the relevant proportion of the maximum-titrate ligand concentration. This is essential if the ligand is chiral. For achiral ligands, this is not necessary as the ligand will be devoid of any CD and the CD spectrum of the baseline will be sufficient for the data processing.
7. Difference in the CD signal at each point is plotted against the respective ligand concentration, fitting with a Hill function for determination of the binding dissociation constant K_d and the stoichiometry of the interaction using a nonlinear regression analysis [46].

3.4 Processing of Circular Dichroism Data Collected at B23

Data is processed using CDApps available at the B23 beamline for use on-site and as a USB downloadable copy for use off-site (found at Computer > Software (W:) > cd_programs > CD Apps > USB). For processing, the following steps are recommended.

3.4.1 Loading CDApps for Processing of Circular Dichroism Data Collected at B23

Below are the initial steps required to load CDApps software:

1. Open the CDApps software by:
   (a) On-site—Computer > Software (W:) > cd_programs > CD Apps > CD Apps.exe. Double click the CD Apps icon.
   (b) Off-site—double-clicking the CD Apps icon.

The following options page enables the User to select the type of analysis appropriate for the data set.

(a) CD Measurement (including Automation)—analysis options available include 1 ton spectra, titration, UV denaturation, thermal melt using both the Quantum Peltier (+5 °C to +95 °C) and Linkam MDS 600 (−150 °C to +350 °C) controllers and 96-well format.

(b) CD Titration.

The analysis using both of these methods shall be discussed.

For all types of analyses, the most important bits of information for subsequent analyses including conversion of units and secondary structure estimations (SSE) are concentration, pathlength, concentration and molecular weight and $\Delta\varepsilon$ (the average amino acid molecular weight).

3.4.2 Analysis Using the "CD Measurement (Including Automation)" Option

(a) Select the "CD Measurement (including Automation)" (highlighted in Fig. 1).

(b) Click on the "Sample Description" section of the sample line to be completed and enter details, e.g., name of the sample (Fig. 2).

(c) To add experimental data to the spreadsheet:

Select the "Experiment Data" tab, selecting the appropriate button for the type of file to be uploaded (.ols or .csv/.txt).

Generally, files collected at the beamline are saved as .ols files. Files collected using both the benchtop Chirascan instrument and the beamline modules can be converted into .csv format.

Attach the appropriate data file to the sample line by clicking on the file to highlight (line becomes blue), then drag and drop the file across to the sample line.

Experiment details (number of scans, wavelength ranges, wavelength increment) are automatically completed for the attached file.

(d) Complete the remaining empty experiment details (concentration, concentration units (mg/mL or μM), volume (μL), $\Delta\varepsilon$ calculated, molecular weight (Da), temperature (°C), pathlength (cm), slit (mm), and integration time(s)).

(e) New rows can be created by selecting "Add blank row" and repeating **steps 3–5**. If all of the details of a row are to be replicated for additional rows, this can easily be done by a single right-click and selecting the "replicate" option. Appropriate data can then be attached using **step (d)**.

Note: For any type of analysis, the correct scan type needs to be selected and appropriate baseline allocated. Baseline is assigned to background or buffer measurements, which will be subtracted from all other measurements selected in the group during the analysis. Depending on the analysis selected, the software will subtract the baseline differently. For 1 to *n* Spectra multiple entries should be used during the analysis, one per sample measured which can each contain multiple scans per loaded file. The 1 to *n* Spectra option creates an

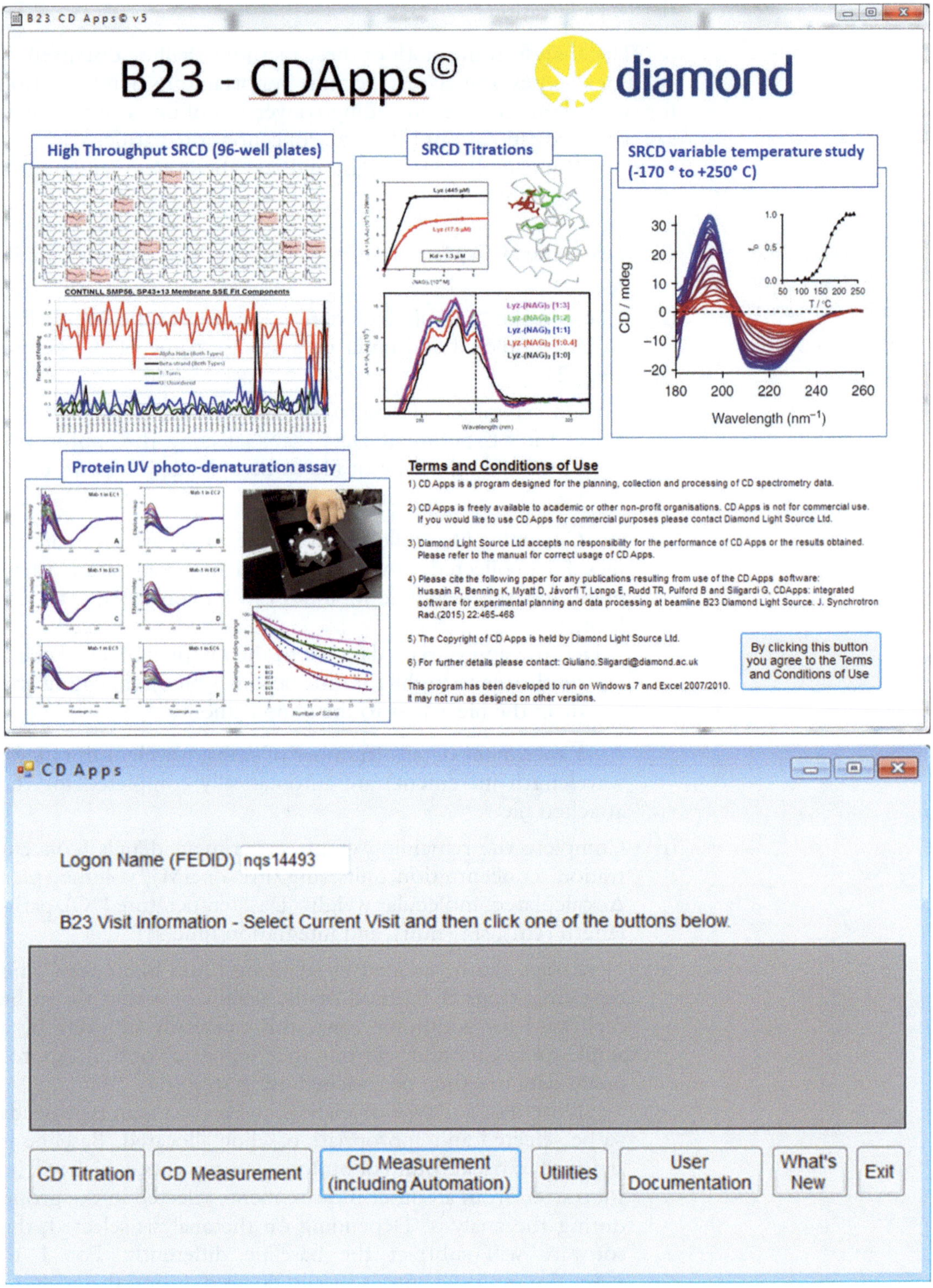

Fig. 1 (Top) Terms and conditions page which appears upon opening. (Bottom) Initial options page which opens upon loading of the program

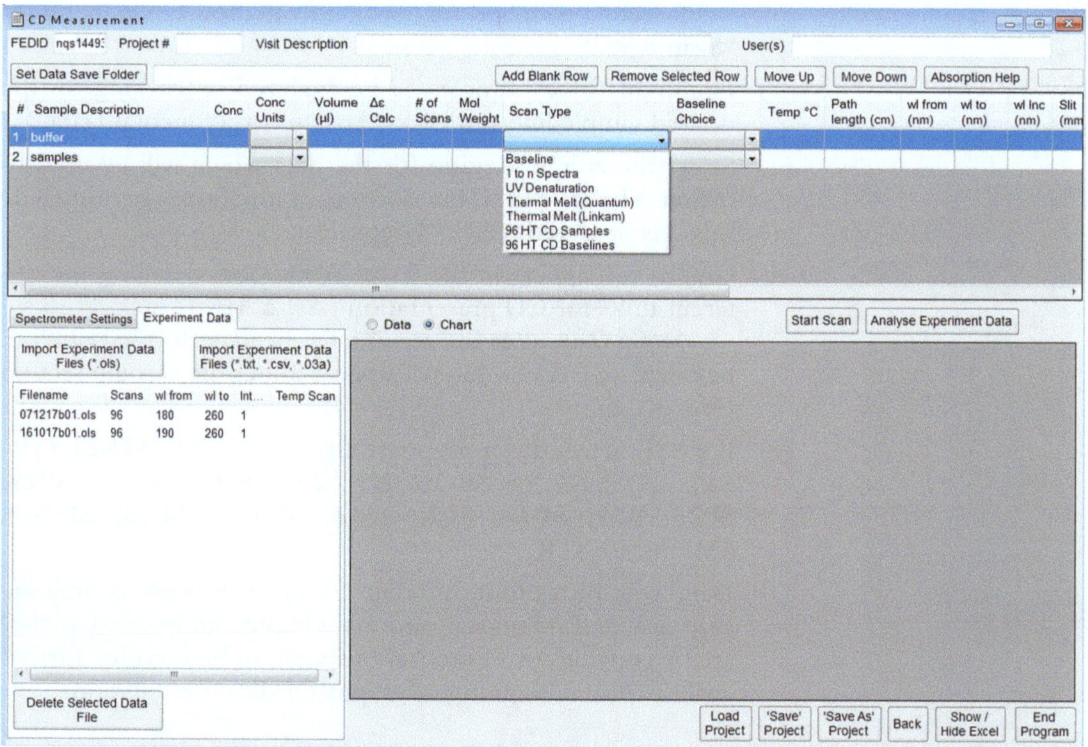

Fig. 2 Opening screen for CDApps. From here all of the experimental details including data type and experimental parameters are added ready for the analysis. From here the initial information is input before progressing through the analysis steps by pressing "Analyse Experiment Data"

average scan for each loaded file and subtracts the baseline from each sample. UV Denaturation analyses subtract the average spectra of the baseline from every individual scan in the file.

(f) From here, a range of analyses are available for different types of data sets. The type of analysis is selected using the "Scan type" drop-down menu for each sample line (*see* **Note 17**). Each analysis option will be discussed in Subheading 3.4.

1 to *n* Analysis

Multiple samples can be analyzed by completing the required number of sample lines as detailed in Subheading 3.3, **step 2**, and highlighting the desired number of samples before selecting the "Analyse Experimental Data."

(a) Complete the workbook with the appropriate number of sample lines and data.

Ensure that the data files to be analyzed are within the same wavelength range and increments. For analyses in the far-UV region (180–260 nm) where secondary structure estimations (SSE) are desired as part of the analysis, endure that loaded files are collected in 1 nm increments.

Select "1 to n Spectra" option from drop-down menu for "Scan type."

(b) Highlight (blue) samples to be analyzed by clicking on the desired sample line and click "Analyse Experimental Data."

(c) Select the zeroing region for the data which will be used to off-set all data sets. This is often in a region where no CD signal is observed. Click "Graph Data."

(d) Graphs will appear in the Excel Workbook, detailing the different units for CD presentation ($\Delta\varepsilon$, ΔA).

Note: Only after the zeroing has been performed can SSE be calculated. A new tab will appear under the zeroing section (Fig. 3b, c).

(e) For SSE, select the appropriate algorithm (CONTINLL [19–21], CDSSTR, or SECLON3 [20]) and database (SP29, SP23, SP1, SP37, SP2, SP42, SMP50, SP43, SDP48, SMP56, CLSTR).

(f) Data will be output into an Excel Workbook containing graphical and numerical outputs. Graphs can be used as they are by copying, or numerical outputs can be used for further analysis and redesigning of graphs for data presentation.

UV Denaturation Analysis

UV denaturation analyses investigate the unfolding process of a protein as a result of exposure to UV light from the synchrotron light source during a CD experiment.

1. Select UV denaturation option from drop-down menu for "Scan type" (Fig. 2).
2. Highlight the Sample and Baseline sample lines. Highlighted lines will appear blue.

   Note: Unlike the 1 to *n* analysis option, only one sample can be analyzed at a time.
3. Click "Analyse Experiment Data."
4. Set the off-set region for the data (where no CD signal is observed).
5. Select the wavelength to monitor for UV denaturation by either clicking on the region of the observed signal in CDApps or selecting the wavelength from the drop-down menu (Fig. 4).
6. Click "Graph Wavelength to Monitor" button.
7. Under the UV Denaturation Fitting tab in CDApps, select the most appropriate fit for the data points (Fig. 4).
8. Alternatively, copy the data from the "Folding Change" tab of the Excel workbook for analysis using alternative software.
9. Under the SSE (Protein) tab, select appropriate algorithm for protein secondary structure estimation (Fig. 5).

A.

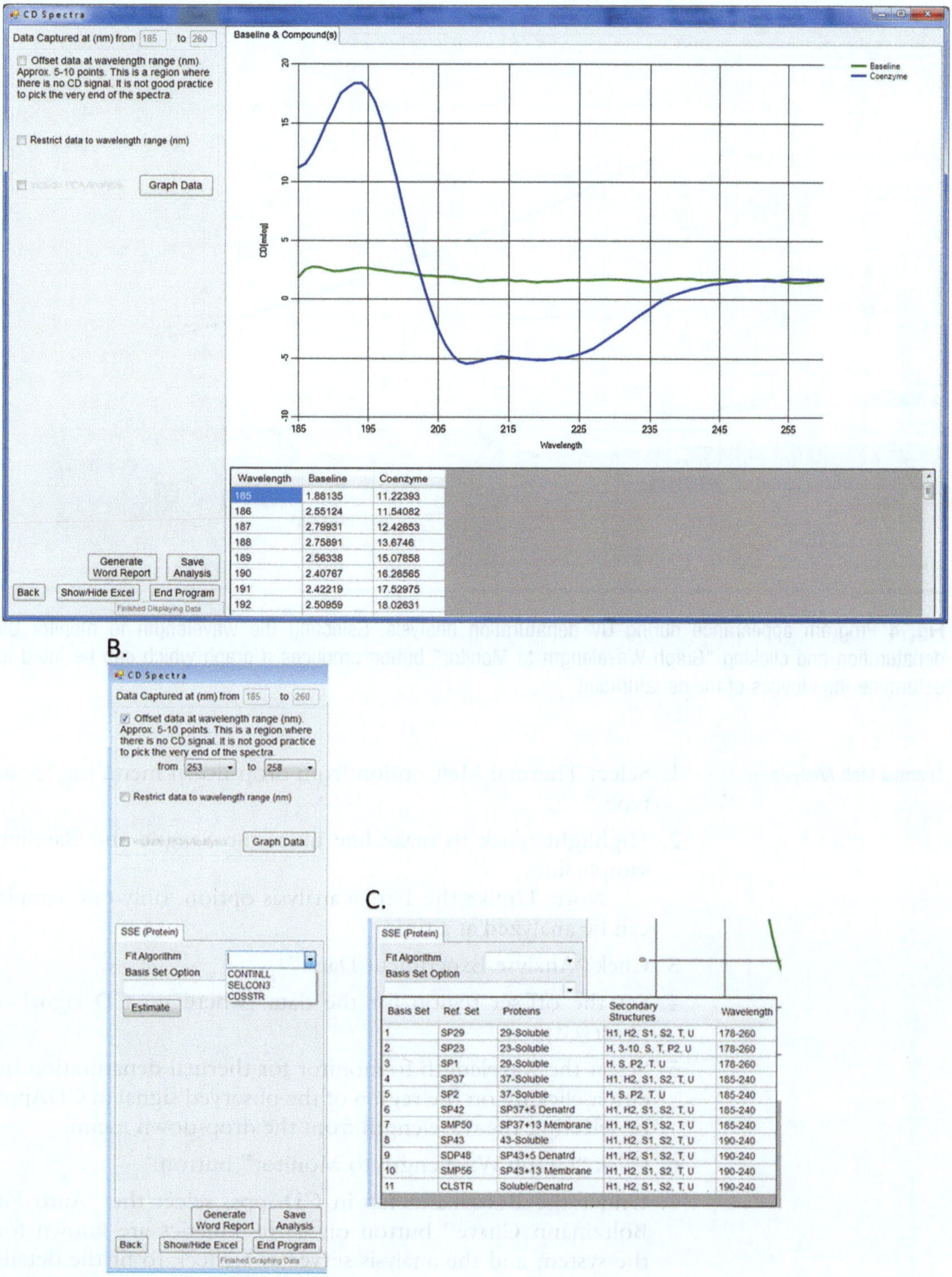

Fig. 3 (**a**) Analysis page of CDApps after pressing "Analyse Experimental Data" button for "1 to *n*" function. (**b**) After setting the data offset a SSE tab appears from which the appropriate algorithm and (**c**) reference dataset is selected

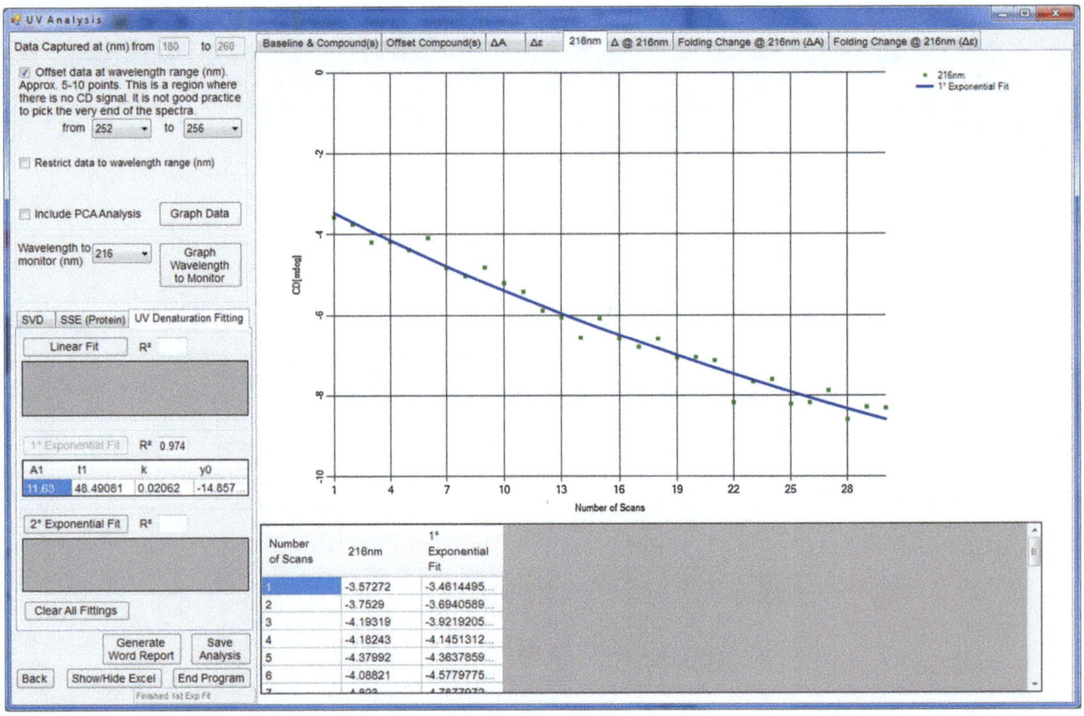

Fig. 4 Program appearance during UV denaturation analysis. Selecting the wavelength to monitor UV denaturation and clicking "Graph Wavelength to Monitor" button produces a graph which can be fitted to determine the kinetics of the denaturation

Thermal Melt Analysis

1. Select Thermal Melt option from drop-down menu for "Scan type."
2. Highlight (click to make line blue) the Sample and Baseline sample lines.

   Note: Unlike the 1 to *n* analysis option, only one sample can be analyzed at a time.
3. Click "Analyse Experiment Data."
4. Set the off-set region for the data (where no CD signal is observed).
5. Select the wavelength to monitor for thermal denaturation by either clicking on the region of the observed signal in CDApps or selecting the wavelength from the drop-down menu.
6. Click "Graph Wavelength to Monitor" button.
7. Under the Boltzmann tab in CDApps, select the "Auto Fit Boltzmann Curve" button or, if the kinetics are known for the system and the analysis serves as a check to fit the details that can be manually inputted, then click "Manual Fit Boltzmann Curve" button (Fig. 6).

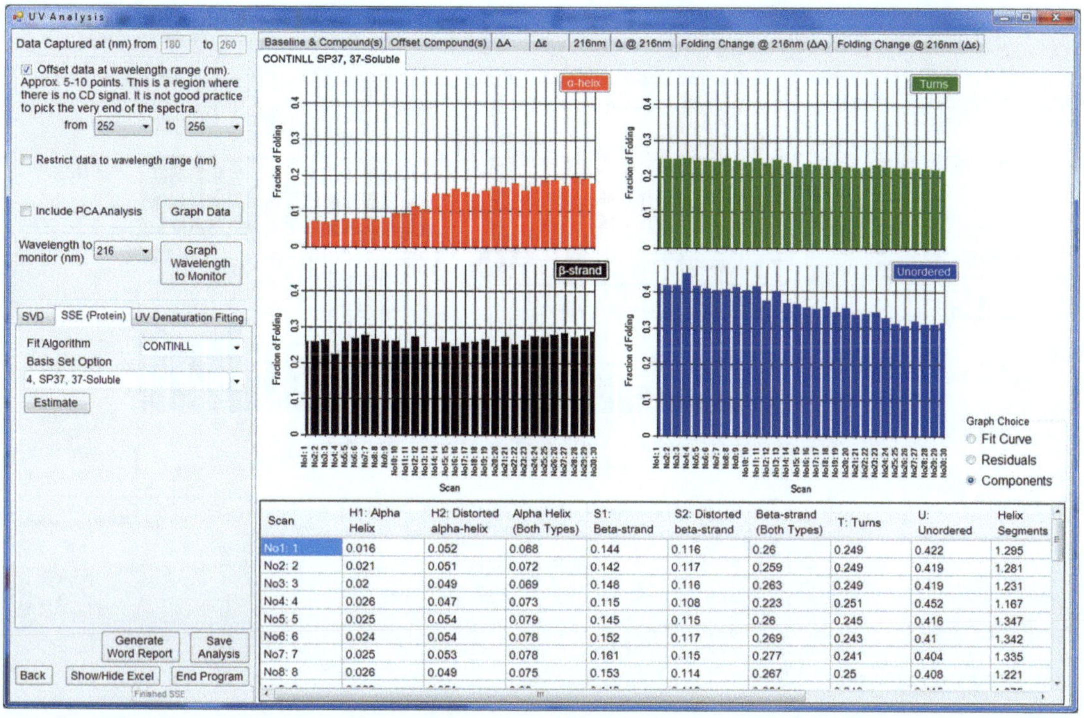

Fig. 5 SSE for each measurement of the denaturation file. SSE can be performed for each scan in the file using the options under the "SSE (protein)" tab

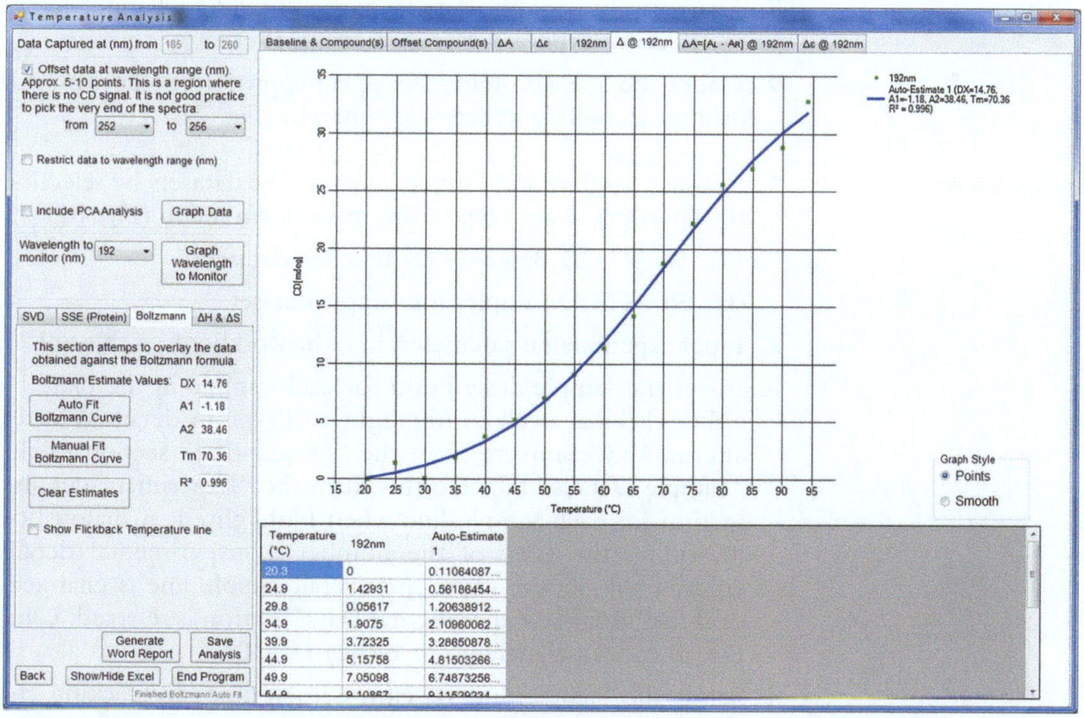

Fig. 6 Analysis window for Thermal Melt analysis. Change in CD can be monitored at a specific wavelength for calculation of denaturation kinetics. Kinetics calculated using the Boltzmann tab

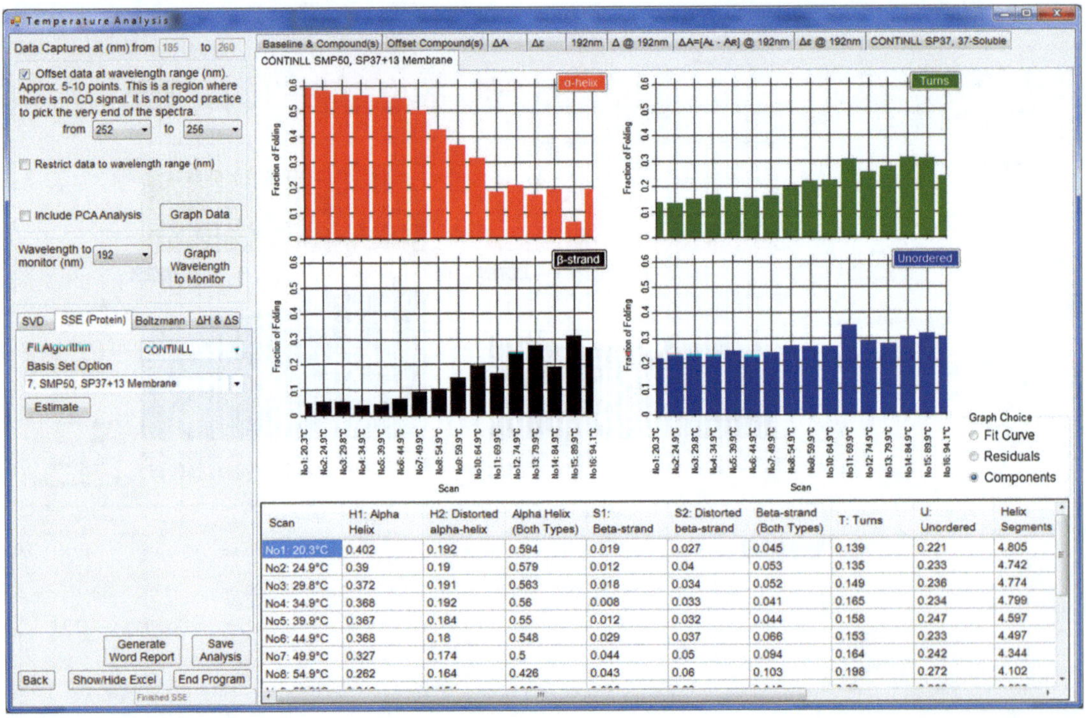

Fig. 7 SSE for scans taken at each temperature of the denaturation. SSE is performed by selecting the parameters under the "SSE (Protein)" tab

8. Alternatively, copy the data from the "Folding Change" tab of the Excel workbook for analysis using alternative software.
9. Under the SSE (Protein) tab, select appropriate algorithm for protein secondary structure estimation (Fig. 7).

96-Well Analysis

1. Allocate baseline and sample labels to the datasets by selecting the appropriate scan type from drop-down menu (Fig. 8a):
   (a) 96 HTCD Baselines for baseline dataset.
   (b) 96 HTCD Samples for sample dataset.
2. Input experimental parameters into the workbook (*see* **Note 17**).
3. Insert the sample description for each sample in the plate by either clicking a cell to highlight and type in details directly, alternatively complete from the "Description" section of the "Sample Details" tab. Do this from the "Experiment details" section for each sample line when highlighted. A dialog box informing the User of the number of experimental details entered will appear when the overall sample line is changed, or if the "Analyse Experimental Data" button is selected. Click OK to continue analysis (*see* **Notes 18–20**).
4. Select the Sample and Baseline sample lines by clicking the desired line (the line becomes blue).

A.

B.

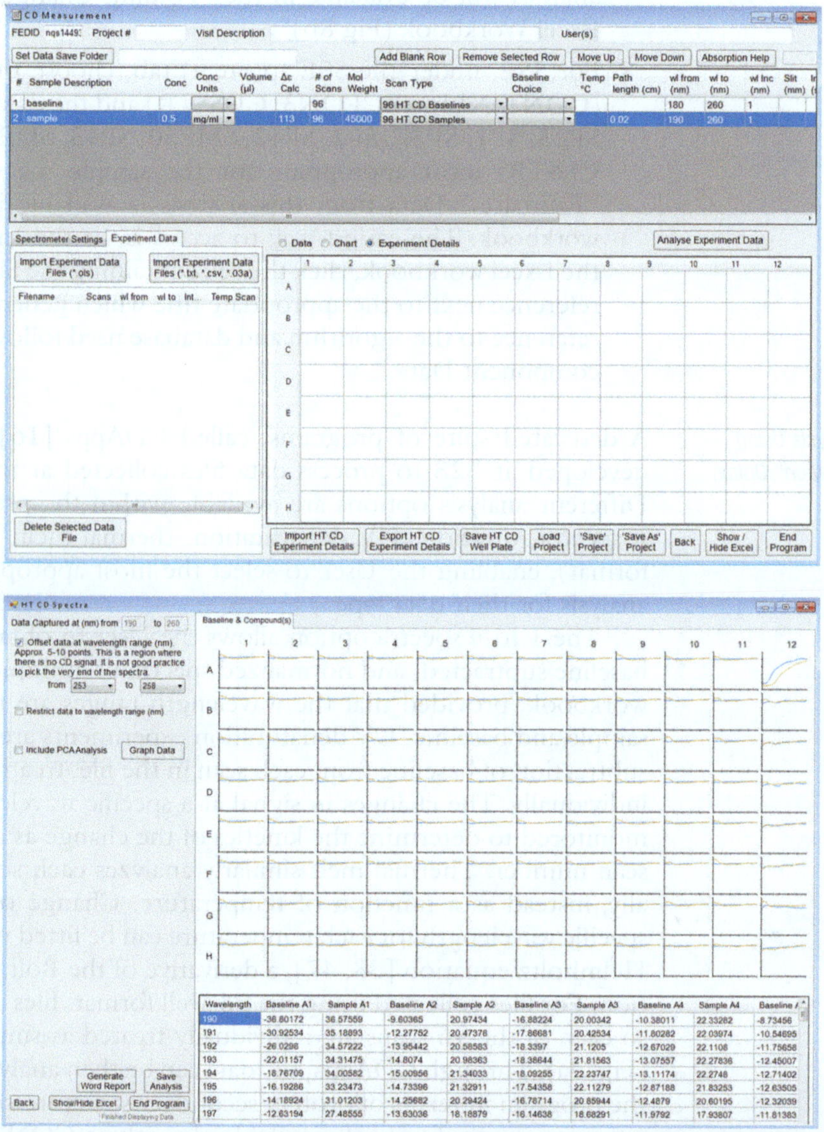

Fig. 8 Windows during the analysis for HTCD data. (**a**) Initial window where experimental details are input for the analysis under the "Experimental Details" section of the "Experiment Data" tab (**b**) off-set region for the data where no CD signal is observed

5. Click "Analyse Experiment Data" (*see* **Note 17**).
6. Set the off-set region for the data (where no CD signal is observed). Click "Graph Data" button (Fig. 8b).
7. If a particular wavelength is to be monitored across the plate for a change in signal, this can be selected from the "Wavelength to monitor (nm) section of CDApps. Data for this

analysis will be available in the "Folding Change" tab of the Excel Workbook (Fig 8b)".

8. For SSE, under the SSE (protein) tab, choose the algorithm (CONTINLL, SECLON3, CDSSTR) and the database (SP29, SP23, SP1, SP37, SP2, SP42, SMP50, SP43, SDP48, SMP56, CLSTR) most appropriate for the sample measured. Click "Estimate." Data from this analysis is available in the Excel workbook. The easiest way to access is via the Index page of the Excel workbook, click the cell containing the Sheet number reference next to the appropriate title which generally contains reference to the algorithm and database used followed by "SSE component Data."

3.4.3 Analysis Using the "CD Titration" Option

A dedicated suite of programs, called CDApps [16], have been developed at B23 to process data files collected at the beamline. Different analysis options are available within the program (1 to *n* spectra, titration, UV denaturation, thermal melt, and 96-well format), enabling the User to select the most appropriate type of analysis for their data type.

The 1 to *n* spectra option allows the User to obtain averaged, baseline subtracted, and normalized spectra of multiple files in one workbook, provided that the wavelength ranges are the same for sample and baseline. UV denaturation experiments are analyzed by subtraction of baseline from each scan in the file, treating each scan individually. The changes in signal at a specific wavelength can be monitored to determine the kinetics of the change as a function of scan number. Thermal melt similarly analyzes each scan individually, instead as a function of temperature. Change of signal at a specific wavelength for each temperature can be fitted with a Gibbs-Helmholtz equation [38, 47], a derivative of the Boltzmann equation. For files collected using the 96-well format, files are analyzed, so each of the 96 scans are individually treated as single spectrum (Fig. 3a). Through CDApps, the data can be then analyzed through the program in terms of protein secondary structure content using either or three of publicly available algorithms CONTINLL [19–21], CDSSTR, and SECLON3 [20] (Fig. 3b).

1. Load CDApps and click the "CD Titration" button on the option window (Fig. 9).

Setting Up the Analysis Workbook of "Fixed" Volume Titrations in CDApps

For the fixed volume titration option, calculations will be made for each titration step to determine the volume of each component to combine with a defined fixed volume. Each titration step will be a new, individual experiment. Therefore, this option is best for cases where an abundance of material is available as it can be costly in material determined on the total volume for each titration step and the number of molar ratio points to be investigated. Downstream analysis does not require a dilution factor because all titrations are performed in the same volume; therefore, points after ligand

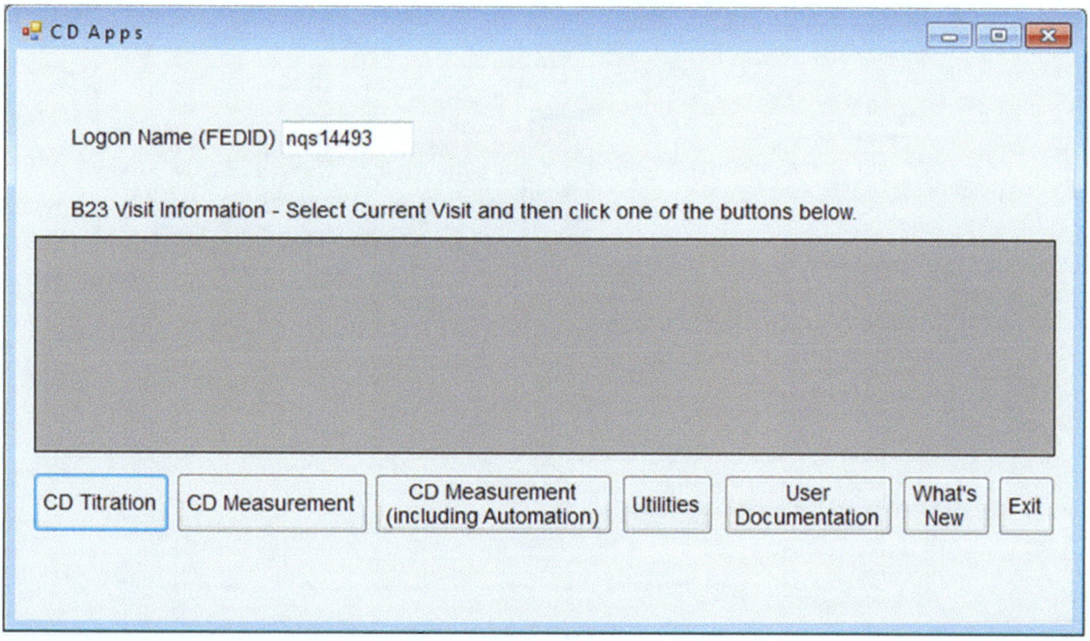

Fig. 9 Option window in CDApps

subtraction and zeroing can be taken directly for analysis of the binding constants.

1. Select the "Fixed" volume option at the top of the workbook.
2. Under the "Experiment Settings" tab, complete the details for the cell parameters, Host and Ligand. The information provided here will be used to determine the volumes required at each step of the titration (Fig. 10a).
3. To determine the Ligand stock concentration to be used and the volumes to be used at each titration point, in the spreadsheet section of the workbook enter the titration points under the "Molar Ratio" column (Fig. 10a).
4. Add the number of the Ligand stock (1, 2, or 3) which will be used at each point (Fig. 10a) to the "Stock Number" column. Ensure aliquot volumes are feasible within the error of the pipettes. For ease of the User, each Ligand stock is color coded to quickly identify which stock to use.
5. Complete experimental parameters under the "Spectrometer Settings" tab (Fig. 10b).
6. Import data files into the workbook under the "Experiment Data" tab by selecting the appropriate button for the data to be imported (either .ols or .csv/.txt files) and locate the files. When multiple files are to be imported, highlight all appropriate files (Ctrl + Click/Shift + Click). Click Open. The files will be imported into CDApps in the "Imported Data Files" column (Fig. 10c).

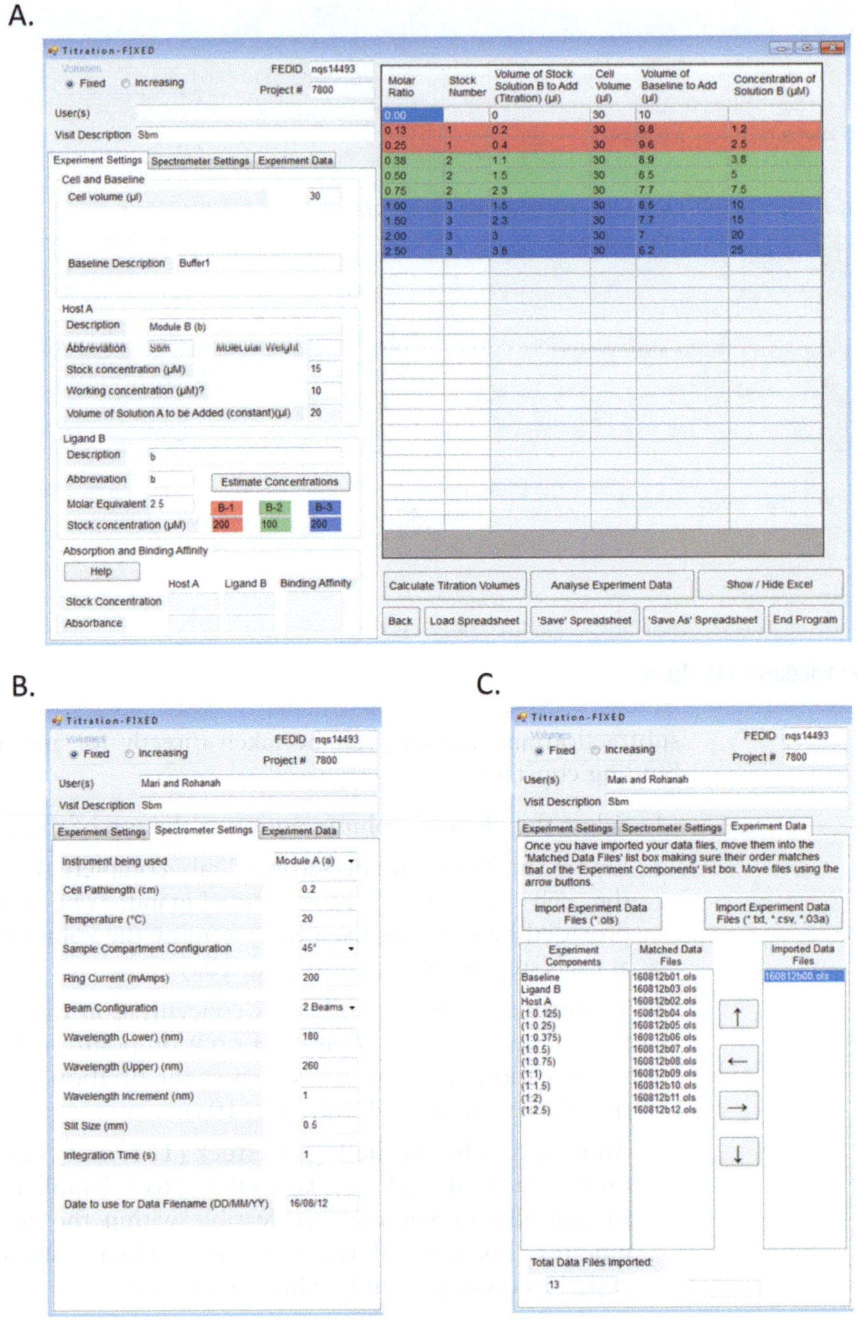

Fig. 10 Analysis workbook windows in CDApps for titrations in fixed volume. (**a**) Experiment setting, (**b**) Spectrometer setting, (**c**) Experiment data

7. Match the appropriate files to the "Experiment Component." This can be done one by one or in bulk. Four arrows are used to control the order and matching of files (Fig. 10c).
   (a) ↑ and ↓ control the position of the file in the "Imported Data Files" column.

(b) ← and → control the transfer of files between the "Imported Data Files" and "Matched Data Files" columns.

8. Only when in the "Imported Data Files" column can the order of the files be changed. Once they have been moved to the "Matched Data Files" column they cannot be reordered. To rearrange the order, files will have to be moved to the "Imported Data Files" column, ordered, and then transferred to the "Matched Data Files" column.

9. Files can be transferred between columns either individually by selecting individual files or in bulk by highlighting multiple files (Ctrl + Click/Shift + Click) before using ← and →.

10. Once the workbook has been completed, save the spreadsheet by clicking ""Save As" Spreadsheet." This will enable the spreadsheet to be loaded at another time to continue the analysis and to ensure that a copy has been saved in case of any downstream erroneous analysis.

11. The Workbook is saved as an Excel workbook, and the experimental plan can be printed from the "Titration" tab of the workbook for the convenience of the User to perform the experiment. The plan details the volumes required at each step, and ligand stocks are color coded for quick identification by the User (Fig. 11).

12. Instruction on how to perform the analysis is provided in Subheading 3.4.3.3.

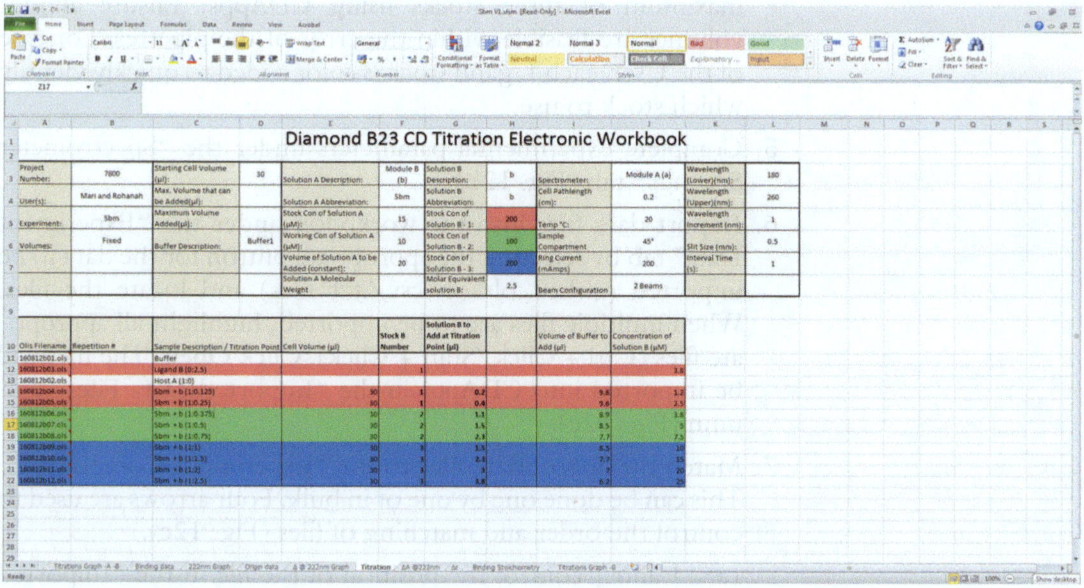

Fig. 11 Titration tab of the workbook detailing the experiment plan can be printed for a convenient plan to use for the experiment in the lab

Setting Up the Analysis Workbook of "Increasing" Volume Titrations in CDApps

For titrations performed in an increasing volume, a single sample of the Host is made, and aliquots of Ligand stock are added at each titration step so that a cumulative concentration of ligand is achieved at each step. For these types of experiments, during the analysis a dilution factor will have to be applied to the different spectra of the Host at each titration point to account for the reduced signal intensity resulting from the dilution caused by the addition of each ligand aliquot.

1. Select the "Increasing" volume option at the top of the workbook.
2. Under the "Experiment Settings" tab, complete the details for the cell parameters, Host and Ligand. The information provided here will be used to determine the volumes required at each step of the titration (Fig. 12a). CDApps will automatically calculate the maximum volume of ligand which can be added within acceptable dilution limits.
3. To determine the Ligand stock concentration to be used and the volumes to be used at each titration point, in the spreadsheet section of the workbook enter the titration points under the "Molar Ratio" column (Fig. 12a).
4. Add the number of the Ligand stock (1, 2, or 3) which will be used at each point (Fig. 12a) to the "Stock Number" column. CDApps will notify you if the maximum volume of added ligand is exceeded. If this happens, adjust the concentrations of the ligand stocks to reduce the total added volume, using a maximum of three stocks using CDApps. Ensure aliquot volumes are feasible within the error of the pipettes. For ease of the User, each Ligand stock is color coded to quickly identify which stock to use.
5. Complete experimental parameters under the "Spectrometer Settings" tab (Fig. 12b).
6. Import data files into the workbook under the "Experiment Data" tab by selecting the appropriate button for the data to be imported (either .ols or .csv/.txt files) and locate the files. When multiple files are to be imported, highlight all appropriate files (Ctrl + Click/Shift + Click). Click Open. The files will be imported into CDApps in the "Imported Data Files" column (Fig. 12c).
7. Match the appropriate files to the "Experiment Component." This can be done one by one or in bulk. Four arrows are used to control the order and matching of files (Fig. 12c).
   (a) ↑ and ↓ control the position of the file in the "Imported Data Files" column.
   (b) ← and → control the transfer of files between the "Imported Data Files" and "Matched Data Files" columns.

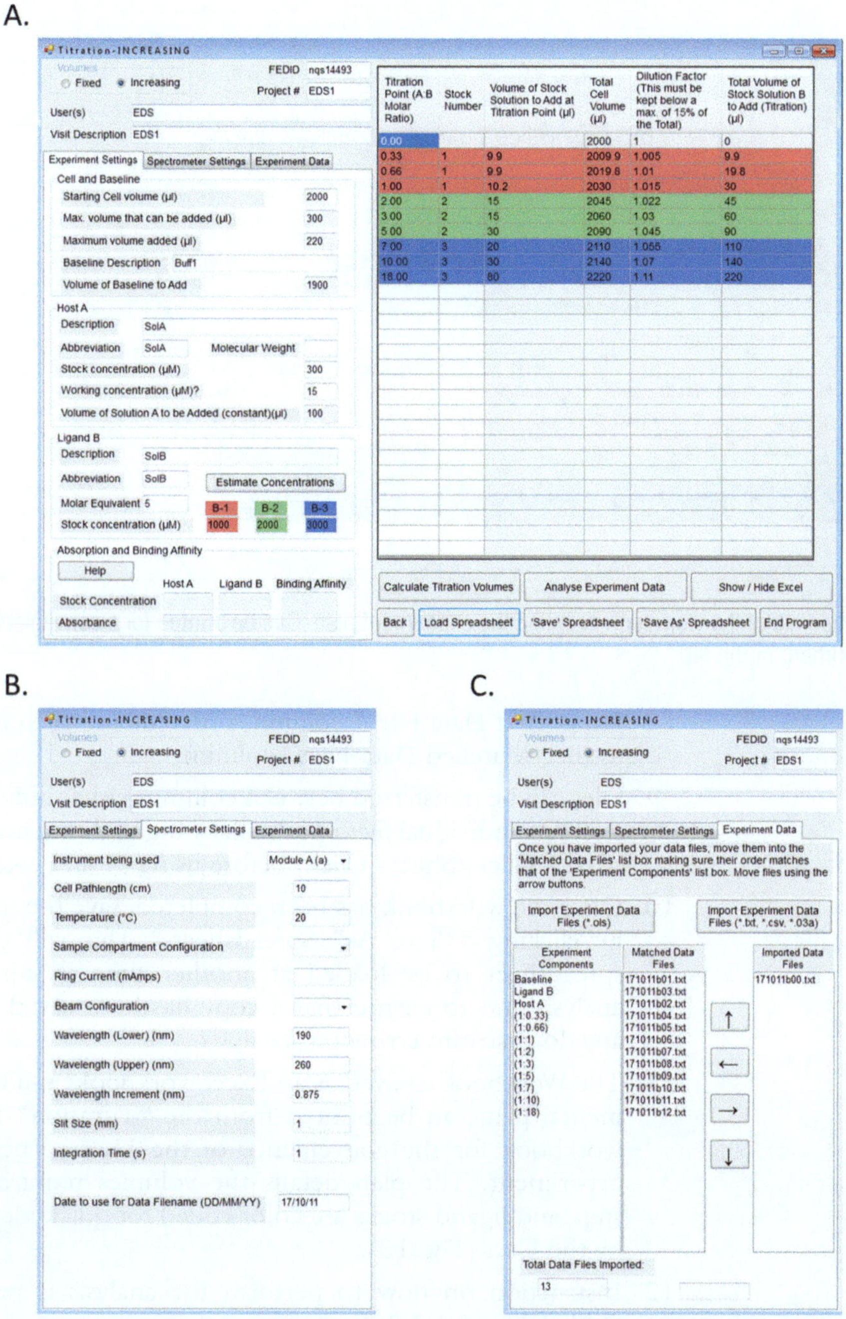

Fig. 12 Analysis workbook windows in CDApps for titrations in increasing volume. (**a**) Experiment setting, (**b**) Spectrometer setting, (**c**) Experiment data

8. Only when in the "Imported Data Files" column can the order of the files be changed. Once they have been moved to the "Matched Data Files" column they cannot be reordered. To rearrange the order, files will have to be moved to the

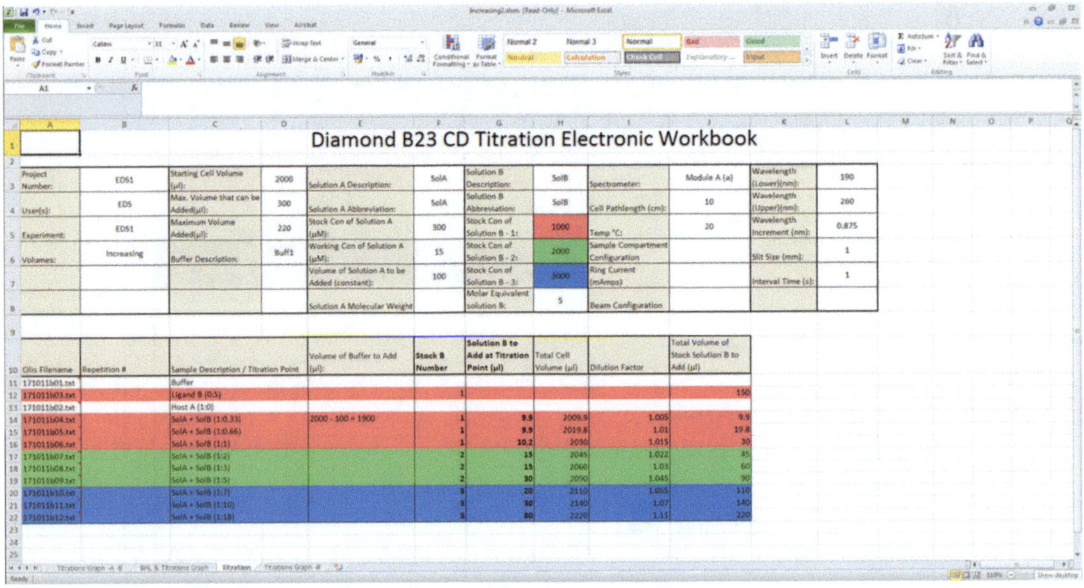

Fig. 13 Titration tab of the workbook detailing the experiment plan. Can be printed for a convenient plan to use for the experiment in the lab

"Imported Data Files" column, ordered, and then transferred to the "Matched Data Files" column.

9. Files can be transferred between columns either individually by selecting individual files or in bulk by highlighting multiple files (Ctrl + Click/Shift + Click) before using ← and →.
10. Once the workbook has been completed, save the spreadsheet by clicking ""Save As" Spreadsheet." This will enable the spreadsheet to be loaded at another time to continue the analysis and to ensure that a copy has been saved in case of any downstream erroneous analysis.
11. The Workbook is saved as an Excel workbook, and the experimental plan can be printed from the "Titration" tab of the workbook for the convenience of the User to perform the experiment. The plan details the volumes required at each step, and ligand stocks are color coded for quick identification by the User (Fig. 13).
12. Instruction on how to perform the analysis is provided in Subheading 3.4.3.3.

Determination of the Binding Constants

1. Load CDApps, select CD Titration button, and load the appropriate Spreadsheet. All details and
2. Click "Analyse Experiment Data" button.
3. Off set data to a region where no CD signal is detected.

4. If necessary, restrict the wavelength range of the data to remove noise. For measurements in the far-UV region requiring SSE, ensure restriction does not compromise the wavelength ranges required by the algorithms and datasets, e.g., 190 nm and above.

5. From the "Titrations –A –B" tab or the "Wavelength to monitor (nm)" section, select the wavelength from which binding constant calculations will be made. This is usually the region on the spectra which shows the biggest change in magnitude over the course of the titration. Click "Graph Wavelength to monitor" button (Fig. 14).

A.

B.

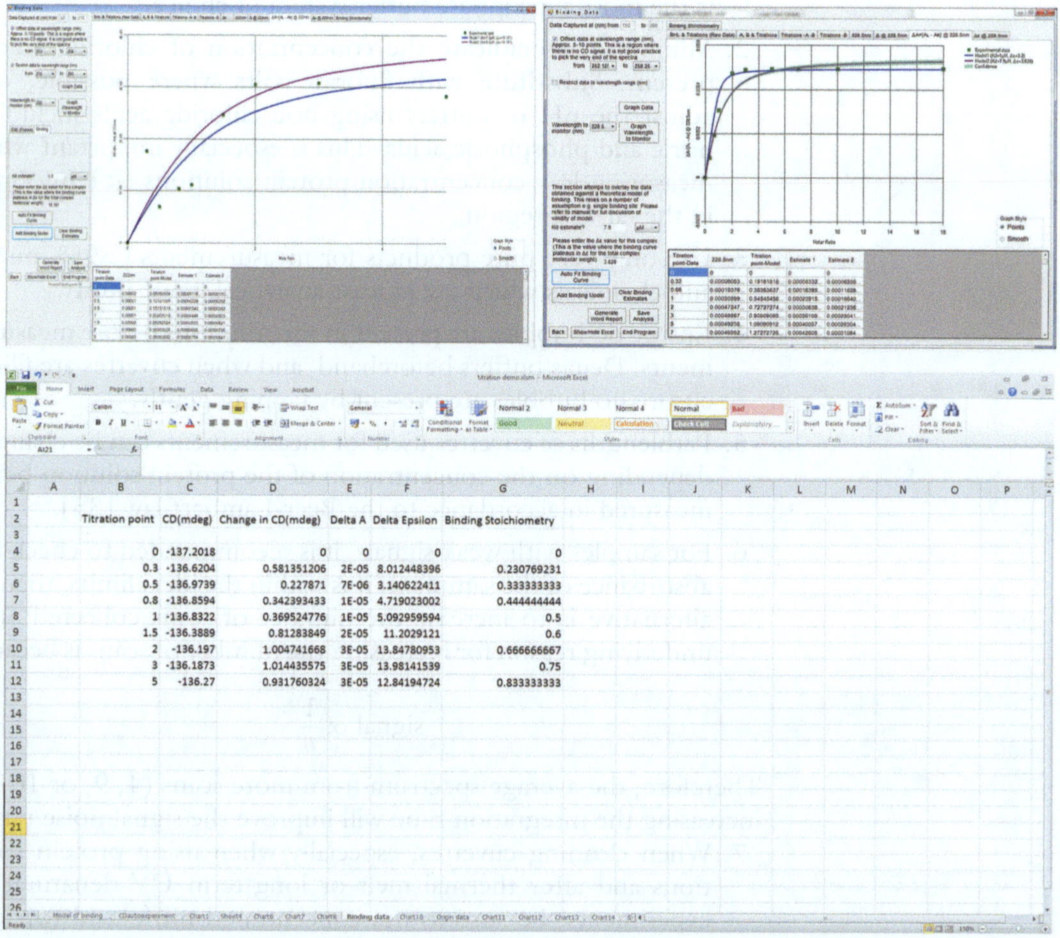

Fig. 14 Analysis of the binding constants from titration data. (**a**) Analysis window in CDApps. (**b**) Binding Data tab of the analysis workbook. Data can be exported to alternative programs for further analysis and display. The example in Fig. 14a left has been chosen deliberately to show the results of a preliminary CD titration. The experimental data appear to be a bit scattered in the plot intensity versus ligand concentration indicating that for the membrane protein, the incubation time might require to be increased and also that extra intermediate points might confirm that subtle change between the first two point and also between the 5th and 6th point

6. A "Binding" tab will then appear. From here fitting can be performed for the points to determine the K_d of the binding interaction. Fitting can either be automatically fitted through CDApps using the "Auto Fit Binding Curve" button, or manually by entering $\Delta\varepsilon$ values and clicking "Add Binding Model" button.

4 Notes

1. Conduct measurements in the "most UV-transparent" buffer possible, achieved using buffering systems at their lowest working concentration for the system of interest. Phosphate buffer is the most appropriate buffer system of choice.
2. Minimize or eradicate the concentration of chloride anions present. Substitute with fluoride salts where possible, and adjust the pH of buffers using non-chloride acids including nitric and phosphoric acids. This is especially important when measuring low-concentration protein solutions (<1 mg/mL) in the far-UV region.
3. Use only very pure products for measurements (>95% pure), and chemicals which are at least analytical grade purity.
4. Ensure no bubbles are present in the solutions before measurements. Degas buffers beforehand, and when cuvettes are filled, ensure no bubbles are present by tapping gently.
5. Pathlengths of cuvettes used for measurements can be changed depending on the concentration of the protein solution being measured in accordance to the Beer-Lambert law [35].
6. For samples with weak signals, it is recommended to check the absorbance of the sample. If it is within the ideal limits, then an alternative is to increase the number of scans collected. The underlying reason for increasing the number of scans is because

$$\text{signal} \propto \frac{1}{n}$$

Therefore, the average spectrum from more scans (4, 9, or 16) or increasing the integration time will improve the signal:noise ratio.

7. When cleaning cuvettes, especially when using protein solutions and after thermal melt or long-term UV denaturation experiments, be sure to use concentrated nitric acid to clean thoroughly wet cells from wash of 96% ethanol. Also thoroughly rinse and dry the cells before reuse to remove residual acid which can change the pH affecting the protein folding.
8. Remove all debris and solid particulates by centrifugation beforehand and using only the supernatant, or filtering the solution through a 0.2 μm filter.

9. Be sure that the HV of the PMT detector does not exceed the recommended voltage from the CD instrument manufacturer (for B23 is about 600 V, which is common with other benchtop instruments using the PMT detector) as this is a sign of over-saturation of the detector. If this occurs, remove the portion of spectra for which this occurs before analysis or decrease the sample concentration, or reduce the cell pathlength accordingly.
10. Observe the absorbance and discount measurements which are above 1.5. Reduce the concentration to reduce the absorbance, or alternatively decrease the cell pathlength. Also consider the absorbance contribution from the ligand as this should be factored in for ligand-binding studies, especially titration experiments.
11. In cases where the ligand absorbs at 280 nm, a lower starting protein concentration can be used (0.4 mg/mL) to allow for the absorbance contribution of the ligand throughout the titration.
12. When conducting titration experiments in increasing volume, ensure the total added ligand volume does not exceed 20% of the original volume (CDApps will warn when a 15% dilution has been reached). This ensures any changes observed are solely as a result of ligand binding rather than any potential "concentration effects" influencing the proteins conformation or local tertiary structure. To aid this, use ligand stocks at the highest possible concentration (usually in the region of 10–20 times the final concentration of ligand required for the titration).
13. When conducting ligand-binding studies ensure thorough and gentle mixing in the solution before measurement, avoiding the formation of bubbles (*see* **Note 4**). Check for a homogenous appearance. Look for signs of insufficient mixing, e.g., streaking of solutions, and rectify before measurement. This is incredibly important when conducting titration experiment, and care should be taken to thoroughly mix after the addition of each aliquot.
14. Determine the incubation time required for the system to stabilize after the addition of ligand by determining the time required for consecutively measured spectra to overlap and no longer change. The determined time can be used as standard for subsequent measurements.
15. Care should be taken to monitor for precipitation during the titration, a sign of unsuitable conditions for the protein (e.g., though changes to the pH of the solution upon addition of the ligand) and the possibility of reduced protein concentration which should be accounted for during the analysis for accuracy

especially for measurements in the far-UV region and SSE. Solid particulates should be removed to prevent light scattering.

16. The extent of UV denaturation is specific to a protein. Use experience of the proteins behavior and stability to judge the number of scans to collect.
17. When using CDApps for analyses of data sets more complicated than standard 1 to *n* measurements (UV denaturation, Thermal Melt, 96 HT format), only one dataset can be analyzed at a time.
18. When analyzing data for the 96-HT format, be sure to complete the experimental details and parameters for all of the scans in the file.
19. Ensure the number of scans in the workbook match the number of scans in the uploaded data files.
20. After selection of the HTCD analysis options, experimental parameters (e.g., concentration and units, $\Delta\varepsilon$, molecular weight, and pathlength) will have to be re-entered. A dialog box will ask if you want the value to be applied to all of the scans in the file, click OK if this is the case, otherwise they can be manually inserted for each individual scan under the "Experiment details" section in CDApps.

Acknowledgement

We like to thank Diamond Light Source for access to B23 beamline (CM12182, CM14484, CM16778, CM19680).

References

1. Javorfi T, Hussain R, Myatt D, Siligardi G (2010) Measuring circular dichroism in a capillary cell using the B23 synchrotron radiation CD beamline at Diamond Light Source. Chirality 22(1E):E149–E153
2. Hussain R, Javorfi T, Siligardi G (2012) Circular dichroism beamline B23 at the Diamond Light Source. J Synchrotron Radiat 19 (1):132–135
3. Hussain R, Jávorfi T, Rudd TR, Siligardi G (2016) High-throughput SRCD using multiwell plates and its applications. Sci Rep 6 (1):38028
4. Hussain R, Jávorfi T, Siligardi G (2012) Spectroscopic analysis: synchrotron radiation circular dichroism. In: Carreira EM, Yamamoto H (eds) Comprehensive chirality, vol 8. Elsevier, Amsterdam, pp 438–448
5. Hussain R, Longo E, Siligardi G, Hussain R, Longo E, Siligardi G (2018) UV-denaturation assay to assess protein photostability and ligand-binding interactions using the high photon flux of diamond B23 beamline for SRCD. Molecules 23(8):1906
6. Hughes CS, Longo E, Phillips-Jones MK, Hussain R (2017) Characterisation of the selective binding of antibiotics vancomycin and teicoplanin by the VanS receptor regulating type A vancomycin resistance in the enterococci. Biochim Biophys Acta Gen Subj 1861 (8):1951–1959
7. Phillips-Jones MK, Patching SG, Edara S, Nakayama J, Hussain R, Siligardi G (2013) Interactions of the intact FsrC membrane histidine kinase with the tricyclic peptide inhibitor siamycin I revealed through synchrotron

radiation circular dichroism. Phys Chem Chem Phys 15(2):444–447
8. Siligardi G, Hussain R, Patching SG, Phillips-Jones MK (2014) Ligand- and drug-binding studies of membrane proteins revealed through circular dichroism spectroscopy. Biochim Biophys Acta 1838(1 Pt A):34–42
9. Hussain R, Harding SE, Hughes CS, Ma P, Patching SG, Edara S et al (2016) Purification of bacterial membrane sensor kinases and biophysical methods for determination of their ligand and inhibitor interactions. Biochem Soc Trans 44(3):810–823
10. Bettaney KE, Sukumar P, Hussain R, Siligardi G, Henderson PJF, Patching SG (2013) A systematic approach to the amplified expression, functional characterization and purification of inositol transporters from Bacillus subtilis. Mol Membr Biol 30(1):3–14
11. Runti G, Lopez Ruiz Mdel C, Stoilova T, Hussain R, Jennions M, Choudhury HG et al (2013) Functional characterization of SbmA, a bacterial inner membrane transporter required for importing the antimicrobial peptide Bac7 (1-35). J Bacteriol 195(23):5343–5351
12. Patching S, Edara S, Ma P (2012) Interactions of the intact FsrC membrane histidine kinase with its pheromone ligand GBAP revealed through synchrotron radiation circular dichroism. Biochem Biophys Acta 1818 (7):1595–1602
13. Hassan KA, Jackson SM, Penesyan A, Patching SG, Tetu SG, Eijkelkamp BA et al (2013) Transcriptomic and biochemical analyses identify a family of chlorhexidine efflux proteins. Proc Natl Acad Sci U S A 110(50):20254–20259
14. Kalverda AP, Gowdy J, Thompson GS, Homans SW, Henderson PJF, Patching SG (2014) TROSY NMR with a 52 kDa sugar transport protein and the binding of a small-molecule inhibitor. Mol Membr Biol 31 (4):131–140
15. Hughes CS, Longo E, Phillips-Jones MK, Hussain R (2017) Quality control and biophysical characterisation data of VanSA. Data Br 1(14):41–47
16. Hussain R, Benning K, Javorfi T, Longo E, Rudd TR, Pulford B et al (2015) CDApps: integrated software for experimental planning and data processing at beamline B23, Diamond Light Source. J Synchrotron Radiat 22 (2):465–468
17. Kelly SM, Jess TJ, Price NC (2005) How to study proteins by circular dichroism. Biochim Biophys Acta 1751(2):119–139
18. Hennessey JP, Johnson WC (1981) Information content in the circular dichroism of proteins. Biochemistry 20(5):1085–1094
19. Provencher SW, Glöckner J (1981) Estimation of globular protein secondary structure from circular dichroism. Biochemistry 20(1):33–37
20. Sreerama N, Woody RW (2000) Estimation of protein secondary structure from circular dichroism spectra: comparison of CONTIN, SELCON, and CDSSTR methods with an expanded reference set. Anal Biochem 287 (2):252–260
21. van Stokkum IHM, Spoelder HJW, Bloemendal M, van Grondelle R, Groen FCA (1990) Estimation of protein secondary structure and error analysis from circular dichroism spectra. Anal Biochem 191(1):110–118
22. Greenfield NJ (2006) Using circular dichroism spectra to estimate protein secondary structure. Nat Protoc 1(6):2876–2890
23. Micsonai A, Wien F, Kernya L, Lee Y-H, Goto Y, Réfrégiers M, Kardos J (2015) Accurate secondary structure prediction and fold recognition for circular dichroism spectroscopy. Proc Natl Acad Sci U S A 112(24): E3095–E3103
24. Lobley A, Whitmore L, Wallace BA (2002) DICHROWEB: an interactive website for the analysis of protein secondary structure from circular dichroism spectra. Bioinformatics 18 (1):211–212
25. Siligardi G, Hussain R (1998) Biomolecules interactions and competitions by non-immobilised ligand interaction assay by circular dichroism. Enantiomer 3(2):77–87
26. Kumar S (2006) Organic chemistry spectroscopy of organic compounds. Guru Nanak Dev University, Amritsar
27. Iyer KS, Klee WA (1973) Direct spectrophotometric measurement of the rate of reduction of disulfide bonds. The reactivity of the disulfide bonds of bovine-lactalbumin. J Biol Chem 248 (2):707–710
28. Seo A, Jackson JL, Schuster JV, Vardar-Ulu D (2013) Using UV-absorbance of intrinsic dithiothreitol (DTT) during RP-HPLC as a measure of experimental redox potential in vitro. Anal Bioanal Chem 405 (19):6379–6384
29. Han JC, Han GY (1994) A procedure for quantitative determination of tris(2-carboxyethyl)phosphine, an odorless reducing agent more stable and effective than dithiothreitol. Anal Biochem 220(1):5–10
30. Wingfield PT (2001) Use of protein folding reagents. Curr Protoc Protein Sci. Appendix 3:Appendix 3A
31. Knubovets T, Osterhout JJ, Klibanov AM (1999) Structure of lysozyme dissolved in neat organic solvents as assessed by NMR and

CD spectroscopies. Biotechnol Bioeng 63 (2):242–248

32. Lide DR (2010) In: Haynes WM (ed) CRC handbook of chemistry and physics, 91th edn. CRC Press (Taylor and Francis Group), Boca Raton, FL
33. Zinna F, Resta C, Gorecki M, Pescitelli G, Di Bari L, Javorfi T, Hussain R, Siligardi G (2017) Circular dichroism imaging: mapping the local supramolecular order in thin films of chiral functional polymers. Macromolecules 50 (5):2054–2060
34. Johnson WC (1985) Circular dichroism and its empirical application to biopolymers. Methods Biochem Anal 31:61–163
35. Lindon JC, Tranter GE, Koppenaal D (2016) Encyclopedia of spectroscopy and spectrometry. 3rd ed. 287 p
36. Damon AJH, Kresheck GC (1982) Influence of surfactants on the conformation of β-lactoglobulin B using circular dichroism. Biopolymers 21(5):895–908
37. Plangger H, Scheibenzuber M, Blümelhuber G, Meyer-Pittroff R (2003) Influence of high pressure on the secondary structure of poly-L-lysine. In: Winter R (ed) Advances in high pressure bioscience and biotechnology II. Springer, Berlin, Heidelberg
38. Greenfield NJ (2007) Using circular dichroism collected as a function of temperature to determine the thermodynamics of protein unfolding and binding interactions. Nat Protoc 1 (6):2527–2535
39. Gutiérrez-Mejía FA, van IJzendoorn LJ, Prins MWJ (2015) Surfactants modify the torsion properties of proteins: a single molecule study. New Biotechnol 32(5):441–449
40. Sun C, Yang J, Wu X, Huang X, Wang F, Liu S (2005) Unfolding and refolding of bovine serum albumin induced by cetylpyridinium bromide. Biophys J 88(5):3518–3524
41. Hayashi R, Kakehi Y, Kato M, Tanimizu N, Ozawa S, Matsumoto M (2002) Circular dichroism under high pressure. Prog Biotechnol 19:583–590
42. Brown EM, Groves ML (1985) Effect of temperature on the circular dichroism spectra of-β_2-microglobulins. FEBS Lett 184 (1):36–39
43. Hussain R, Jávorfi T, Siligardi G (2012) Circular dichroism beamline B23 at the Diamond Light Source. J Synchrotron Radiat 19 (1):132–135
44. Clarke DT, Jones G (2004) CD12: a new high-flux beamline for ultraviolet and vacuum-ultraviolet circular dichroism on the SRS, Daresbury. J Synchrotron Radiat 11(2):142–149
45. Tanaka M, Yagi-Watanabe K, Kaneko F, Nakagawa K (2009) First observation of natural circular dichroism spectra in the extreme ultraviolet region using a polarizing undulator-based optical system and its polarization characteristics. J Synchrotron Radiat 16 (4):455–462
46. Prodromou C, Siligardi G, O'Brien R, Woolfson DN, Regan L, Panaretou B, Ladbury JE, Piper PW, Pearl LH (1999) Regulation of Hsp90 ATPase activity by tetratricopeptide repeat (TPR)-domain co-chaperones. EMBO J 18(3):754–762
47. Singleton DG, Hussain R, Siligardi G, Kumar P, Hrdlicka PJ, Berova N et al (2016) Increased duplex stabilization in porphyrin-LNA zipper arrays with structure dependent exciton coupling. Org Biomol Chem 14 (1):149–157

Chapter 7

Measuring Small Molecule Binding to *Escherichia coli* AcrB by Surface Plasmon Resonance

Steven W. Polyak, Rumana Mowla, and Henrietta Venter

Abstract

Antimicrobial resistance (AMR) is rapidly becoming one of the great healthcare challenges. A common mechanism employed by pathogenic bacteria to avoid the action of certain antibiotics is to overexpress efflux pumps that can extrude these drugs from the cell rendering them ineffective. Small molecule inhibitors that target bacterial efflux pumps provide a route toward reversing AMR. Here, we describe the application of surface plasmon resonance (SPR) technology to characterize protein:small molecule interactions between the inner membrane protein AcrB subunit of the *Escherichia coli* AcrA-AcrB-TolC efflux pump and its substrates and novel inhibitors. The SPR assay provides quantitative data about the kinetics of binding that can help guide the development of new chemotherapies to combat AMR.

Key words Surface plasmon resonance, Binding kinetics, Small molecule, Efflux pumps, Antimicrobial resistance, Antibiotic discovery, AcrA-AcrB-TolC, Efflux pump inhibitors

1 Introduction

The rise of antimicrobial resistance has rendered many classes of antibiotics ineffective for the treatment of bacterial infections [1, 2]. New approaches are desperately required to combat antibiotic resistance. One promising strategy under investigation is the reversal of resistance through the pharmacological inhibition of bacterial efflux pumps [3–5]. These transmembrane transport proteins are responsible for the extrusion of antibacterial agents from the bacteria, thereby preventing high, cytotoxic concentrations from accumulating inside the cell [6, 7]. The overexpression of efflux pumps is a common mechanism employed by pathogenic bacteria to avoid the antimicrobial activity of multiple, unrelated classes of chemotherapeutics [8, 9]. The clinically relevant antibiotic efflux pumps from Gram-negative bacteria are tripartite assemblies consisting of an inner membrane protein of the resistance nodulation division (RND) family that are responsible for the recognition and export of antibiotics, an outer membrane pore,

Nikolaos E. Labrou (ed.), *Targeting Enzymes for Pharmaceutical Development: Methods and Protocols*, Methods in Molecular Biology, vol. 2089, https://doi.org/10.1007/978-1-0716-0163-1_7,

and a periplasmic adapter protein. The best studied Gram-negative bacterial efflux pumps are the tripartite AcrA-ArcB-TolC transporter present in *Escherichia coli* and the MexA-MexB-OprM homolog found in *Pseudomonas aeruginosa* [6, 10]. Their ability to extrude a wide range of discrete compounds with unrelated chemical structures explains their crucial role in multidrug resistance [5, 11, 12]. Therefore, efflux pumps are highly attractive targets for the development of new products to reverse antimicrobial resistance [3].

Targeting AcrA-AcrB-TolC is an active area of antibiotic drug discovery with several groups identifying small molecule inhibitors through screening of natural product libraries and rational design [13–23]. Biological characterization of efflux pump inhibitors has been performed primarily using cell-based assays to either measure synergy with antibacterial agents or disruption of the extrusion of fluorescent dyes such as Hoechst 33342, TMA-DPH, ethidium bromide or Nile Red. Chemical optimization of these promising hits requires reliable biochemical and structural biology tools to facilitate the design, synthesis, and biological evaluation of the next generation analogs. However, obtaining quantitative structure-activity relationship data has been hampered by the lack of a facile, homogenous biochemical assay to directly measure ligand binding. Likewise, the crystallization of integral membrane proteins, such as the AcrB subunit, are notoriously difficult which is why only a handful of co-complexes of AcrB bound to substrates [20, 24–28] or inhibitors [20, 29] have been solved using X-ray crystallography. Recent advances to facilitate structural studies include engineering a soluble, readily crystallizable version of AcrB where the hydrophobic transmembrane helices have been removed [20] and the development of cryo-electron microscopy to understand the molecular mechanisms governing the function of the efflux pump and ligand binding [30]. The available structural data reveal that antibiotics translocate through a hydrophobic channel, encased by the side chains of phenylalanine residues at positions 136, 178, 610, 615, and 628, that runs through the periplasmic domain of AcrB. Substrates first enter a proximal multidrug binding site then translocate into a deeper, distal binding site through a functional rotating mechanism before extrusion through the TolC outer membrane pore. Inhibitors that bind into a pocket adjacent into to the deep binding pocket in AcrB, known as the hydrophobic trap, sterically hinder conformation changes that are required for the rotating mechanism to extrude the substrates [20, 29]. Accordingly, small molecules that move through this hydrophobic channel are also lipophilic often making them challenging to study in biochemical assays, where they are prone to aggregation and nonspecific binding in aqueous biological media.

Here, we present a method to measure the direct binding of small molecules to AcrB using surface plasmon resonance (SPR) technology. SPR is a powerful technique to measure the direct binding between two molecules and has been used extensively to quantitate bimolecular protein:protein, protein:DNA, and protein: small molecule interactions [31], including the binding of substrates and inhibitors to *E. coli* AcrB [32–34]. SPR has the advantage over other binding assays that neither binding partner needs to be modified, for example with fluorescent tags or radiolabels. Instead, one binding partner (the ligand) is chemically immobilized onto the surface of a sensor chip. Various concentrations of the second binding partner (the analyte) dissolved in a running buffer are then injected across the sensor. When binding occurs between the two molecules at the sensor surface, a change in refractive index occurs that can be monitored in real time. These binding responses are measured in resonance units (RU). Two types of sensorgrams can be generated from SPR experiments depending upon the kinetics of the bimolecular interaction. If the kinetic association (k_a) and dissociation (k_d) rates are fast and outside the parameters of quantitation for the software, sensorgrams measuring equilibrium binding responses at each concentration of analyte can be generated (Fig. 1a). This data can be plotted as a binding isotherm allowing calculation of maximum binding (B_{max}) and an equilibrium binding constant ($K_d = 1/2B_{max}$). This equilibrium binding analysis is typical for protein:small molecule interactions and has been observed for AcrB binding substrates Hoechst-33342, TMA-DPH, doxorubicin, minocycline, and novobiocin as well as efflux pump

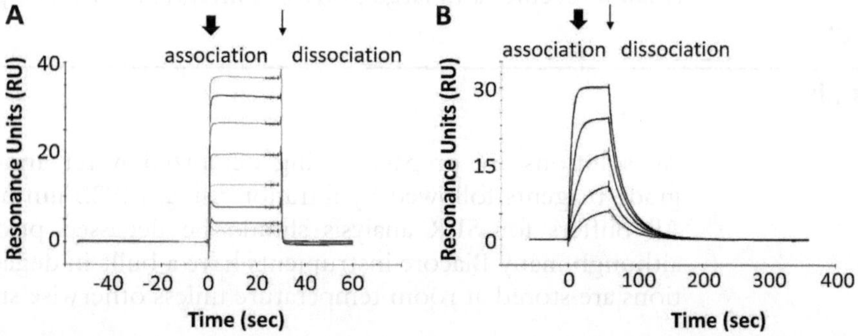

Fig. 1 Alternate models of SPR analysis. Typical sensorgrams for bimolecular interactions are shown. Sensorgrams are plots showing changes in the surface resonance, measured in resonance units, as a function of time. Heavy arrows show the time when an analyte is first injected across the sensor surface, and the light arrow shows the end of the injection step. The association phase measures the binding of the analyte to the immobilized ligand. The dissociation step is an injection of buffer and disassembly of the complex. The binding response is normalized by subtraction of nonspecific binding of analyte on a control surface. (**a**) When the interaction has very fast association and dissociation rates, the "Affinity" model is applied for the data analysis. The data is plotted as a binding isotherm to define maximum binding (B_{max}). An equilibrium binding constant K_d can be calculated from the curve ($K_d = 1/2B_{max}$). This data is common for small molecules binding to AcrB. (**b**) The "Kinctic" model can be used when the association and dissociation rates can be calculated by the software

inhibitors [32, 33]. Alternatively, when the binding kinetics are slower (Fig. 1b), kinetic constants k_a and k_d can be quantitated and equilibrium binding constants calculated ($K_a = k_a/k_d$; $K_d = 1/K_a$). This data is common for strong protein:protein interactions but may also be observed for high-affinity protein:small molecule interactions that are characterized by slow dissociation rates.

In this report, we present protocols for the immobilization of the AcrB subunit of the *E. coli* AcrA-AcrB-TolC efflux pump and assays to determine the equilibrium binding constants for substrates and inhibitors. We also describe our experimental design and data analysis that addresses nonspecific binding of hydrophobic analytes to the sensor surfaces and non-specific binding to the large membrane domain of AcrB. One approach commonly employed to address nonspecific binding is to perform the SPR binding assays simultaneously with both wild-type protein and a mutant engineered such that it is incapable of ligand binding. However, this strategy is not satisfactory for efflux pumps due to lack of an appropriate mutant that effectively abolishes ligand binding. Here, the sugar transport protein GalP was included as a negative control to account for nonspecific binding. This membrane protein was selected as it bears certain structural similarities with AcrB, including its homo-trimeric assembly where each subunit contains 12 transmembrane helices [35]. However, as GalP has high selectivity for binding sugars it presents as an attractive negative control. The protocols outlined here may be adaptable for researchers likewise wanting to measure binding interactions involving lipophilic small molecules and large polytopic integral membrane proteins.

2 Materials

All solutions are prepared using deionized water and analytical grade reagents followed by filtration using a 0.22 μm membrane. All buffers for SPR analysis should be degassed prior to use, although many Biacore instruments have a built-in degasser. Solutions are stored at room temperature unless otherwise stated.

2.1 Reagents and Buffers

1. Biacore T100/T200/S200.
2. Series S CM5 sensorchips for Biacore T100/T200/S200. Store sensorchips at 4 °C but equilibrate to room temperature before loading into the Biacore.
3. 10× Phosphate buffer saline (PBS): Dissolve 80 g NaCl, 2 g KCl, 14.4 g Na_2HPO_4, 2.4 g KH_2PO_4 in 800 mL of MilliQ water. Adjust pH to 7.4 with HCl and make final volume to 1 L.
4. Running Buffer A: 1.05× PBS, 0.05% Tween 20 (*see* **Notes 1** and **2**).

5. Running Buffer B: 1× PBS, 0.05% Tween 20, 5% DMSO.
6. Amine Coupling Buffer: 10 mM sodium acetate in MilliQ water, pH 4.5 (*see* **Note 3**).
7. Purified AcrB: His-tagged AcrB was expressed and purified as previously described [33]. The protein was first desalted in PBS by application to a 5 mL Whatman filtration unit (Millipore) with a cutoff of 100 kDa before being concentrated to 0.5 mg/mL. AcrB was then diluted tenfold in Amine Coupling Buffer.
8. Purified GalP: His-tagged GalP was purified as previously described [36, 37]. The protein was first desalted in PBS by application to a 5 mL Whatman filtration unit (Millipore) with a cutoff of 100 kDa before being concentrated to 0.5 mg/mL. GalP was then diluted tenfold in Amine Coupling buffer (*see* **Note 4**).
9. 0.4 M 1-ethyl-3-(3-dimethylaminopropyl)-carbodiimide hydrochloride (EDC) in MilliQ water. Store 100 μL aliquots at −20 °C. Thaw one aliquot per flow cell just prior to use.
10. 0.1 M *N*-hydroxysuccinimide (NHS) in MilliQ water. Store 100 μL aliquots at −20 °C. Thaw one aliquot per flow cell just prior to use.
11. EDC/NHS: 1:1 (v/v) mixture of 0.4 M EDC and 0.1 M NHS, prepared just prior to injection onto Biacore (*see* **Note 5**).
12. Blocking Buffer: 50 mM Tris in MilliQ water, pH 7.4 (*see* **Note 6**).
13. Dimethyl sulphoxide (DMSO) (*see* **Note 7**).
14. Solvent correction stock solution 1 (4.5% DMSO): 9.5 mL Running Buffer A + 0.45 mL DMSO (*see* **Notes 8** and **9**).
15. Solvent correction stock solution 2 (5.8% DMSO): 9.5 mL Running Buffer A + 0.58 mL DMSO.
16. 50% DMSO for washing injection needle.

3 Methods

3.1 Protein Immobilization Through Amine Coupling

Each Series S CM5 sensorchip has four individual flow cells. The experiment will be performed as follows:

(a) Flow cell 1: Blank immobilization.

(b) Flow cell 2: Immobilized AcrB.

(c) Flow cell 3: Blank Immobilization.

(d) Flow cell 4: Immobilized GalP.

1. Prime the Biacore T100/T200/S200 and equilibrate a CM5 sensorchip with Running Buffer A.
2. For the preparation of all flow cells use the Manual function in the Biacore software (Tools –> Manual). Each flow cell should be prepared individually before commencing the preparation of the next.
3. Activate flow cell 1 by injecting EDC/NHS for 15 min at a flow rate of 10 μL/min, then quench the reaction with Blocking buffer for 5 min at 30 μL/min.
4. Immobilize AcrB onto flow cell 2. Activate surface using EDC/NHS mix for 15 min at a flow rate of 10 μL/min. Inject AcrB diluted in the Amine Coupling Buffer for 25 min at a flow rate of 10 μL/min. Note the final level of response obtained.
5. If the target RU is achieved for flow cell 2 (typically 20,000 RU), quench the coupling reaction with Blocking buffer for 5 min at 30 μL/min. If not, continue injection of AcrB until this target is achieved. The sensorchip surface will be remaining chemically reactive for approximately 45 min after activation with EDC/NHS treated.
6. Activate flow cell 3 by injecting EDC/NHS for 15 min at a flow rate of 10 μL/min, then quench the reaction with Blocking buffer for 5 min at 30 μL/min.
7. Immobilize GalP onto flow cell 4. Activate surface using EDC/NHS mix for 15 min at a flow rate of 10 μL/min. Inject GalP diluted in Amine Coupling Buffer for 25 min at a flow rate of 10 μL/min. Note the final level of response obtained—aim to immobilize the same number of RU obtained in **step 4**. If this level is not reached, continue to use the Manual tool to inject more GalP across the activated surface.
8. Once the target RU is achieved for flow cell 4, quench the coupling reaction with Blocking buffer for 5 min at 30 μL/min.

3.2 Preparation of Small Molecule Analytes

This method can be used to prepare a two-fold dilution series of eight compounds in a single 96-well plate. The concentration range of the analytes assayed here was 500 μM to 488 nM. However, this can be adjusted depending upon the affinity of the interaction. Ideally the highest concentration being tested should be 10× the K_d.

1. Accurately weigh <5 mg of dry solid compound into a glass vial using an analytical balance (*see* **Note 10**).
2. Add an appropriate volume of DMSO such that the final concentration of compound is 10 mM.
3. Vortex sample until all solid has completely dissolved. Gently warm the solution in a water bath if required.

4. In a polypropylene microtube dilute 50 μL of 10 mM compound (100% DMSO) into 950 μL of Running Buffer A. This will yield an analyte concentration of 500 μM in 1× PBS and 5% DMSO.
5. One 96-well plate is used to prepare 11 serial dilutions of 8 separate compounds (*see* **Note 11**).
6. Transfer 200 μL of compound 1 from **step 4** (500 μM solution) into Well A1 of the 96-well plate.
7. Repeat **step 6** for another seven compounds by applying each sample to a different position in the first well of column 1.
8. Add 100 μL of Running Buffer B to all unused wells in columns 2–12 of the 96-well plate.
9. Using an 8-barrel multichannel pipette, withdraw 100 μL of sample from column 1 and add this to buffer in column 2.
10. If the multichannel pipettor has a "mix" function, mix the samples in column 2. Otherwise, mix sample by withdrawing sample and dispensing back into same well several times.
11. Repeat the serial dilution across the 96-well plate for all samples in columns 3–11.
12. The wells in column 12 containing Running Buffer B (i.e., no compound) represent zero analyte.
13. Seal the 96-well plate with a pierceable plate seal.

3.3 Preparation of Solutions for the Solvent Correction Curve

In eight separate vials, combine Solvent correction stock solutions 1 and 2 in the volumes shown in Table 1.

Table 1
Preparing solvent correction solutions

Vial no.	Solvent correction stock solution 1 (4.5% DMSO) (μL)	Solvent correction stock solution 2 (5.8% DMSO) (μL)
1	0	1400
2	200	1200
3	400	1000
4	600	800
5	800	600
6	1000	400
7	1200	200
8	1400	0

3.4 Assay Small Molecule Binding to AcrB by SPR

The bimolecular interaction between AcrB and analytes displays fast association and dissociation kinetics. Hence, it is appropriate to perform equilibrium binding analysis. As the AcrB:analyte complexes dissociate quickly there is no need to include a regeneration step nor use long dissociation times.

1. Open the "*Multi Cycle kinetics*" protocol in the Biacore software. Set the following parameters:

   Collect data across all four flow cells using the "multi" data collection mode.

   Flow rate at 30 μL/min (*see* **Note 12**).

   Five start-up cycles to stabilize the baseline.

   The association time was 60 s and dissociation time 120 s.

   Wash the injection needle with 50% DMSO between samples.

   The solvent correction solutions were injected every 30 cycles.
2. Load 96-well plate, solvent correction solutions, running buffer, and 50% injection needle wash solution into the sample rack.
3. Enter details about all samples, and their positions in the sample rack, into the software.
4. Commence the binding assay. Ensure there is an adequate amount of Running Buffer B for the run.

3.5 Curve Fitting and Statistical Analysis

1. Open the BiaEvaluation software.
2. Analyze the data using the *Affinity Binding* protocol.
3. The binding response at each analyte concentration is set to 5 s before the end of the injection.
4. Export the RU values for all four flow cells as a rich text file.
5. Open the rtf in Excel.
6. Calculate analyte binding to AcrB ($AcrB_{corr}$) by subtracting background (i.e., Fc 2-1).
7. Calculate analyte binding to GalP ($GalP_{corr}$) by subtracting background (i.e., Fc 4-3).
8. Calculate specific analyte binding to AcrB ($AcrB_{sp} = AcrB_{corr} - GalP_{corr}$).
9. Enter the data into GraphPad Prism.
10. Plot $AcrB_{sp}$ against analyte concentration.
11. Determine the binding constants B_{max} and K_d using the relationship:

$$B_{max}[S]/(K_d + [S])$$

where analyte binding is represented by B, the drug concentration by $[S]$, the maximal binding by B_{max}, and the analyte concentration yielding $1/2B_{max}$ by the dissociation constant K_d (*see* **Note 13**).

4 Notes

1. To prevent contamination of the Biacore surfaces and tubing, all buffers should be degassed and filtered through 0.2 μm filters before use.
2. Tween-20 is used as a surfactant to prevent nonspecific binding to the tubing and microfluidics of the Biacore. It is recommended to use the highest grade of Tween-20 available. Certain commercial suppliers sell P20, which is a high grade of Tween-20.
3. Use a low ionic strength buffer for optimal immobilization. The pH of the Amine Coupling Buffer should be below the isoelectric point (pI) of the target protein. The pI of a protein can be calculated using an online isoelectric point calculator and the sequence of your protein including any tags. This provides a starting point for scouting experiments using buffer at least 0.5 pH units below the pI. We used 10 mM sodium acetate buffer at pH 4.8 for both AcrB and GalP.
4. Although GalP is only 52 kDa, the size of the protein oligomer-detergent micelle complex is larger than 100 kDa. A filtration unit with a size of 100 kDa has to be used to prevent concentration of empty n-dodecyl-β-maltoside (DDM) detergent micelles which have an approximate size of 70 kDa.
5. Aliquots of NHS and EDC should be thawed and premixed as soon as possible before their use. When preparing multiple flow cells using the Manual function it is recommended to complete the activation, immobilizing, and blocking of one surface before preparing the next flow cell.
6. Blocking of the unreacted groups on the dextran surface can be performed using a compound containing a primary amine group. GE supplies ethanolamine as a blocking reagent in their Amine Coupling kit. However, this may disrupt the biological activity of certain proteins. We suggest 50 mM Tris pH 7.0 as a milder blocking reagent.
7. The quality of DMSO can vary between different suppliers. Use analytical grade or higher.
8. DMSO is commonly used for dissolving and assaying organic compounds. However, DMSO contributes significantly to the bulk refractive index of samples and buffers (i.e., 1% DMSO contributes approximately 1200 RU to the response level). Therefore, it is important that the measured responses are accurately compensated for any variations in DMSO concentration. This is known as solvent correction.
9. It is important that the concentration of DMSO in the running buffer and analytes are matched as closely as possible. Running

buffer B, Solvent correction stock solutions 1 and 2 and solubilized compounds (analytes) should all be prepared using same aliquot of DMSO. Prepare all solutions on the same day as their use.

10. Avoid storing materials dissolved in DMSO in plastic vials for prolonged periods as DMSO will dissolve certain plastics over time.
11. Use 96-well plates that are made from plastics that are resistant to DMSO, such as polypropylene.
12. Use flow rates of 30–50 μL/min to help avoid rebinding of analytes and mass transport effects.
13. Each binding assay should be performed at least three separate times using at least two separate protein immobilizations.

Acknowledgments

This research was supported financially by the National Natural Science Foundation of China (81673284), National Health and Medical Research Council of Australia (GN1147538), China–Australia Centre for Health Sciences Research and Australian Postgraduate Award (RM). All experiments were performed on a BiaCore 200 which from part of the Biophysical Characterisation Suite at The University of South Australia.

References

1. Theuretzbacher U (2017) Global antimicrobial resistance in Gram-negative pathogens and clinical need. Curr Opin Microbiol 39:106–112
2. Butler MS, Blaskovich MA, Cooper MA (2017) Antibiotics in the clinical pipeline at the end of 2015. J Antibiot (Tokyo) 70:3–24
3. Wang Y, Venter H, Ma S (2016) Efflux pump inhibitors: a novel approach to combat efflux-mediated drug resistance in bacteria. Curr Drug Targets 17:702–719
4. Spengler G, Kincses A, Gajdács M, Amaral L (2017) New roads leading to old destinations: efflux pumps as targets to reverse multidrug resistance in bacteria. Molecules 22:468
5. Ruggerone P, Murakami S, Pos KM, Vargiu AV (2013) RND efflux pumps: structural information translated into function and inhibition mechanisms. Curr Top Med Chem 13:3079–3100
6. Venter H, Mowla R, Ohene-Agyei T, Ma S (2015) RND-type drug efflux pumps from Gram-negative bacteria: molecular mechanism and inhibition. Front Microbiol 6:377
7. Nikaido H, Pagès JM (2012) Broad-specificity efflux pumps and their role in multidrug resistance of Gram-negative bacteria. FEMS Microbiol Rev 36:340–363
8. Arzanlou M, Chai WC, Venter H (2017) Intrinsic, adaptive and acquired antimicrobial resistance in Gram-negative bacteria. Essays Biochem 61:49–59
9. Masi M, Réfregiers M, Pos KM, Pagès JM (2017) Mechanisms of envelope permeability and antibiotic influx and efflux in Gram-negative bacteria. Nat Microbiol 2:17001
10. Zwama M, Yamaguchi A (2018) Molecular mechanisms of AcrB-mediated multidrug export. Res Microbiol 169:372–383
11. Opperman TJ, Nguyen ST (2015) Recent advances toward a molecular mechanism of efflux pump inhibition. Front Microbiol 6:421
12. Blair JM, Bavro VN, Ricci V, Modi N, Cacciotto P, Kleinekathöfer U, Ruggerone P,

Vargiu AV, Baylay AJ, Smith HE et al (2015) AcrB drug-binding pocket substitution confers clinically relevant resistance and altered substrate specificity. Proc Natl Acad Sci U S A 112:3511–3516

13. Wang Y, Mowla R, Guo L, Ogunniyi AD, Rahman T, De Barros Lopes MA, Ma S, Venter H (2017) Evaluation of a series of 2-napthamide derivatives as inhibitors of the drug efflux pump AcrB for the reversal of antimicrobial resistance. Bioorg Med Chem Lett 27:733–739
14. Wang Y, Mowla R, Ji S, Guo L, De Barros Lopes MA, Jin C, Song D, Ma S, Venter H (2018) Design, synthesis and biological activity evaluation of novel 4-subtituted 2-naphthamide derivatives as AcrB inhibitors. Eur J Med Chem 143:699–709
15. Vargiu AV, Ruggerone P, Opperman TJ, Nguyen ST, Nikaido H (2014) Molecular mechanism of MBX2319 inhibition of Escherichia coli AcrB multidrug efflux pump and comparison with other inhibitors. Antimicrob Agents Chemother 58:6224–6234
16. Aparna V, Dineshkumar K, Mohanalakshmi N, Velmurugan D, Hopper W (2014) Identification of natural compound inhibitors for multidrug efflux pumps of Escherichia coli and Pseudomonas aeruginosa using in silico high-throughput virtual screening and in vitro validation. PLoS One 9:e101840
17. Handzlik J, Szymańska E, Alibert S, Chevalier J, Otrębska E, Pękala E, Pagès JM, Kieć-Kononowicz K (2013) Search for new tools to combat Gram-negative resistant bacteria among amine derivatives of 5-arylidenehydantoin. Bioorg Med Chem 21:135–145
18. Opperman TJ, Kwasny SM, Kim HS, Nguyen ST, Houseweart C, D'Souza S, Walker GC, Peet NP, Nikaido H, Bowlin TL (2014) Characterization of a novel pyranopyridine inhibitor of the AcrAB efflux pump of Escherichia coli. Antimicrob Agents Chemother 58:722–733
19. Nguyen ST, Kwasny SM, Ding X, Cardinale SC, McCarthy CT, Kim HS, Nikaido H, Peet NP, Williams JD, Bowlin TL et al (2015) Structure-activity relationships of a novel pyranopyridine series of Gram-negative bacterial efflux pump inhibitors. Bioorg Med Chem 23:2024–2034
20. Sjuts H, Vargiu AV, Kwasny SM, Nguyen ST, Kim HS, Ding X, Ornik AR, Ruggerone P, Bowlin TL, Nikaido H et al (2016) Molecular basis for inhibition of AcrB multidrug efflux pump by novel and powerful pyranopyridine derivatives. Proc Natl Acad Sci U S A 113:3509–3514
21. Choudhury D, Talukdar AD, Chetia P, Bhattacharjee A, Choudhury MD (2016) Screening of natural products and derivatives for the identification of RND efflux pump inhibitors. Comb Chem High Throughput Screen 19:705–713
22. Ohene-Agyei T, Mowla R, Rahman T, Venter H (2014) Phytochemicals increase the antibacterial activity of antibiotics by acting on a drug efflux pump. Microbiology 3:885–896
23. Lomovskaya O, Warren MS, Lee A, Galazzo J, Fronko R, Lee M, Blais J, Cho D, Chamberland S, Renau T et al (2001) Identification and characterization of inhibitors of multidrug resistance efflux pumps in Pseudomonas aeruginosa: novel agents for combination therapy. Antimicrob Agents Chemother 45:105–116
24. Nakashima R, Sakurai K, Yamasaki S, Nishino K, Yamaguchi A (2011) Structures of the multidrug exporter AcrB reveal a proximal multisite drug-binding pocket. Nature 480:565–569
25. Hung LW, Kim HB, Murakami S, Gupta G, Kim CY, Terwilliger TC (2013) Crystal structure of AcrB complexed with linezolid at 3.5 Å resolution. J Struct Funct Genom 14:71–75
26. Ababou A, Koronakis V (2016) Structures of gate loop variants of the AcrB drug efflux pump bound by erythromycin substrate. PLoS One 11:e0159154
27. Eicher T, Cha HJ, Seeger MA, Brandstätter L, El-Delik J, Bohnert JA, Kern WV, Verrey F, Grütter MG, Diederichs K et al (2012) Transport of drugs by the multidrug transporter AcrB involves an access and a deep binding pocket that are separated by a switch-loop. Proc Natl Acad Sci U S A 109:5687–5692
28. Murakami S, Nakashima R, Yamashita E, Matsumoto T, Yamaguchi A (2006) Crystal structures of a multidrug transporter reveal a functionally rotating mechanism. Nature 443:173–179
29. Nakashima R, Sakurai K, Yamasaki S, Hayashi K, Nagata C, Hoshino K, Onodera Y, Nishino K, Yamaguchi A (2013) Structural basis for the inhibition of bacterial multidrug exporters. Nature 500:102–106
30. Wang Z, Fan G, Hryc CF, Blaza JN, Serysheva II, Schmid MF, Chiu W, Luisi BF, Du D (2017) An allosteric transport mechanism for the AcrAB-TolC multidrug efflux pump. Elife 6:e24905
31. Damborský P, Švitel J, Katrlík J (2016) Optical biosensors. Essays Biochem 60:91–100
32. Tikhonova EB, Yamada Y, Zgurskaya HI (2011) Sequential mechanism of assembly of

multidrug efflux pump AcrAB-TolC. Chem Biol 18:454–463

33. Mowla R, Wang Y, Ma S, Venter H (2018) Kinetic analysis of the inhibition of the drug efflux protein AcrB using surface plasmon resonance. Biochim Biophys Acta Biomembr 1860:878–886
34. Tikhonova EB, Zgurskaya HI (2013) Assessment of multidrug efflux assemblies by surface plasmon resonance. Methods Mol Biol 966:133–155
35. Zheng H, Taraska J, Merz AJ, Gonen T (2010) The prototypical H+/galactose symporter GalP assembles into functional trimers. J Mol Biol 396:593–601
36. Venter H, Ashcroft AE, Keen JN, Henderson PJ, Herbert RB (2002) Molecular dissection of membrane-transport proteins: mass spectrometry and sequence determination of the galactose-H+ symport protein, GalP, of Escherichia coli and quantitative assay of the incorporation of [ring-2-13C]histidine and (15)NH(3). Biochem J 363:243–252
37. Ward A, Hoyle C, Palmer S, O'Reilly J, Griffith J, Pos M, Morrison S, Poolman B, Gwynne M, Henderson P (2001) Prokaryote multidrug efflux proteins of the major facilitator superfamily: amplified expression, purification and characterisation. J Mol Microbiol Biotechnol 3:193–200

Chapter 8

Systematic Screening of Viral Entry Inhibitors Using Surface Plasmon Resonance

Penmetcha K. R. Kumar

Abstract

Surface plasmon resonance (SPR) analytical method was initially used as biosensor for analyzing diverse biomolecular interactions and recently gained important place in the drug discovery. Here, I describe the procedures for screening of inhibitors against the viral proteins using the SPR. Using the described procedures, in the past, we were able to identify several antiviral products that interfere viral-host receptor proteins interactions.

Key words Surface plasmon resonance, Viral antigens, Biacore, Inhibitors, RNA, Receptors

1 Introduction

Surface plasmon resonance (SPR) is a powerful analytical method commonly explored in the analyses of wide range of biomolecular interactions. Compared to different biophysical techniques, such as nuclear magnetic resonance [1, 2], X-ray crystallography [3], mass spectroscopy [4], isothermal titration calorimetry [5], protein thermal shift [6], affinity capillary electrophoresis [7], other affinity chromatography [8], and label-free biosensor methods [9], the SPR allows the real-time monitoring of biomolecular interactions, requires very small amounts of sample for analyses (usually micrograms), allows the simultaneous use of reference ligands, and more importantly, requires no modifications or ligand labeling. The use of SPR in drug discovery has been reviewed [10–12] and its applications in the primary and secondary screening of drug leads against various targets have been described [13–17]. These studies suggest the usefulness of SPR-based approaches to the screening and identification of effective inhibitors.

In an SPR system, the interacting biomolecule is typically immobilized on the surface of a sensor using a matrix such as dextran. Beneath the dextran surface, a thin gold layer is attached

Nikolaos E. Labrou (ed.), *Targeting Enzymes for Pharmaceutical Development: Methods and Protocols*, Methods in Molecular Biology, vol. 2089, https://doi.org/10.1007/978-1-0716-0163-1_8, © Springer Science+Business Media, LLC 2020

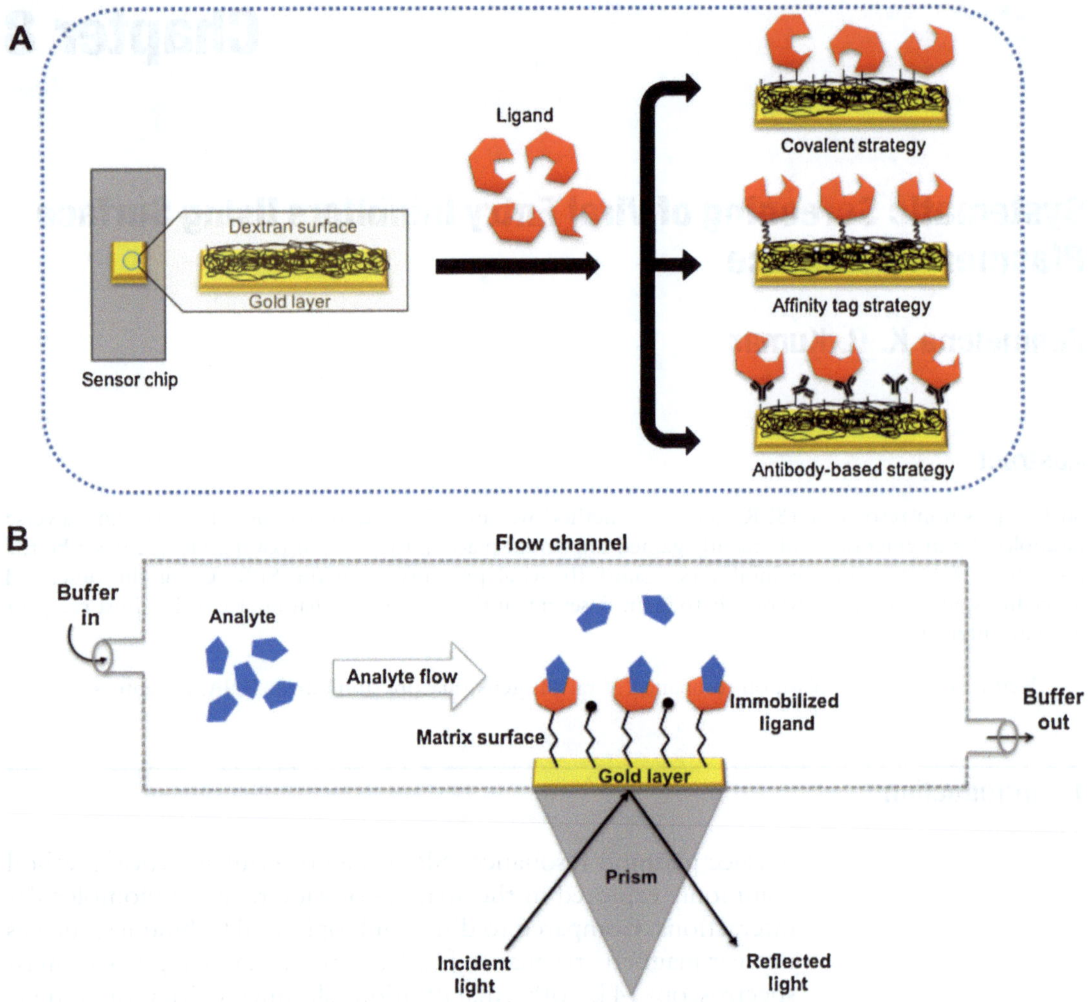

Fig. 1 Surface plasmon resonance system. (**a**) Immobilization strategies for attaching ligand over the sensor surface. (**b**) Schematic representation of Biomolecular interactions in the surface plasmon resonance

to a glass surface (Fig. 1). Currently, two common approaches are used for stably immobilizing ligands on the sensor chip surface: a covalent coupling and non-covalent capturing method. In the former case, the sensor surface is coated with different functional groups that react with an appropriate functional group from the ligand to form stable linkages. In the latter case, capture involves the use of high-affinity recombinant tags or antibodies (Fig. 1a). In the covalent method of immobilization, different functional groups of ligand explored for bonding to the matrix that includes amino, carboxyl, and sulfhydryl groups (Fig. 1a). It is possible that some of these functional groups are important for the ligand activity. Targeting these functional groups for bonding to the matrix may not only inactivate the ligand upon bonding to the matrix but also allow the ligand to orient in a random fashion. The non-covalent method for immobilizing the ligand uses either fusion protein (His-tag or GST-tag proteins) or ligand antibodies (Fig. 1a). This method has the advantage of achieving properly oriented immobilized ligands and the means to efficiently remove immobilized ligands and their binding partners to regenerate a clean surface for the next round of screening. However, this strategy faces other issues such as high ligand consumption and the need to prepare tagged ligands. Therefore, an immobilization method must be carefully selected with consideration of the strategy selected for the screening of drug-like compounds. In addition to the immobilization methods used, assay format is also important when considering the primary or secondary screening of drug-like compounds.

Once the ligand is immobilized on the surface, the composite surface is placed within a flow cell, and a cognate biomolecule (analyte) that interacts with an immobilized biomolecule is injected into the flow cell in the presence of a suitable buffer. Surface plasmons are generated when an incident beam of light is directed toward the gold surface at a critical angle. The critical angle depends on the refractive index near the surface, which changes when an analyte binds to the immobilized biomolecule (Fig. 1b). The resulting change in refractive index at the surface is measured, and the extent of that change correlates with the mass concentration near the surface. All these changes are displayed as a plot (sensogram) of resonance units (RU) (Fig. 1b). Three distinct phases are observed in the sensogram when an analyte is injected on the immobilized protein surface: (1) an association phase, records analyte binds to the immobilized protein during the injection process; (2) an equilibrium phase, observed at the end of the analyte injection process where the analyte binding rate reaches a steady state; and (3) a dissociation phase is viewed when the analyte dissociates from the immobilized protein. This occurs during the continuous flow of buffer over the immobilized surface (Fig. 2). Importantly, all of these events can be monitored in real-time fashion. The residual analyte remaining on the sensor surface can be removed and the surface can be regenerated without damaging the activity of the immobilized protein. The regeneration of an active ligand surface, including the removal of bound analytes especially in high-throughput screens, as the ligand surface must be used multiple times in each successive round of screening previous screens and the maintenance of ligand activity, is crucial. In the past, several ligands have been analyzed using the SPR [18–23].

We adopted some of these procedures and modified them in order to screen inhibitors efficiently against the viral antigens which interfere in their cognate interaction with human receptors. The procedures reported, here, allowed to screen inhibitors efficiently in a continuous fashion [18–20].

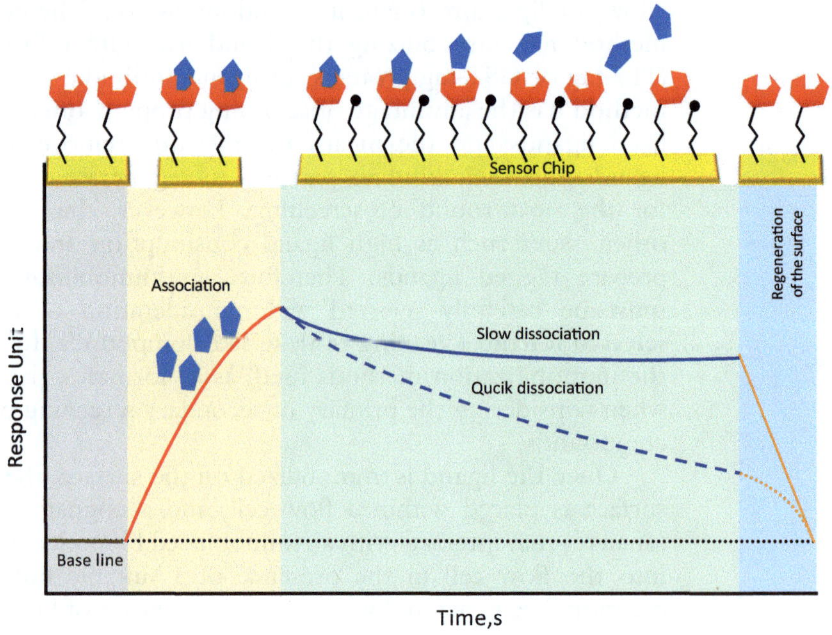

Fig. 2 Sensogram response observed upon analyte binding with immobilized ligand

2 Materials

All solution must be prepared using miliQ water and stored at room temperature (25 °C). Follow all waste disposal regulations for disposing waste solutions.

2.1 Reagents, Buffers, Sensor Chips, and Machine

1. Amine Coupling kit type 2 (BR-100-50, GE Healthcare, USA).
2. Surfactant P20 (GE Healthcare, USA).
3. 10× HBS-P+ buffer (GE Healthcare, USA).
4. CM5 chips (GE Healthcare, USA).
5. BiacoreT100.
6. Acetate Buffer 100 mM pH (4.0, 4.5, 5.0, and 5.5).
7. NaOH 50 mM.
8. HEPES 10 mM (pH 7.4).
9. NaCl (1 M, 150 mM).
10. Polysorbate 20 (0.005%).
11. Regeneration solution (SDS, 0.1%).

2.2 Proteins and Inhibitors

1. Recombinant HSV-1 gD was expressed as a C-terminal His-tag fusion protein (1–285 amino acids) in the baculovirus system and purified as described [21].
2. HVEM was purchased from Sino Biologicals, Inc., China.
3. Heparin sulfate biotin sodium salt.

3 Methods

3.1 Immobilization of Ligand on the Sensor Chip

1. Dock the new CM5 sensor chip in the instrument BiacoreT100 using chip port.
2. For the ligand immobilization, the optimum pH condition for immobilization needs to be evaluated by scouting. This was carried out either by manual injection or by using wizard template. Dilute the ligand (HVEM) in 10 mM acetate buffer pH ranging from 4.0, 4.5, 5.0, and 5.5 to get 100 nM final concentration. Inject the ligand sample (100 μL) into the flow cell of CM5 Chip at 25 °C.
3. The wizard template is used as follows: select Flow path 1 and enter the buffer name and the pH values to be tested.
4. Enter the test sample name in solution column, followed by entering 120 and 10 values in the columns contact time and flow rate, respectively. Prepare 50 mM NaOH solution (100 μL) and enter this information in the surface washing solution column. Insert these tubes at appropriate positions and run the wizard program. The resulted sensogram is shown in Fig. 3a.
5. Once the best conditions for the HVEM immobilization from the pH scouting was determined, based on the desired amount of immobilization required, the HVEM was immobilized using the following protocol. In our case, based on the above pH scouting, pH 5.0 was selected for immobilization of HVEM [19]. Select flow cells 1 and 2 as control and sample flow cell, respectively. Activate the surface of flow cell 1 and 2 by EDC/NHS followed by injection of HVEM (100 nM in acetate buffer pH 5.0 solution) over the activated surface of only flow cell 2. Block the remaining unreacted activated surface of flow cell 2 and the control flow cell 1 with ethanolamine solution (supplied in the kit). The entire process of immobilization of HVEM into the flow cell 2 is shown in Fig. 3b.
6. The flow cell 1 serves as a control flow cell in evaluating the HVEM-gD protein interactions.
7. Wash both flow cells (1 and 2), once blocked, with binding buffer (10 mM HEPES, pH 7.4, 150 mM NaCl, and 0.005% polysorbate 20) for another 35 min.
8. The amount of HVEM immobilized in the flow cell 2 was estimated to be, in this case, 1725 response unit (Fig. 3b).

3.2 Monitoring the Analyte Binding to the Ligand

1. For analyzing the interaction with gD protein of HSV-1, use the above prepared HVEM immobilized sensor chip.
2. To determine the HVEM-gD protein interactions and rate kinetics, use two flow cells (flow cell 1 and 2) in the CM5

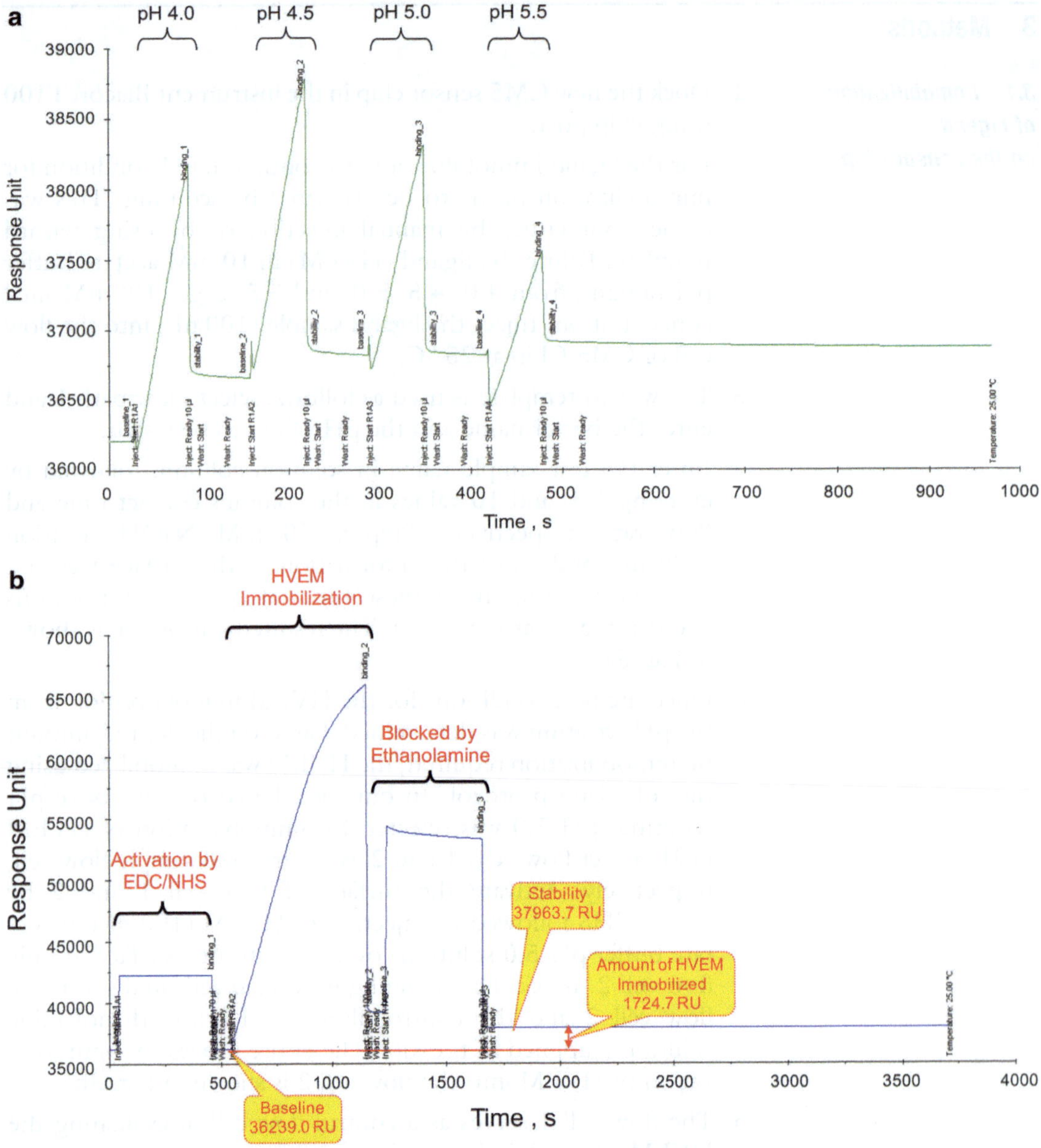

Fig. 3 (**a**) Evaluation of HVEM ligand immobilization amounts by pH scouting. (**b**) Immobilization of HVEM ligand on the CM5 chip by amino-coupling reaction at pH 5.0

chip. Here, a single-cycle kinetics example is described, where different concentrations of purified gD protein (20, 40, 80, 160, 320 nM) flown over the surface of immobilized HVEM with a flow rate 30 μL/min into flow cells (1 and 2), one concentration after the other, in a continuous fashion at 25 °C without regenerating the surface of the sensor chip.

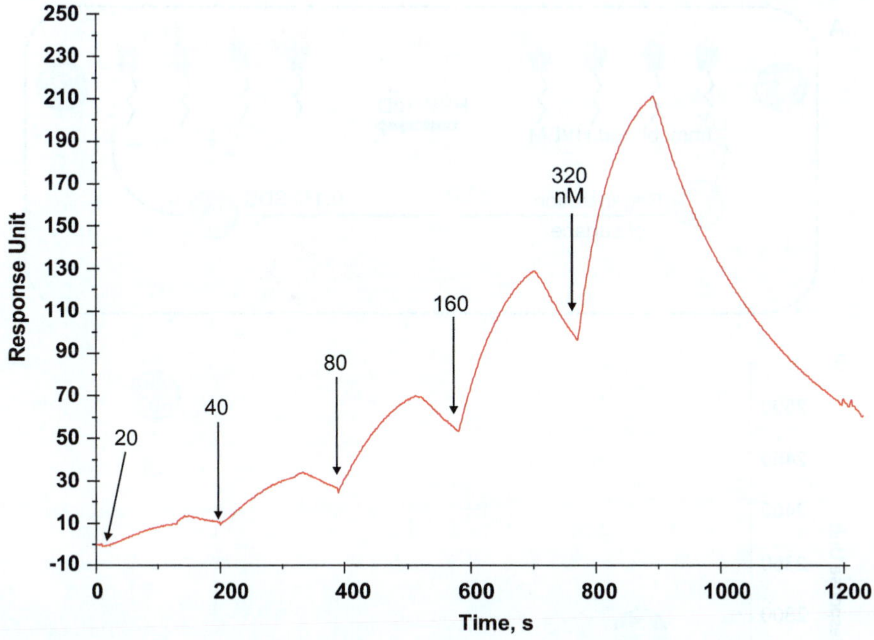

Fig. 4 Single-cycle kinetics for analyzing HVEM-gD protein of HSV-1 interactions

3. The observed sensogram responses after subtracting from the background responses in the flow cell 1 are shown in Fig. 4.
4. Determine the binding rates of association and dissociation from the sensogram after data fitted to a 1:1 binding model. Single-cycle kinetics experiments have been found to be superior compared to the multi-injection analyses [22, 23].

3.3 Removing the Analyte from the Sensor Surface

1. For multiple analyses on the same chip, it is important not only to remove the bound analyte (gD protein of HSV-1) from the ligand but also the ligand binding affinity should retain (Fig. 5a). Toward this, a few regeneration reagents are recommended by supplier for regenerating the sensor surface [24]. From these, we tested two regeneration reagents, 1 M NaCl and 0.1% SDS, and found that 0.1% SDS efficiently removed the gD protein from the immobilized ligand sensor surface.
2. Regeneration of HVEM surface: For the demonstration of entire process of analyte binding to the ligand followed by stripping of analyte from the ligand, the following sequential protocol is used. Inject 100 μL of 300 nM of gD of HSV-1 (in binding buffer) into the flow cell 2 with a flow rate of 30 μL/min over the HVEM surface for 8 min. Inject 0.1% SDS regeneration solution at a flow rate of 10 μL/min for 2 min at 25 °C. Entire process of a representative sensogram is shown in Fig. 5b: the immobilized HVEM surface (step 1),

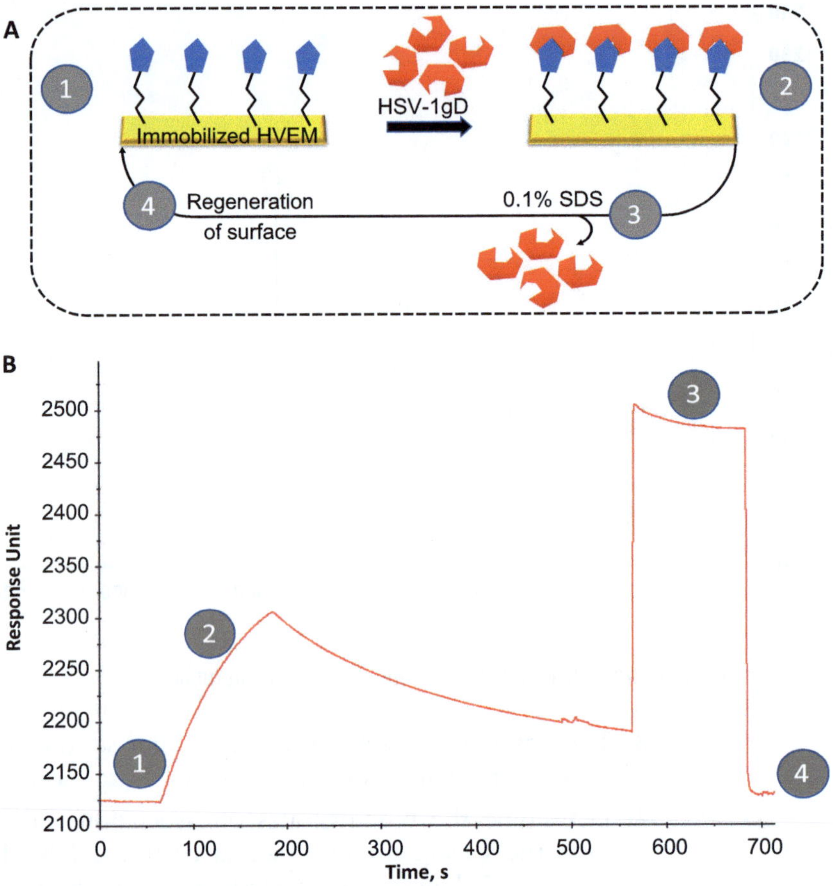

Fig. 5 Regeneration of sensor surface for the next round of analyses of HVEM-gD interactions. (**a**) Schematic representation of regeneration of sensor surface. (**b**) The observed sensogram during the entire binding and regeneration processes

gD protein binding to the HVEM (step 2), upon injection of regeneration solution (step 3), and regenerated HVEM surface for the next round of analysis (step 4). The sensogram reached the baseline to the original level of response value observed for immobilized HVEM (step 1, Fig. 5b), suggesting that the gD protein is removed from the HVEM surface.

3.4 Regeneration of Sensor Surface for the Multiple Round of Analyses Using SPR

1. The regeneration reagent in principle should not only facilitate in removing the previously bound gD protein from the HVEM but also make sure that the HVEM binding affinity should retain the binding affinity to the gD protein. If these two conditions are achieved, it reduces using multiple sensor chips for the analyses and further reduces the time required for screening as well as the cost (scheme in Fig. 5a).
2. To evaluate whether the 0.1% SDS regeneration solution affect the HVEM binding to gD protein, two cycles of analysis of

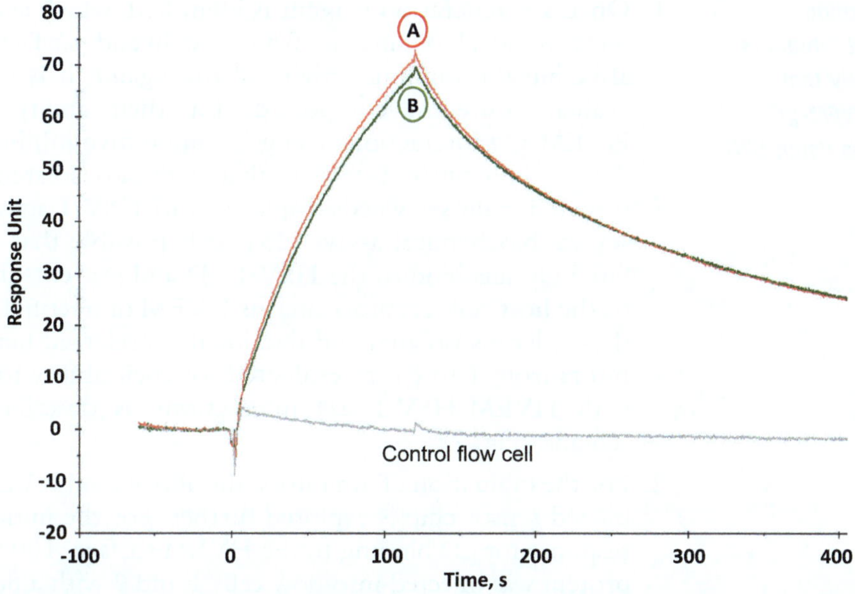

Fig. 6 Sensograms observed when gD protein injected on HVEM immobilized surface before and after treating with regeneration solution

HVEM-gD interactions and regeneration of sensor surface are performed (similar to scheme shown in Fig. 5a). For this, the above used HVEM immobilized sensor chip is explored as described: Inject 100 μL of 100 nM of gD of HSV-1 (in binding buffer) with a flow rate of 30 μL/min for 7 min at 25 °C and observe a sensogram response (Fig. 6a). To remove gD protein from the HVEM surface, as described above, inject regeneration solution with a flow rate of 10 μL/min for 2 min at 25 °C followed by washing with binding buffer for 1 min with a flow rate of 30 μL/min at 25 °C. Next, to test whether the HVEM surface retained the same affinity after the treatment with regeneration solution, the binding analyses is repeated by injecting the same amount (100 nM) of gD of HSV-1 (in binding buffer) over the regenerated surface with a flow rate of 30 μL/min for 7 min at 25 °C. The resulting sensogram response signal was observed (Fig. 6b). For direct comparison, the two analyzed sensogram responses were overlaid, i.e., before and after the treatment with regenerating solution and found that the sensogram responses were identical, suggesting that the HVEM binding activity toward gD protein was not altered the HVEM surface by the regeneration solution. Further, we repeated the same process for more than 20 times and found no significant effect on HVEM binding ability to the gD protein. However, in the control flow cell 1 do not show any response signal, suggesting that gD protein does not bind to the flow cell surface (blank surface without HVEM).

3.5 Screening of Different Inhibitors Continuously that Interfere HVEM-gD Interactions Using SPR

1. Once a regeneration reagent is identified, which allows complete removal of analyte from the ligand surface without affecting the binding activity of the ligand, it is possible to evaluate different compounds for their ability to block HVEM-gD interactions using a competitive inhibition assay format. Previously, different sulfated glycans isolated and fractionated from sea weeds displayed anti-HSV-1 activity-based several biochemical assays [25]. It is possible that these sulfated glycans bind to the HSV-1 gD and prevents its binding to the host cell receptors such as HVEM or nectin-1. Some of these glycans isolated and fractionated (referred here as inhibitors from 1 to 8) are evaluated for their ability to interfere with HVEM-HSV-1 gD interactions as described in the scheme (Fig. 7).
2. For the evaluation of inhibitors, the above used HVEM immobilized sensor chip is explored further. For the initial binding response for gD binding to the HVEM surface, 100 nM of gD protein was injected into flow cells 1 and 2 with a flow rate of 30 μL/min for 7 min at 25 °C. Similarly, binding responses for sulfate glycans and heparin to the HVEM surface were evaluated by the injection of 310 nM of each glycans into flow cells 1 and 2 with a flow rate of 30 μL/min for 7 min at 25 °C. To determine the interference of different sulfated glycans (inhibitors ranging from 1 to 8), and heparin on the gD-HVEM interactions, each glycan was diluted in binding buffer and mixed with HSV-1 gD protein to a final concentration of 100 nM. The final concentration of each glycan and heparin was 310 nM. The mixed reaction solution was incubated further for 10 min at 25 °C and inject into flow cells 1 and 2 for 2 min with a flow rate of 30 μL/min. The dissociation time was set for 2 min. Each inhibitor binding sensogram responses were collected from flow cell 2 after deducting the responses from flow cell 1. For each inhibitor flow of continuous analysis was as follows: (i) gD-HVEM analysis (red sensogram), (ii) regeneration of sensor surface, (iii) then analysis of inhibitor binding to HVEM surface (green sensogram), (iv) regeneration of sensor surface, and (v) evaluation of gD-HVEM interactions in the presence of sulfated glycans (blue) (Fig. 8).
3. All inhibitor responses were compared and found that inhibitors 4 and 2, 3 interfere HVEM-gD interactions efficiently and moderately, respectively. Whereas inhibitors 1, 5, 6, 7, 8, and heparin were found to be inefficient in blocking the HVEM-gD interactions at least at 310 nM concentration level (Fig. 8). Importantly, the inhibitors (1–8 and heparin) tested here are

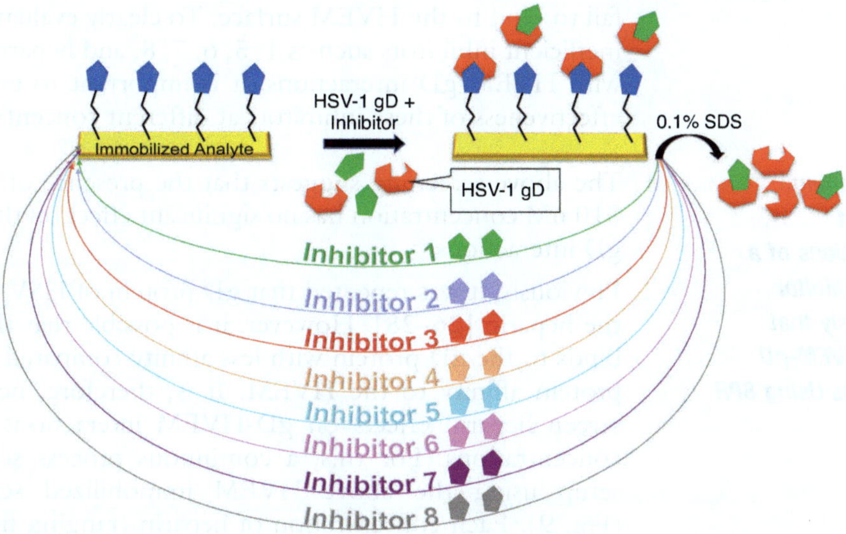

Fig. 7 Schematic representation of screening different inhibitors continuously that interfere HVEM-HSV-1 gD protein interactions using SPR

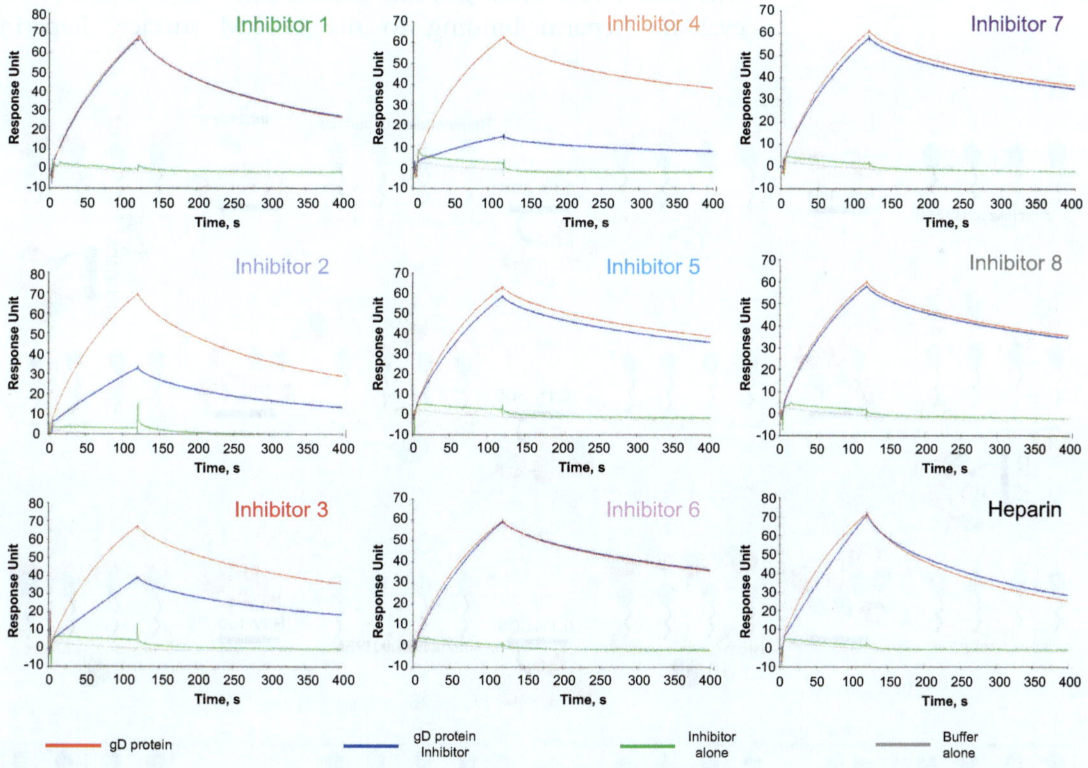

Fig. 8 Sensograms observed for HVEM-HSV-1 gD interactions in the presence of different inhibitors by using SPR

fail to bind to the HVEM surface. To clearly evaluate whether inefficient inhibitors such as 1, 5, 6, 7, 8, and heparin interfere with HVEM-gD interactions, it is important to evaluate the effectiveness of these inhibitors at different concentrations.

3.6 Screening of Different Concentrations of a Specific Inhibitor Continuously that Interfere HVEM-gD Interactions Using SPR

1. The above screening suggests that the presence of heparin at 310 nM concentration has no significant effect on the HVEM-gD interactions.
2. Previously, it was reported that gD protein of HSV-1 binds to the heparin [26–28]. However, it is possible that the heparin binds to the gD protein with less affinity compared to the gD protein affinity to the HVEM. It is, therefore, necessary to screen heparin effects on gD-HVEM interactions at higher concentrations. For this, a continuous process scheme was setup using the above HVEM immobilized sensor chip (Fig. 9). Each concentration of heparin (ranging from 0.630 to 5.0 μM final concentration) was diluted in binding buffer and mixed with 100 nM of HSV-1 gD protein and the reaction mixture was incubated for 10 min at 25 °C before injecting with a flow rate of 30 μL/min over the flow cells 1 and 2. To evaluate heparin binding to the HVEM surface, heparin

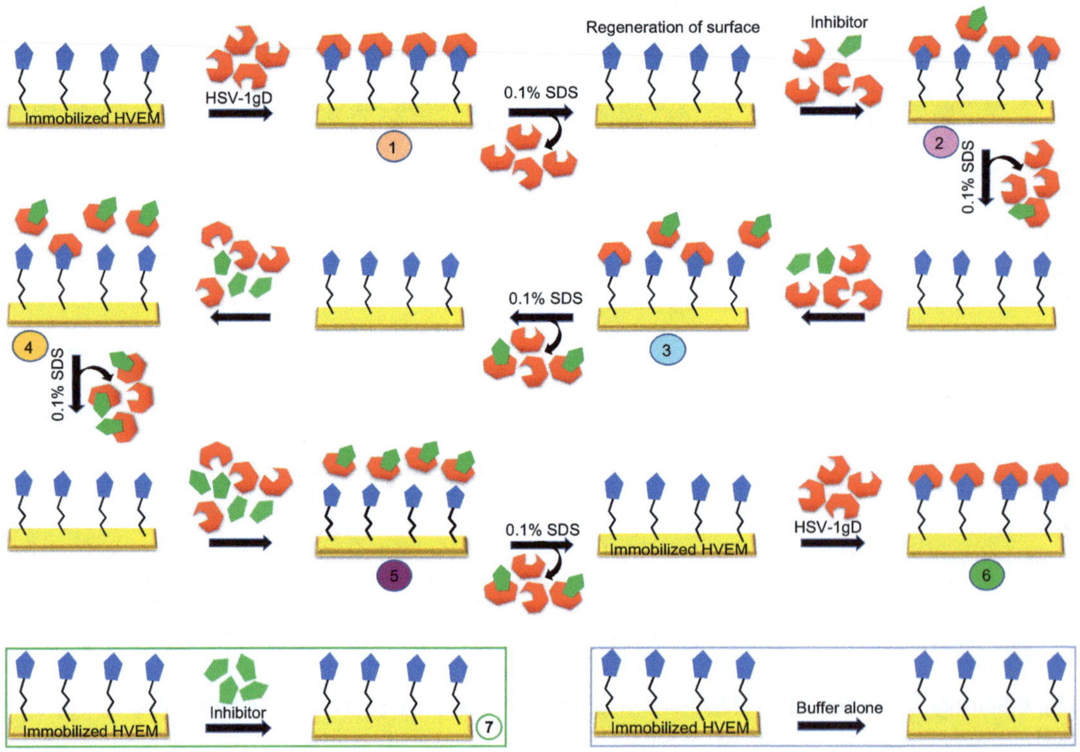

Fig. 9 Continuous analyses of HVEM-HSV-1 gD interactions in the presence of different concentrations of heparin by using SPR

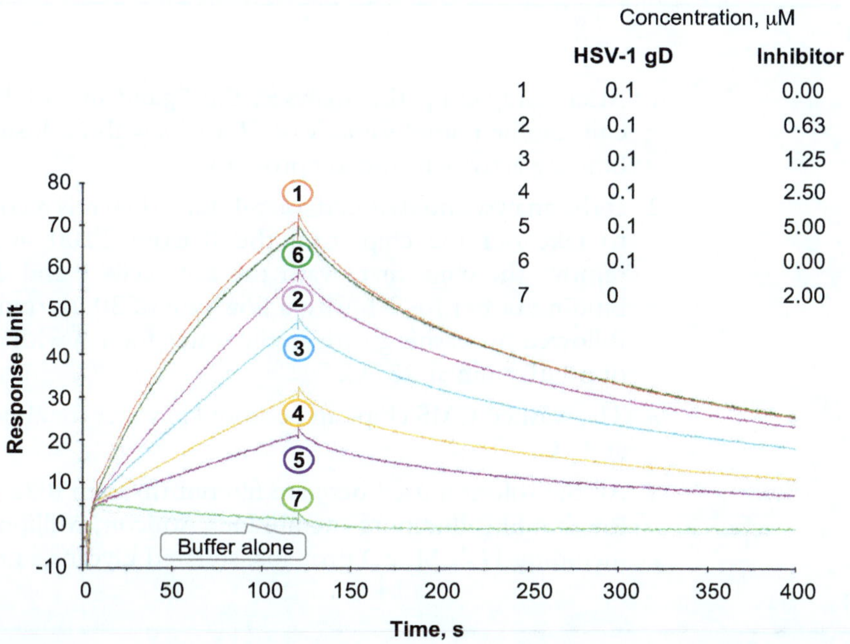

Fig. 10 Sensograms observed for HVEM-HSV-1 gD interactions in the presence of different concentrations of a heparin by using SPR

(2.0 μM) was also injected into the flow cells 1 and 2, similar to the other samples. Each concentration of heparin binding sensogram responses were collected from flow cell 2 after deducting the responses from flow cell 1.

3. After collection of the sensogram response, for each concentration of heparin, the HVEM surface regenerated using 0.1% SDS with a flow rate of 10 μL/min and repeat the screening for the next concentration as shown in Fig. 7. Upon comparison of sensor responses for different heparin concentrations (from 0.630 to 5.0 μM), it was observed that the heparin interferes the HVEM-gD interactions at higher concentrations (Fig. 10). Whereas, the heparin alone (at 2.0 μM) had no binding ability to the HVEM surface suggesting that heparin interact specifically with gD protein and prevents binding to the HVEM surface.
4. Integrating these responses at different concentrations of heparin calculated IC_{50}, 50% inhibition required for inhibiting the HVEM-gD interactions. It was observed that IC_{50} value for heparin was about 2.0 μM.

4 Notes

1. After completing the analyses, the ligand immobilized CM5 chip can be reused for at least 30 times without losing HVEM binding activity to the gD proteins.
2. If the analyses interrupted for >4 days, then it is recommended to take out the chip from the Biacore T100 machine. To remove the chip, first, wash the flow cells 1 and 2 with the binding buffer for 1 h with a flow rate of 30 μL/min at 25 °C followed by washing with sterile water for 1 h with a flow rate of 60 μL/min at 25 °C.
3. The washed CM5 chip can be stored in water for about 3 weeks at 4 °C.
4. All the solution used here are filtered through 0.22 μm filters, low binding durapore membrane (Amicon, Millipore, USA), including HVEM, gD proteins, sulfated glycans, and heparin.

Acknowledgment

This work was supported by grants from JST and AIST to P.K.R.K.

References

1. Shuker SB, Hajduk PJ, Meadows RP et al (1996) Discovering high-affinity ligands for proteins: SAR by NMR. Science 274:1531–1534
2. Mayer M, Meyer B (1999) Characterization of ligand binding by saturation transfer difference NMR spectroscopy. Angew Chem Int Ed 38:1784–1788
3. Blundell TL, Jhoti H, Abell C (2002) High-throughput crystallography for lead discovery in drug design. Nat Rev Drug Discov 1:45–54
4. Duong-Thi MD, Bergstrom M, Fex T et al (2013) High-throughput fragment screening by affinity LC-MS. J Biomol Screen 18:160–171
5. Ladbury JE, Klebe G, Freire E (2010) Adding calorimetric data to decision making in lead discovery: a hot tip. Nat Rev Drug Discov 9:23–27
6. Kranz JK, SChalk-Hihi C (2011) Protein thermal shifts to identify low molecular weight fragments. Methods Enzymol 493:277–298
7. Lewis LM, Engle LJ, Pierceall WE et al (2004) Affinity capillary electrophoresis for the screening of novel antimicrobial targets. J Biomol Screen 9:303–308
8. Duong-Thi MD, Meiby E, Bergstrom M et al (2011) Weak affinity chromatography as a new approach for fragment screening in drug discovery. Anal Biochem 414:138–146
9. Proll F, Fechner P, Proll G (2001) Direct optical detection in fragment-based screening. Anal Bioanal Chem 393:1557–1562
10. Myszka DG, Rich RL (2000) Implementing surface plasmon resonance biosensors in drug discovery. Pharm Sci Technol Today 3:310–317
11. Cooper MA (2003) Label-free screening of biomolecular interactions. Anal Bioanal Chem 377:834–842
12. Lofas S (2004) Optimizing the hit-to-lead process using SPR analysis. Assay Drug Dev Technol 2:407–416
13. Huber W (2005) A new strategy for improved secondary screening and lead optimization using high-resolution SPR characterization of compound-target interactions. J Mol Recognit 18:273–281
14. Geschwindner S, Olsson LL, Albert JS et al (2007) Discovery of a novel warhead β-secretase through fragment-based lead generation. J Med Chem 50:5903–5911

15. Godemann R, Madden J, Kramer J et al (2009) Fragment-based discovery of BACE1 inhibitors using functional assays. Biochemistry 48:10743–10751
16. Cole DC, Olland AM, Jacob J et al (2010) Identification and characterization of acidic mammalian chitinase inhibitors. J Med Chem 53:6122–6128
17. Giannetti AM (2011) From experimental design to validation hits a comprehensive walk-through of fragment lead identification using surface plasmon resonance. Methods Enzymol 493:169–218
18. Gopinath SC, Hayashi K, Kumar PKR (2012) Aptamer that binds to the gD protein of herpes simplex virus 1 and efficiently inhibits viral entry. J Virol 86:6732–6744
19. Gopinath SC, Hayashi K, Lee JB et al (2013) Analysis of compounds that interfere with herpes simplex virus-host receptor interactions using surface plasmon resonance. Anal Chem 85:10455–10462
20. Suenaga E, Kumar PKR (2014) An aptamer that binds efficiently to the hemagglutinins of highly pathogenic avian influenza viruses (H5N1 and H7N7) and inhibits hemagglutinin-glycan interactions. Acta Biomater 10:1314–1323
21. Zhang N, Yan J, Lu G et al (2011) Binding of herpes simplex virus glycoprotein D to nectin-1 exploits host cell adhesion. Nat Commun 2:577. https://doi.org/10.1038/ncomms1571
22. Karlesson R, Katsamba PS, Nordin H et al (2006) Analyzing a kinetic titration series using affinity biosensors. Anal Biochem 349:136–147
23. Suenaga E, Mizuno H, Kumar PKR (2012) Monitoring influenza hemagglutinin and glycan interactions using surface plasmon resonance. Biosens Bioelectron 32:195–201
24. Biacore: Concentration analysis Handbook BR-1005-12 Edition AB (2007) GE Healthcare, Supplied with T100 Biacore machine. 53–54
25. Hayashi K, Kawauchi M, Nakai C et al (2001) Characterization of inhibitory action of concanamycins against herpes simplex virus. Antivir Chem Chemother 12:51–59
26. Srisomporn P, Hayashi K, Lee JB et al (2001) Effects of structural modification of calcium spirulan, a sulfated polysaccharide from spirulina platensis, on antiviral activity. Chem Pharm Bull 49:484–485
27. Shukla D, Liu J, Blaiklock P et al (1999) A novel role for 3-O-sulfated heparan sulfate in herpes simplex virus 1 entry. Cell 99:13–22
28. Liu J, Shriver Z, Pope RM (2002) Characterization of a heparan sulfate octasaccharide that binds to herpes simplex virus type 1 glycoprotein D. J Biol Chem 277:33456–33467

Chapter 9

Screening of Beta-Secretase Inhibitors by Capillary Electrophoresis-Mass Spectrometry

Jan Schejbal, Roman Řemínek, and Zdeněk Glatz

Abstract

Alzheimer's disease is the most common cause of dementia, currently afflicting almost 40 million patients worldwide. According to the amyloid cascade hypothesis, the pathogenesis of the disease could be slowed down or even stopped by the inhibition of beta-secretase, making this aspartic acid protease a potentially important drug target site. Capillary electrophoresis is a promising technique for screening putative enzyme inhibitors due to highly effective separations, minuscule sample and other chemicals consumption, compatibility with a variety of detection techniques, and high throughput via automation. This chapter presents a method based on capillary electrophoresis coupled to mass spectrometry detection for kinetic and inhibition assays of the beta-secretase reaction with a decapeptide derived from an amyloid precursor protein.

Key words Capillary electrophoresis, Mass spectrometry, Alzheimer's disease, Beta-secretase, Enzyme inhibition, Screening

1 Introduction

Alzheimer's disease (AD) is a severe neurodegenerative disorder characterized by progressive memory and cognitive function impairment, with development into dementia and eventually death [1]. Several prescription drugs treating symptoms of AD are currently available; however, none of them stop the progression of the disease itself. Since increased levels of the 42-amino-acid residue form of amyloid β peptide (Aβ1–42) were found in the brain tissue of most patients, according to the amyloid cascade hypothesis AD pathogenesis is started by an imbalance between Aβ1–42 production and its clearance [2]. Aβ1–42 subsequently aggregates into amyloid-derived diffusible ligands and amyloid plaques causing a degradation of neurons [3, 4]. The specific inhibition of beta-secretase (beta-site amyloid precursor protein cleaving enzyme 1, BACE1), the aspartic acid protease responsible

Nikolaos E. Labrou (ed.), *Targeting Enzymes for Pharmaceutical Development: Methods and Protocols*, Methods in Molecular Biology, vol. 2089, https://doi.org/10.1007/978-1-0716-0163-1_9, © Springer Science+Business Media, LLC 2020

for Aβ1–42 production, thus appears to be a promising way to slow down or even stop the progression of the disease [5–7].

Capillary electrophoresis (CE) represents a suitable analytical technique in this field due to its high separation efficiency, minuscule sample and other reagents consumption, high throughput via automation and very limited need for sample pretreatment, as well as its straightforward coupling with a variety of optical, electrochemical, or mass-spectrometric (MS) detection systems [8–11]. For these reasons, a protocol of a CE method with MS detection for the screening of BACE inhibitors is presented in this chapter. A decapeptide derived from an amyloid precursor protein BACE1 site with the "Swedish mutation" [12], primary sequence Ser-Glu-Val-Asn-Leu-Asp-Ala-Glu-Phe-Arg, further referred to as SEVNLDAEFR, is used as the substrate cleaved by BACE1 into two proteolytic products—the first with the sequence Ser-Glu-Val-Asn-Leu, further referred to as SEVNL; and the second with the sequence Asp-Ala-Glu-Phe-Arg, further referred to as DAEFR [13, 14]. Using the unlabeled substrate eliminates the drawbacks of other commonly applied BACE1 kinetic and inhibition assays, which are almost exclusively based on Förster resonance energy transfer (FRET) technology [15, 16], such as the high background signal, very limited solubility of fluorescently labeled substrates, and risk of false positive/negative results [13, 16]. The method can thus be favorably applied for the rapid screening of putative BACE1 inhibitors and/or confirmation of the results obtained using conventional FRET assays.

2 Materials

Prepare all solutions using ultrapure deionized water (resistivity at 25 °C > 18 MΩ cm) and analytical grade reagents, unless otherwise instructed. The solutions can be stored for 1 week at room temperature unless indicated otherwise. Consistently comply with all applicable waste disposal regulations.

2.1 BACE1 Enzyme Assay

1. Incubation buffer (IB): 50 mM sodium acetate (NaAc), pH 4.25. Weigh 0.41015 g NaAc and quantitatively transfer to a 100 mL volumetric flask using a funnel and water. Fill the flask up to four-fifths with water and mix until the crystals dissolve (*see* **Note 1**). Fill up the flask and mix the solution. Transfer the solution to a suitable reclosable jar that can accommodate a pH electrode. Adjust pH using acetic acid (HAc) to pH 4.25 (*see* **Note 2**). Close and store at 4 °C.
2. Ice-cold acetonitrile (ACN) for reaction termination (*see* **Note 3**).
3. Stock solutions of tested inhibitor: 3 mM stock solution in dimethyl sulfoxide (DMSO) (*see* **Note 4**).

4. Stock solutions of peptide substrate (SEVNLDAEFR) and product standards (SEVNL and DAEFR), all available from BACHEM (Bubendorf, Switzerland): dilute in DMSO to a final peptide concentration of 20 mM and store at −20 °C (*see* **Note 5**).
5. Stock solution of internal standard: Prepare 1 mM human angiotensin II (At II) stock solution in DMSO.
6. Working solution of BACE1: Dilute the BACE1 solution available from Sigma Aldrich (St. Louis, MO, USA) with IB to obtain a 100 U mL^{-1} BACE1 concentration (*see* **Note 6**).
7. Working solutions of substrate and inhibitor: Prepare fresh every day from stock solutions by dilution in IB. The specific concentrations of working solutions are given by the specific assay setting.
8. Solution of internal standard: Prepare a 5 μM working solution of At II in IB.

2.2 CE-MS Analysis

1. Background electrolyte (BGE): 12.5% (v/v) HAc. Prepare 100 mL calibrated cylinder and fill partially with water. Use a 25 mL calibrated cylinder to prepare 12.5 mL HAc and transfer it quantitatively to the 100 mL cylinder containing water. Fill up to 100 mL with water. Mix and transfer the BGE to a reclosable jar.
2. Bare silica capillary: Prepare a 45 cm-long bare silica capillary (75 μm inner diameter) with rectangularly cut ends. Use a diamond capillary cutter to precisely cut the capillary (*see* **Note 7**).
3. Conditioning and washing solutions: Every day prepare fresh 50% HAc and 6% NH_4OH in water (v/v) for washing and conditioning procedures. These solutions can be prepared directly in 1.5 mL glass vials using micropipettes.
4. Sheath liquid (SL): Mix MS-grade MeOH and water (1:1, v/v) in a suitable container for your SL delivery system (*see* **Note 8**).

3 Methods

3.1 CE-MS Analysis

This protocol is designed for an Agilent 7100 CE system coupled to a Bruker maXis Impact QTOF MS by the coaxial sheath-liquid flow CE-MS ESI interface by Agilent. The coupling of CE to MS should be performed according to the respective manufacturer's instructions.

1. Before the final setup of the instrument, remove 5 cm of the capillary polyimide coating at the capillary outlet and 5 mm at the capillary inlet (*see* **Note 9**).

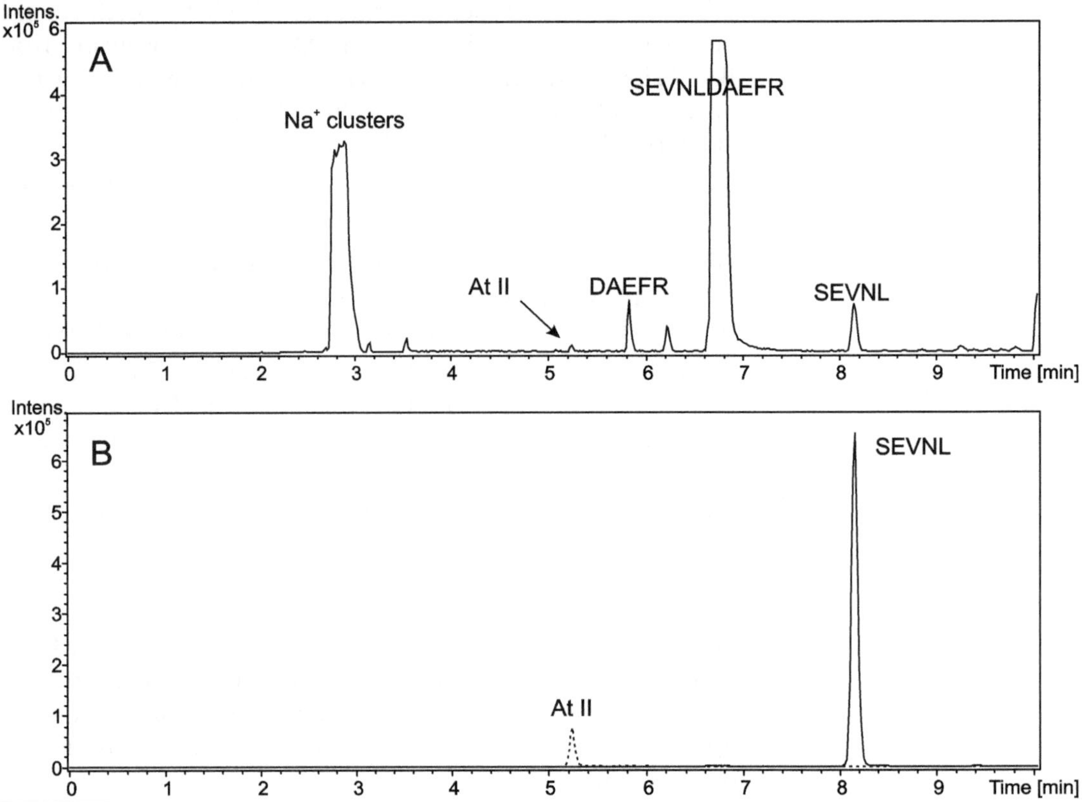

Fig. 1 Base peak electropherogram of CE-MS analysis of sample acquired during kinetics assay using 600 μM substrate (**a**) and extracted ion electropherograms of internal standard At II (dashed line ---) and quantified product SEVNL (solid line —). (Adapted with permission from ref. [13])

Table 1
Dilution table for calibration curve

Sample (SEVNL concentration) (μM)	IB (μL)	Calibration working solution (μL)	Calibration working solution 10× diluted in IB (μL)
0.5	37.5	–	12.5
1	25	–	25
2	–	–	50
4	40	10	–
8	30	20	–
12	20	30	–
16	10	40	–
20	–	50	–

2. Every capillary needs to be conditioned prior to its first use and then every day before use according to the following conditioning sequence: 50% HAc for 2 min, water for 2 min, 6% NH_4OH for 5 min, water for 2 min, and BGE for 5 min at 25 °C.
3. Set the MS detection method using electrospray ionization (ESI) in positive mode. Ground the ESI needle and set the ESI voltage at the MS sampling capillary entrance to −5 kV. Set the flow rate of the SL to 4 μL min^{-1} (final flow through CE-MS interface). Set the drying gas flow rate to 5 L min^{-1}, the drying temperature to 180 °C, and nebulization gas pressure to 0.4 bar (40 kPa) (*see* **Note 10**).
4. Set the acquisition mass range from 50 to 1600 *m/z* at a spectra rate of 1 Hz. Respective analytes can be detected as follows: peptide substrate SEVNLDAEFR as the protonated molecular ion $[M + 2H]^{2+}$ $m/z = 590.286 \pm 0.005$; proteolytic product SEVNL as $[M + H]^{1+}$ $m/z = 561.288 \pm 0.005$; proteolytic product DAEFR as both $[M + H]^{1+}$ $m/z = 637.294 \pm 0.005$ and $[M + 2H]^{2+}$ $m/z = 319.151 \pm 0.005$ and the internal standard At II as $[M + 2H]^{2+}$ $m/z = 523.775 \pm 0.005$ (*see* **Note 11**).
5. To achieve a high *m/z* accuracy, commonly occurring sodium salt clusters $[Na(CH_3COONa)_n]^{1+}$ originating from the HAc present in the BGE and the sodium ions in samples can be used for internal mass calibration for each analysis. Automated internal mass calibration using these clusters is available in the software Bruker Data Analysis. External mass calibration should be performed at least once every 2 weeks.
6. Set up the CE method as follows: preconditioning by rinsing the capillary with 50% HAc for 1 min, water for 1 min, 6% NH_4OH for 2 min, water for 1 min, and BGE for 4 min. Set the sample injection to 25 mbar (kPa) for 5 s. Set the separation voltage to 15 kV (333.3 V cm^{-1}, positive polarity) with a linear voltage gradient from 0 to 15 kV during the first minute.
7. Sample electropherograms are available in Fig. 1.
8. Prepare samples for the calibration curve as follows: prepare the working solution including 380 μM substrate and 20 μM concentrations of both reaction products in IB (*see* **Note 12**). Dilute this working solution in IB to acquire samples including the 0.5–20 μM product according to Table 1. Transfer 40 μL of the sample to 50 μL of ice-cold ACN, add 10 μL of 5 μM internal standard working solution, and vortex thoroughly.
9. Analyze the samples according to the CE-MS method and use the ratio of SEVNL peak area/internal standard peak area plotted against the concentration of SEVNL to acquire the calibration curve (*see* **Note 13**).

Table 2
Dilution table for BACE1 kinetic assay

Sample (substrate concentration) (μM)	IB (μL)	Substrate working solution (μL)	BACE1 working solution (μL)
50	60.8	3.2	16
100	57.6	6.4	16
200	51.2	12.8	16
300	44.8	19.2	16
400	38.4	25.6	16
600	25.6	38.4	16
800	12.8	51.2	16
1000	–	64	16

3.2 BACE1 Kinetic Assay

1. Prepare all required working solutions—BACE1, substrate, and internal standard—in IB. The concentration of the substrate working solution should be set to 1.25 mM.
2. Prepare a set of twenty-four 500 μL self-lock tubes and label them for eight samples with different substrate concentrations, each of them in triplicate.
3. Set the thermomixer to 1500 RPM, 60-min incubation time, and heat it to 37 °C.
4. Fill the tubes according to Table 2, while adding the BACE1 working solution last (*see* **Note 14**).
5. Vortex all the tubes briefly, transfer them into the preheated thermomixer, and start the incubation.
6. Prepare twenty-four 500 μL self-lock tubes and label them. Shortly before the incubation is finished, put 50 μL of ice-cold ACN into all tubes. Once the incubation is completed, add 40 μL of reaction mixture to the correspondingly labeled tubes. Vortex all tubes to terminate the enzymatic reaction and add 10 μL of internal standard working solution (*see* **Note 14**).
7. Vortex all tubes and transfer 70 μL of the prepared sample to the labeled 250 μL polypropylene snap vial and close with a polyethylene olefin snap cap.
8. Analyze the samples as soon as possible according to the CE-MS analytical method. Short-term storage of the samples before analysis is possible at 4 °C.
9. Calculate the reaction rate from the acquired concentrations of reaction product in the respective samples for the incubation

time of 60 min and 20 U mL^{-1} enzyme concentration. Plot the reaction rate against substrate concentration and calculate the Michaelis-Menten kinetics parameters using regression analysis.

3.3 BACE1 Inhibition Assay

1. Prepare all required working solutions—BACE1, substrate, internal standard, and inhibitor—in IB. The concentration of the substrate working solution should be set to 2 mM.
2. Prepare several working solutions of inhibitor in order to be able to cover a concentration range of at least six orders of magnitude (*see* **Note 15**).
3. Prepare a set of twenty-four 500 μL self-lock tubes and label them for eight samples of different inhibitor concentrations, each of them in triplicate.
4. Set the thermomixer to 1500 RPM, 60-min incubation time, and heat it to 37 °C.
5. Fill the tubes according to Table 3, while adding the substrate working solution last (*see* **Note 16**). The final volume is 80 μL and final concentrations are 200 μM substrate and 20 U mL^{-1} BACE1; the final concentration of inhibitor is given in Table 3 (*see* **Note 14**).
6. Vortex all the tubes briefly, transfer them into the preheated thermomixer, and start the incubation.
7. Prepare twenty-four 500 μL self-lock tubes and label them. Shortly before the incubation is finished, put 50 μL of ice-cold ACN into all tubes. Once the incubation is completed, add 40 μL of incubated sample to the correspondingly labeled tubes. Vortex all tubes to terminate the enzymatic reaction and add 10 μL of internal standard working solution (*see* **Note 14**).
8. Vortex all tubes and transfer 70 μL of the prepared sample to the labeled 250 μL polypropylene snap vial and close with a polyethylene olefin snap cap.
9. Analyze the samples as soon as possible according to the CE-MS analytical method. Short-term storage of the samples is possible at 4 °C.
10. Calculate the reaction rate from the acquired concentrations of reaction product in the respective samples and known incubation time and enzyme concentration. Plot the reaction rate against the logarithmic concentration of the inhibitor and calculate the half-maximal inhibitory concentration (IC_{50}) value. Nonlinear regression analysis according to the single-site competition model equation $Y = \text{Bottom} + (\text{Top} - \text{Bottom})/1 + 10^{X-\log IC50}$ can be used for this purpose. Y represents the reaction rate, X is the logarithm of the inhibitor concentration, and Bottom and Top are the lower and upper

Table 3
Dilution table for inhibition assay (*see* Note 17)

Sample (inhibitor conc.) (nM)	IB (μL)	0.2 nM inhibitor solution (μL)	20 nM inhibitor solution (μL)	200 nM inhibitor solution (μL)	20 μM inhibitor solution (μL)	BACE1 working solution (μL)	Substrate working solution (μL)
0.1	16	40	–	–	–	16	8
1	52	–	4	–	–	16	8
10	16	–	40	–	–	16	8
20	48	–	–	8	–	16	8
40	40	–	–	16	–	16	8
100	16	–	–	40	–	16	8
1000	52	–	–	–	4	16	8
10,000	16	–	–	–	40	16	8

limits of the curve, corresponding to the SEVNL formation rate with the highest and the lowest inhibitor concentration in the reaction mixture.

4 Notes

1. NaAc is readily soluble at this concentration. If facing problems with solubility, recheck compliance with the protocol.
2. Make sure the pH meter is properly calibrated before starting. Concentrated HAc can be used to precisely adjust pH at this concentration of NaAc. Using a 200 μL micropipette proved to be the most convenient for this purpose. The use of concentrated HAc also minimizes the changes in total volume of the solution and thus keeps the changes in final NaAc concentration below a significant level.
3. The cooling of ACN provides a more convenient and precise handling of the liquid. Room-temperature ACN cannot be precisely handled with micropipettes. Using ACN to terminate the enzymatic reaction is beneficial for subsequent CE-MS analysis, as it provides sample stacking within the CE, which is beneficial in terms of method sensitivity.
4. In this protocol, the reference synthetic inhibitor LY2886721 is used. In general, a 3 mM stock solution can be recommended for all potent inhibitors. Stock solutions of inhibitors with IC_{50} above 1 μM should be prepared at a higher concentration with respect to their respective potency and total DMSO content in the incubation mixture, which must be limited to 5% so as to

have no significant effect on BACE1 activity. Please note that the addition of substrate into the incubation mixture will also increase the DMSO content.

5. This concentration of stock solution allows for the preparation of samples containing up to 1 mM substrate, while ensuring that the concentration of DMSO does not exceed 5% (v/v) so as to have no significant effect on BACE1 activity. The preparation of stock solution with a higher concentration of substrate is not recommended, as issues with solubility of the substrate upon dilution in IB can be observed.
6. Before dilution, check the specific activity of each BACE1 batch. The differences can be very high and severely affect the outcome of the assay.
7. Take a bare silica capillary, 75 μm inner diameter, with a clean end. If the cut is not clean, use a diamond capillary cutter to clean the end, by cutting off approximately 5 mm of the capillary. Then, cut 45.5 cm of the capillary using a ceramic cutter and then use the diamond cutter to cut off the surplus 0.5 cm. Recheck that both ends of the capillary to be used have a clean cut orthogonal to the capillary axis. A rough capillary inlet may cause inconsistency in sample injection, while a rough capillary end coupled to MS will negatively affect electrospray ionization.
8. In our case, 80 mL of SL was sufficient for comfortable maintenance of the system. Regular checking of SL level is necessary to avoid running experiments without SL.
9. In our case—the coupling of an Agilent 7100 CE system to a Bruker MaXis Impact QTOF MS via an Agilent coaxial ESI interface (part no. G1607-60002)—the optimal procedure was to remove the capillary coating after its insertion into the interface. The interface was positioned so that approximately 10 cm of the capillary protruded from the ESI needle, and 5 cm of the polyimide coating at the capillary outlet could be safely removed by burning it off. The de-coated part of the capillary was cleaned with MS-grade methanol on Kimtech precision wipes. After this procedure, the MS interface was setup according to the manufacturer instructions. The coating at the inlet part of the capillary was burned off after inserting the capillary into the CE cassette. The inlet part was then cleaned in the same way.
10. When using different CE-MS instrumentation, MS coupling and detection parameters should undergo optimization.
11. As the ionization of the two observed DAEFR protonated molecular ions occurs in an unforeseeable ratio, using the SEVNL proteolytic product for quantification is strongly recommended.

12. The concentrations of both SEVNL and DAEFR should be 20 μM in the working solution. This substrate:products ratio represents a conversion of 5% of the substrate. The inclusion of BACE1 into the calibration samples is not recommended.
13. The calibration range can be extended, as the relation is linear over the concentration range from at least 20 nM to 20 μM.
14. In this protocol, multiple small-volume samples are handled manually, while the incubation time is an important variable and thus must be known and uniform for all samples. Therefore, precise and at the same time-efficient sample handling must be ensured. Proficiency in micropipette handling and the use of dilution tables is highly recommended.
15. Serial dilution is a suitable option for inhibitors with unknown potency; in such cases, a concentration range of eight orders of magnitude is recommended. For inhibitors of known potency, additional samples for data points at and closely around the anticipated IC_{50} should be used, *see* Table 3 for a sample dataset.
16. The addition of substrate as the final part of the incubation mixture allows for the preincubation of the inhibitor with BACE1.
17. This is a sample table for the synthetic BACE1 inhibitor LY2886721 with a previously published IC_{50} value of 20.3 nM [17]. In studies for the elucidation of an inhibition mechanism, include two more sample sets with different substrate concentrations.

Acknowledgments

This work was supported by grant no. GA16-06106S from the Czech Science Foundation.

References

1. Alzheimer's Association (2015) Alzheimer's disease facts and figures. Alzheimers Dement 11:332–384
2. Hardy J, Selkoe DJ (2002) The amyloid hypothesis of Alzheimer's disease: progress and problems on the road to therapeutics. Science 297:353–356
3. Selkoe DJ (2001) Alzheimer's disease: genes, proteins, and therapy. Physiol Rev 81:741–766
4. Eikelenboom P, Veerhuis R, Scheper W, Rozemuller AJM, van Gool WA, Hoozemans JJM (2006) The significance of neuroinflammation in understanding Alzheimer's disease. J Neural Transm 113:1685–1695
5. Vassar R (2001) The beta-secretase, BACE: a prime drug target for Alzheimer's disease. J Mol Neurosci 17:157–170
6. Vassar R, Kuhn PH, Haass C, Kennedy ME, Rajendran L, Wong PC, Lichtenthaler SF (2014) Function, therapeutic potential and cell biology of BACE proteases: current status and future prospects. J Neurochem 130:4–28
7. Querfurth HW, LaFerla FM (2010) Alzheimer's disease. N Engl J Med 362:329–344

8. Glatz Z (2006) Determination of enzymatic activity by capillary electrophoresis. J Chromatogr B 841:23–37
9. Fan Y, Scriba GK (2010) Advances in-capillary electrophoretic enzyme assays. J Pharm Biomed Anal 53:1076–1090
10. Scriba GK, Abromeit H, Hense M, Fan Y (2013) Capillary electrophoretic enzyme assays. Methods Mol Biol 984:285–308
11. Ouimet CM, D'Amico CI, Kennedy RT (2017) Advances in capillary electrophoresis and the implications for drug discovery. Expert Opin Drug Discov 12:213–224
12. Haass C, Lemere CA, Capell A, Citron M, Seubert P, Schenk D, Lannfelt L, Selkoe DJ (1995) The Swedish mutation causes early-onset Alzheimer's disease by β-secretase cleavage within the secretory pathway. Nat Med 1:1291–1296
13. Schejbal J, Slezáčková L, Řemínek R, Glatz Z (2017) A capillary electrophoresis-mass spectrometry based method for the screening of β-secretase inhibitors as potential Alzheimer's disease therapeutics. J Chromatogr A 1487:235–241
14. Řemínek R, Slezáčková L, Schejbal J, Glatz Z (2017) Development and comprehensive comparison of two on-line capillary electrophoretic methods for β-secretase inhibitor screening. J Chromatogr A 1518:89–96
15. Pietrak BL, Crouthamel M-C, Tugusheva K, Lineberger JE, Xu M, DiMuzio JM, Steele T, Espeseth AS, Stachel SJ, Coburn CA, Graham SL, Vacca JP, Shi X-P, Simon AJ, Hazuda DJ, Lai M-T (2005) Biochemical and cell-based assays for characterization of BACE-1 inhibitors. Anal Biochem 342:144–151
16. Mancini F, De Simone A, Andrisano V (2011) Beta-secretase as a target for Alzheimer's disease drug discovery: an overview of in vitro methods for characterization of inhibitors. Anal Bioanal Chem 400:1979–1996
17. May PC, Willis BA, Lowe SL, Dean RA, Monk SA, Cocke PJ, Audia JE, Boggs LN, Borders AR, Brier RA (2015) The potent BACE1 inhibitor LY2886721 elicits robust central Aβ pharmacodynamic responses in mice, dogs, and humans. J Neurosci 35:1199–1210

Chapter 10

Electrophoretic Mobility Shift Assays with GFP-Tagged Proteins (GFP-EMSA)

Alanna E. Sorenson and Patrick M. Schaeffer

Abstract

The electrophoretic mobility shift assay (EMSA) is commonly used for the study of nucleic acid-binding proteins. The technique can be used to demonstrate that a protein is binding to RNA or DNA through visualization of a shift in electrophoretic mobility of the nucleic acid band. A major disadvantage of the EMSA is that it does not always provide an absolute certitude that the band shift is due to the protein under scrutiny, as contaminants in the sample could also cause the band shift. Here we describe a variation of the standard EMSA allowing to visualize with added certitude, the co-localized band shifts of a GFP-tagged protein binding to its cognate nucleic acid target sequence stained with an intercalator, such as GelRed. Herein, we present an illustrative protocol of this useful technique called GFP-EMSA along with specific notes on its advantages and limitations.

Key words Green fluorescent protein, Nucleic acid binding, DNA binding, RNA binding, Electrophoretic mobility shift assay, Band shift assay, Gel shift assay

1 Introduction

The interaction of proteins with nucleic acids is fundamental to transcription, posttranscriptional modification, translation, DNA replication, repair, and recombination. Due to their central function in such a wide variety of essential cellular processes and their role in disease development, nucleic acid-binding proteins have become desired therapeutic targets [1, 2]. Characterization of essential protein:nucleic acid interactions is important for both fundamental research purposes, and as a basis for drug development. The EMSA has long been an important staple tool in the detection and characterization of nucleic acid-binding proteins and their target sequences [3]. In a typical EMSA, target nucleic acid sequences can be labeled with a fluorescent probe, radioactive isotopes nuclei, or stained for detection. The presence of the cognate-binding protein is indirectly visualized through its effect on the mobility of its target sequence [3]. GFP-EMSA is a unique

Nikolaos E. Labrou (ed.), *Targeting Enzymes for Pharmaceutical Development: Methods and Protocols*, Methods in Molecular Biology, vol. 2089, https://doi.org/10.1007/978-1-0716-0163-1_10,

technique that allows the concomitant detection of the protein and its target sequence. GFP-EMSA can be performed with relatively safe nucleic acid-binding dyes post-electrophoresis. This technique is particularly useful for the definitive characterization of protein: nucleic acid interactions [4–8].

2 Materials

2.1 Reagents

1. GFP-tagged nucleic acid-binding protein (*see* **Note 1**).
2. Oligonucleotides (*see* **Note 2**).
3. Appropriate fluorescent label/intercalating dye (*see* **Note 3**).
4. Annealing buffer: 20 mM Tris–HCl pH 8.0, 100 mM NaCl.
5. Binding buffer (dependent on cofactors and strength of interaction if known, *see* **Note 4**; if unknown: 10 mM Tris-HCl pH 8.0, 1% glycerol, 50 mM NaCl).
6. 50% glycerol.
7. Loading buffer (any 5–6× loading buffer with dye to indicate progression).

2.2 Gel Electrophoresis

1. Molecular biology grade agarose (*see* **Note 5**).
2. Glass flask or bottle sufficient for volume of gel required (dependent on number of reactions desired and electrophoresis equipment available).
3. Microwave.
4. Gel electrophoresis systems, gel molds, combs, electrophoresis tank, and power pack (*see* **Note 6**).
5. Running buffer: Tris–borate–EDTA (TBE), 89 mM Tris, 89 mM boric acid, and 2 mM EDTA [9].

2.3 Gel Imaging

1. UV transilluminator and camera with adequate filter or gel documentation system capable of emitting UV, blue and green light, with appropriate filters for red/green imaging (e.g., Syngene G:BOX).
2. Nucleic acid stain if required for oligonucleotide detection (e.g., Biotium GelRed).

3 Methods

3.1 Reagent Preparation

The following sample protocol is typically applied for a GFP-tagged protein binding to a double-stranded (ds) DNA target sequence with post-electrophoretic GelRed staining.

1. Combine equal volumes of both oligonucleotides (1 μM each) in annealing buffer and heat for 2 min at 80 °C.
2. Allow to cool slowly to room temperature (RT). The volume to prepare will depend on the number of reactions required. Reactions are typically 10 μL volume containing 1 μM GFP-tagged DNA-binding protein and target DNA sequence at desired ratio.

3.2 Gel Electrophoresis

1. Prepare a thin 1% agarose gel in TBE without nucleic acid stain (*see* **Note 7**).
2. Mix equal volumes (5 μL each) of GFP-tagged protein (1 μM) with ds DNA (1 μM) in binding buffer (*see* **Note 8**).
3. Include two control reactions that contain protein only and ds DNA only for comparison of their respective electrophoretic mobility (*see* **Note 9**).
4. Incubate reactions at RT for 10 min.
5. Add 2.5 μL 50% glycerol to reactions if binding buffer contained less than 5% glycerol (*see* **Note 10**).
6. Load 10 μL samples into the TBE gel with an additional lane of generic colored loading buffer to enable visualization of progression (*see* **Note 11**).
7. Subject gel to electrophoresis at 80 V for sufficient time to separate bands (*see* **Note 12**).

3.3 Gel Imaging

1. The gel can be exposed to UV or blue light to excite the GFP chromophore (*see* **Note 13**). If a GFP fusion protein band shift is apparent, then incubate gel with 1× GelRed in water for 30 min with rocking at RT (*see* **Note 14**).
2. Expose gel with adequate excitation wavelengths (i.e., for GFP and GelRed) and capture the emitted light with a CCD camera using appropriate filters (*see* **Note 15**, Fig. 1).
3. Analyze image results (*see* **Note 16**).

4 Notes

1. The GFP-tagged protein can be produced recombinantly using typical *E. coli* expression systems [14, 15]. Alternatively, the GFP-tagged protein can be purified from prokaryotic or eukaryotic systems where it is being used as a reporter of gene expression [16] and localization in vivo [17]. GFP-tagged DNA-binding proteins have been expressed in vivo in all domains of life [18–21]. Production using in vitro transcription-translation is also an option [22]. We have successfully applied the technique with cycle 3 GFP [10, 11] as

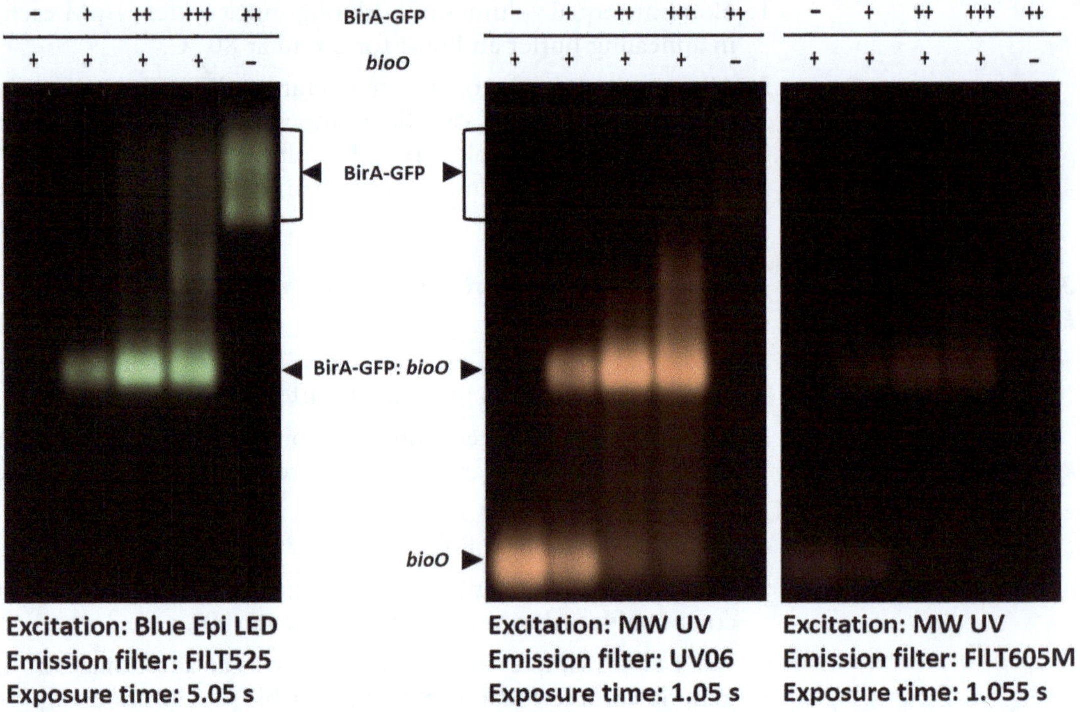

Fig. 1 Example GFP-EMSA. In *E. coli*, the bifunctional BirA protein is responsible for biotinylation of proteins and regulation of biotin synthesis through binding to the biotin operator (*bioO*) [24, 25]. In this example, 5 μL of *bioO* (0.8 μM) was combined with different concentrations of BirA-GFP (0.4, 0.8, and 1.2 μM) in PBS, pH 7.4 with 0.6 mM ATP and biotin and 3 mM $MgCl_2$ to demonstrate complex formation, similarly as described previously [5]. A band shift can be seen with BirA-GFP in the presence of *bioO* (+). Images were captured using a G:BOX Chemi XRQ (excitation, filter, and exposure times detailed below each image) and recolored to highlight differential visualization of GFP (green) and GelRed stained DNA (orange). MW UV denotes medium range UV (302 nm)

well as mCherry [12] fusion proteins and expect it to work with proteins tagged with any fluorescent protein as long as appropriate excitation wavelengths and filters are used, and care has been taken to choose an appropriate DNA fluorescent dye (i.e., the fluorescent protein and DNA dye do not have overlapping emission spectra or excitation maxima). We typically use proteins with a C-terminal GFP tag; however, if this is likely to interfere with binding, an N-terminal tag could be used instead. Dual fluorophore EMSAs can be performed if the imaging system has multiple excitation wavelengths and filters.

2. GFP-EMSA can be performed with DNA and RNA in single or ds form. Many oligonucleotide-binding proteins will require flanking sequences to enable optimal protein binding. We typically use 5 nt flanking sequences both at 5′ and 3′ ends of the target sequence. Nucleic acid-binding proteins have varying affinity and specificity for nucleic acids. The length of

oligonucleotides will be dependent on the binding protein. We have successfully performed GFP-EMSA with oligonucleotides ranging from 20 to 115 bp. It is worth noting that the longer the nucleic acid, the lesser shift in mobility will be observed with a stoichiometric binding mode.

3. Generic fluorescent intercalating dyes can be used such as GelRed (Biotium) or ethidium bromide. Other more specific nucleic acid-binding dyes, such as SYBR Green I or II, can be used if required. Alternatively, an oligonucleotide can be labeled with a fluorescent dye (e.g., Cy3 or Cy5).

4. GFP-EMSA has been performed with a wide range of buffers and salt concentrations. Thus, binding buffer composition will be dependent on the DNA-binding conditions that are to be tested (e.g., salt or pH dependence [13]). A typical binding reaction can be performed in PBS, but other buffer systems such as Tris can be used, and salt concentrations can also be altered depending on application. Some proteins may require the addition of divalent cations such as Mg^{2+} or other cofactors such as ATP.

5. High-quality agarose has very low background fluorescence and is sufficiently transparent for performing GFP-EMSA. However, polyacrylamide gel (8–10%) can be used as an alternative when higher sensitivity is required.

6. It is important to use gel electrophoresis equipment that has not been in contact with intercalating nucleic acid dyes, or it will need to be cleaned thoroughly after previous use to remove any trace of such dyes. This is because intercalating nucleic acid dyes interfere with protein binding.

7. A thin gel improves resolution and reduces background. Pour a gel as thin as possible with sufficient depth to accommodate a 10 μL reaction volume. Polyacrylamide gels can be used (8–12%) in either horizontal or vertical formats. With vertical systems, the unbound protein might not be visible depending on its pI.

8. The amount of protein required can be dependent on the quantum yield of the fluorophore and sensitivity of the imaging equipment. We find that 1 μM GFP-tagged protein is generally sufficient for fluorescence capture. If the stoichiometry of the complex is not known, it can be estimated by trialing different ratios of proteins and oligonucleotides.

9. Control reactions will demonstrate the mobility of each component in isolation and enable visualization of any band shifts produced by the interaction between the protein and nucleic acid (Fig. 1).

10. Glycerol is added to increase the density of the sample to assist with gel loading. It also increases protein stability [23].
11. A colored loading dye is added to a separate well to use as a marker during electrophoresis. It is not added to the samples themselves as it may migrate at the same position as the oligonucleotide(s) and/or protein, making imaging difficult.
12. The time taken for electrophoresis will be dependent on the size of the oligonucleotides as well as percentage of agarose and buffer system. For most proteins 30–60 min will be sufficient. Longer oligonucleotides may require longer electrophoresis for sufficient separation. GFP-EMSA can be paused at any time to check the fluorescent protein band progression using an adequate excitation wavelength. For cycle 3 GFP, simply excite with UV or blue light using a 525 nm filter. A standard "black light" flashlight is often sufficient to check progression during electrophoresis. If using a polyacrylamide gel for electrophoresis, use chilled TBE, keep cool during electrophoresis, and do not exceed 8 V/cm.
13. Cycle 3 GFP has bimodal absorption spectra and can be excited with light ranging from UV to blue (350–480 nm) with an emission maximum at 509 nm.
14. GelRed does not require destaining. The gel can be rinsed with water to reduce background if necessary. However, this has not been an issue with GFP-EMSA using the protocol described. The same procedure can be used for other dyes such as SYBR Green II (use at 1:10,000 in TBE).
15. This protocol is based on the Syngene G:BOX XRQ. Other imaging systems can be used if the light sources and filters are similar. For imaging cycle 3 GFP alone, expose gel to either blue Epi LED (465 nm) or long-wave UV light (365 nm) using a FILT525 filter (516–539 nm) (Fig. 1). For imaging GelRed stained nucleic acid, expose gel to medium-wave UV (MW UV) light (302 nm) using a UV06 filter (572–625 nm) or FILT605M filter (594–610 nm) (Fig. 1). For imaging mCherry, expose gel to green Epi LED (525 nm) using a FILT605M filter (594–610 nm). Possible fluorophores will be dependent on the available equipment.
16. Integrate fluorescent bands with image analysis software (e.g., ImageJ). Overlay images to illustrate and confirm co-localization of GFP and DNA bands in the gel.

References

1. Miyagi T, Shiotani B, Miyoshi R, Yamamoto T, Oka T, Umezawa K, Ochiya T, Takano M, Tahara H (2014) DSE-FRET: a new anticancer drug screening assay for DNA binding proteins. Cancer Sci 105(7):870–874. https://doi.org/10.1111/cas.12420
2. Alonso N, Guillen R, Chambers JW, Leng F (2015) A rapid and sensitive high-throughput screening method to identify compounds targeting protein-nucleic acids interactions. Nucleic Acids Res 43(8):e52. https://doi.org/10.1093/nar/gkv069
3. Hellman LM, Fried MG (2007) Electrophoretic mobility shift assay (EMSA) for detecting protein-nucleic acid interactions. Nat Protoc 2 (8):1849–1861. https://doi.org/10.1038/nprot.2007.249
4. Moreau MJ, Morin I, Schaeffer PM (2010) Quantitative determination of protein stability and ligand binding using a green fluorescent protein reporter system. Mol BioSyst 6 (7):1285–1292. https://doi.org/10.1039/c002001j
5. Askin SP, Bond TEH, Schaeffer PM (2016) Green fluorescent protein-based assays for high-throughput functional characterization and ligand-binding studies of biotin protein ligase. Anal Methods 8(2):418–424
6. Dahdah DB, Morin I, Moreau MJ, Dixon NE, Schaeffer PM (2009) Site-specific covalent attachment of DNA to proteins using a photoactivatable Tus-Ter complex. Chem Commun (Camb) (21):3050–3052. https://doi.org/10.1039/b900905a
7. Johnston EB, Kamath SD, Lopata AL, Schaeffer PM (2014) Tus-Ter-lock immuno-PCR assays for the sensitive detection of tropomyosin-specific IgE antibodies. Bioanalysis 6(4):465–476. https://doi.org/10.4155/bio.13.315
8. Morin I, Schaeffer PM (2012) Combining RNA-DNA swapping and quantitative polymerase chain reaction for the detection of influenza A nucleoprotein. Anal Biochem 420 (2):121–126. https://doi.org/10.1016/j.ab.2011.09.009
9. Green MR, Sambrook J (2012) Molecular cloning: a laboratory manual, vol 3, 4th edn. Cold Springs Hoarbour Laboratory Press, New York
10. Fukuda H, Arai M, Kuwajima K (2000) Folding of green fluorescent protein and the cycle3 mutant. Biochemistry 39(39):12025–12032
11. Wang L, Xie J, Deniz AA, Schultz PG (2003) Unnatural amino acid mutagenesis of green fluorescent protein. J Org Chem 68 (1):174–176. https://doi.org/10.1021/jo026570u
12. Shaner NC, Campbell RE, Steinbach PA, Giepmans BN, Palmer AE, Tsien RY (2004) Improved monomeric red, orange and yellow fluorescent proteins derived from Discosoma sp. red fluorescent protein. Nat Biotechnol 22 (12):1567–1572. https://doi.org/10.1038/nbt1037
13. Moreau MJ, Schaeffer PM (2013) Dissecting the salt dependence of the Tus-Ter protein-DNA complexes by high-throughput differential scanning fluorimetry of a GFP-tagged Tus. Mol BioSyst 9(12):3146–3154. https://doi.org/10.1039/c3mb70426b
14. Tabor S, Richardson CC (1985) A bacteriophage T7 RNA polymerase/promoter system for controlled exclusive expression of specific genes. Proc Natl Acad Sci U S A 82 (4):1074–1078
15. Dubendorff JW, Studier FW (1991) Controlling basal expression in an inducible T7 expression system by blocking the target T7 promoter with lac repressor. J Mol Biol 219(1):45–59
16. Chalfie M (1995) Green fluorescent protein. Photochem Photobiol 62(4):651–656
17. Chudakov DM, Matz MV, Lukyanov S, Lukyanov KA (2010) Fluorescent proteins and their applications in imaging living cells and tissues. Physiol Rev 90(3):1103–1163. https://doi.org/10.1152/physrev.00038.2009
18. Straight AF, Belmont AS, Robinett CC, Murray AW (1996) GFP tagging of budding yeast chromosomes reveals that protein-protein interactions can mediate sister chromatid cohesion. Curr Biol 6(12):1599–1608
19. Mortusewicz O, Schermelleh L, Walter J, Cardoso MC, Leonhardt H (2005) Recruitment of DNA methyltransferase I to DNA repair sites. Proc Natl Acad Sci U S A 102(25):8905–8909. https://doi.org/10.1073/pnas.0501034102
20. Lindhout BI, Fransz P, Tessadori F, Meckel T, Hooykaas PJ, van der Zaal BJ (2007) Live cell imaging of repetitive DNA sequences via GFP-tagged polydactyl zinc finger proteins. Nucleic Acids Res 35(16):e107. https://doi.org/10.1093/nar/gkm618
21. Delpech F, Collien Y, Mahou P, Beaurepaire E, Myllykallio H, Lestini R (2018) Snapshots of archaeal DNA replication and repair in living cells using super-resolution imaging. Nucleic Acids Res. https://doi.org/10.1093/nar/gky829

22. Spirin AS, Swartz JR (2008) Cell-free protein synthesis: methods and protocols. Wiley, Weinheim
23. Vagenende V, Yap MG, Trout BL (2009) Mechanisms of protein stabilization and prevention of protein aggregation by glycerol. Biochemistry 48(46):11084–11096. https://doi.org/10.1021/bi900649t
24. Beckett D, Kovaleva E, Schatz PJ (1999) A minimal peptide substrate in biotin holoenzyme synthetase-catalyzed biotinylation. Protein Sci 8(4):921–929. https://doi.org/10.1110/ps.8.4.921
25. Eisenstein E, Beckett D (1999) Dimerization of the *Escherichia coli* biotin repressor: corepressor function in protein assembly. Biochemistry 38(40):13077–13084

Chapter 11

Online Enantioselective Capillary Electrophoretic Method for Screening Cytochrome P450 3A4 Inhibitors

Roman Řemínek, Zdeněk Glatz, and Wolfgang Thormann

Abstract

The market share of single-enantiomer drugs is steadily increasing. The pharmacodynamics and pharmacokinetics of individual enantiomers can differ considerably. Thus, their characteristics have to be addressed as early as possible in the development process of new pharmaceuticals. Capillary electrophoresis is a promising technique for enantioselective drug metabolism studies due to highly effective separations, minuscule consumption of sample and reagents, compatibility with a variety of detection techniques, high-throughput via automation, and the implementation of online procedures. An online method comprised of the diffusion-based mixing of cytochrome P450 3A4 with racemic ketamine, incubation of the enzyme reaction, separation of the reaction products *S*- and *R*-norketamine, and their quantification is presented in this chapter. Since diffusion is an inherent property of all molecules, the method enables the addition of virtually any compound to the reaction mixture without the need for additional optimization of the mixing conditions, and thus can be favorably used for the rapid screening of putative cytochrome P450 3A4 inhibitors.

Key words Capillary electrophoresis, In-capillary reactor, Transverse diffusion of laminar flow profiles, Enantioselective separation, Drug metabolism, Enzyme inhibition, Cytochrome P450 3A4, Ketamine

1 Introduction

The majority of currently marketed chiral drugs are administered as a racemic mixture of enantiomers in cases where the less-effective enantiomer is benign. However, the rise of single-enantiomer pharmaceuticals is in full swing, as they dominate newly approved substances worldwide [1]. Drug enantiomers may differ in their distribution, rate of metabolism, excretion, potency, and selectivity for target sites. Thus, they have different bioactivity, bioavailability, pharmacologic profiles, and toxicity [2, 3]. A rational drug discovery and development protocol requires that both enantiomers are screened for these characteristics to optimize the likelihood of success in subsequent clinical testing [4]. Biotransformation affects other pharmacokinetic characteristics considerably. Metabolism

Nikolaos E. Labrou (ed.), *Targeting Enzymes for Pharmaceutical Development: Methods and Protocols*, Methods in Molecular Biology, vol. 2089, https://doi.org/10.1007/978-1-0716-0163-1_11, © Springer Science+Business Media, LLC 2020

studies and the assessment of hits/leads in terms of affinity toward drug-metabolizing enzymes thus represent a key part of LADME/Tox (liberation, absorption, distribution, metabolism, excretion, and toxicity) tests. Capillary electrophoresis (CE) represents a promising technique to assess the fate of drug enantiomers via in vitro assays [5]. Besides highly effective separations, the consumption of minuscule amounts of sample and reagents, compatibility with a variety of detection techniques, and high throughput via automation, it offers the possibility of implementing online methods where a fused-silica capillary is used not only as a separation column but also as a nanoliter-scale reaction chamber, integrating the incubation of an enzymatic reaction, separation of its products and their detection and quantitation into a single, fully automated run [6–8].

In this chapter, the protocol of an online enantioselective CE method for the screening of cytochrome P450 3A4 enzyme (CYP3A4) inhibitors, such as ketoconazole, is presented. CYP3A4 is a clinically important enzyme responsible for the biotransformation of more than 50% of commonly prescribed drugs [9] and the principle source of drug-drug interactions defined as the coadministration of a drug, which usually acts as a CYP3A4 inhibitor causing changes in the metabolic rates of other therapeutics [10]. *N*-Demethylation of the chiral phencyclidine derivate ketamine was selected as a model reaction [11–13]. Since the mixing of reaction components inside the capillary based on diffusion is generic, robust, and rapid [14], the principle of transverse diffusion of laminar flow profiles (TDLFP) was adopted [15, 16]. It is based on the alternate introduction of four plugs of the solution comprising ketamine, NADPH and inhibitor, and three plugs of CYP3A4 solution into the capillary by relatively high hydrodynamic pressure [17]. Due to the friction close to the inner wall of the capillary, each injected plug has a parabolic profile and penetrates into the preceding plug in the injection sequence, thereby creating large longitudinal interfaces between them. This is schematically depicted in Fig. 1a. Since even large molecules, such as CYP3A4, are able to diffuse across the short distance of the capillary inner diameter in tens of seconds, the reaction mixture is rapidly formed by transverse diffusion (Fig. 1b). This multiple sandwich arrangement provides a reaction mixture with close to uniform concentration distributions (Fig. 1c) [16, 18], which is necessary for the determination of inhibition parameters such as the half-maximal inhibitory concentration (IC_{50}) and the inhibition constant (K_i). After a sufficient amount of reaction product is formed, the reaction is terminated by the application of voltage and separation of the CYP3A4 and reactants (Fig. 1d). Since diffusion is an inherent property of all molecules, the presented TDLFP-based method enables the addition of virtually any compound to the reaction mixture without the need for additional optimization of the mixing conditions, and thus can

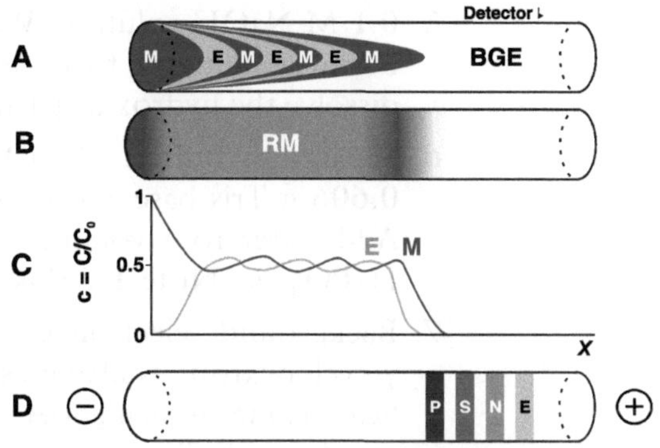

Fig. 1 Injection sequence and principle of TDLFP-based mixing. (**a**) The injection procedure consists of the alternate hydrodynamic introduction of four plugs of ketamine, NADPH, and inhibitor mixture solution (M) and three plugs of CYP3A4 solution (E) into a capillary previously filled with background electrolyte (BGE). The injected plugs have parabolic profiles due to the laminar nature of the flow inside the capillary creating longitudinal interfaces between consecutive plugs. (**b**) Reaction mixture (RM) is quickly formed by transverse diffusion. (**c**) Graphical demonstration of the dependence of dimensionless CYP3A4 and reactants concentrations (*c*) on distance from the capillary inlet (*X*). For the concentration of the injection solutions $c = 1$. (**d**) After a sufficient amount of reaction product (P) is formed, the reaction is terminated by the application of voltage and separation of the E, ketamine (S), and NADPH (N)

be favorably used for the rapid screening of structurally different molecules and their clustering according to their eventual inhibitory effect toward CYP3A4.

2 Materials

Prepare all solutions using ultrapure water (bidistilled or deionized with a resistivity of 18 MΩ cm or higher at 25 °C) and analytical grade reagents. Prepare and store all solutions at room temperature unless indicated otherwise.

2.1 Separation System

1. 2 mL glass vials and caps.
2. Glass vial inserts (microvials).
3. Fused-silica capillary: 50 μm i.d., 375 μm o.d., 64 cm total length, 54 cm effective length, remove polyimide coating at the detection window.
4. 1 M NaOH solution. Weigh 2 g NaOH and transfer to a 50 mL volumetric flask. Add water to a volume of 40 mL and dissolve the hydroxide particles. Make up to 50 mL with water.

5. 0.1 M NaOH solution. Weigh 0.2 g NaOH and transfer to a 50 mL volumetric flask. Add water to a volume of 40 mL and dissolve the hydroxide particles. Make up to 50 mL with water.
6. Separation buffer: 50 mM Tris-phosphate, pH 2.5. Weigh 0.606 g Tris base and transfer to a 100 mL volumetric flask. Add water to a volume of 95 mL and adjust pH with 85% H_3PO_4 (*see* **Note 1**). Make up to 100 mL with water.
7. Background electrolyte (BGE): 3% w/v highly sulfated γ-cyclodextrin, available as stock 20% w/v solution from AB Sciex (Framingham, MA, USA) *see* **Note 2**, prepared in the separation buffer. Transfer 15 mL of separation buffer into a 25 mL volumetric flask, add 3.75 mL of highly sulfated γ-cyclodextrin stock solution, shake vigorously until the solution is homogeneous, and make up to 25 mL with separation buffer.

2.2 Enzyme Reaction

1. Incubation buffer (IB): 100 mM potassium phosphate, pH 7.4. Weigh 1.361 g KH_2PO_4 and transfer to a 100 mL volumetric flask. Add water to a volume of 60 mL and dissolve the phosphate particles. Make up to 100 mL with water. Weigh 1.742 g K_2HPO_4 and prepare solution in the same way as with KH_2PO_4. Titrate K_2HPO_4 solution with KH_2PO_4 solution to reach pH 7.4.
2. Substrate: 100 mM stock solution of racemic ketamine hydrochloride. Weigh 0.01371 g of ketamine using an analytical balance, dissolve with 500 μL methanol, and store at −20 °C. Dilute in the IB to obtain a 1.68 mM working solution. Follow the same procedure to prepare 100 mM *S*-ketamine hydrochloride and 100 mM *R*-ketamine hydrochloride and their 900 μM and 780 μM working solutions, respectively (*see* **Note 3**).
3. Reaction products standard: racemic norketamine hydrochloride, available as 3.84 mM (1 mg/mL of the free base) stock solution in methanol from Cerilliant (Round Rock, TX, USA).
4. Cofactor: 3 mM NADPH. Prepare solution in the IB and keep in a fridge at 5 °C (*see* **Note 4**).
5. CYP 3A4 solution: supersomes containing human CYP3A4, P450 reductase, and cytochrome b_5, available from BD Gentest (Woburn, MA, USA). Divide the CYP3A4 stock solution into 50 μL aliquots after the delivery in order to reduce the number of thaw/freeze circles and store at −80 °C. Dilute in the IB to a working concentration of 400 nM and keep on ice before analysis (*see* **Note 5**).
6. Inhibitor: 10 mM stock solution of ketoconazole. Weigh 0.00266 g of ketoconazole using an analytical balance, dissolve with 500 μL methanol, and store at −20 °C. Dilute in the IB to obtain 15 μM working solution (*see* **Note 3**).

3 Methods

The protocol is designed for a Beckman Coulter Proteomelab PA 800 CE System equipped with a photodiode array UV–VIS detector and operated using the 32 Karat 7.0 software. Conduct all the rinsing steps at a pressure of 20 psi (137.9 kPa).

1. Transfer 1.5 mL of water, rinsing solutions, or buffers into glass vials as follows: 1 vial of 1 M NaOH, 1 vial of 0.1 M NaOH, 1 vial of water, 4 vials of BGE, and 2 vials of IB and place them in the CE system (*see* **Note 6**). Use one empty vial for waste collecting and one vial containing 1.5 mL of water as an outlet waste vial for the injection procedure (*see* **Note 7**).
2. Before its first use, activate the capillary by rinsing with 1 M NaOH for 20 min, 0.1 M NaOH for 10 min, and water for 10 min at 50 °C.
3. Rinse the capillary with 0.1 M NaOH for 10 min, water for 10 min, and BGE for 10 min at 37 °C at the beginning of each day or set of analyses.
4. Set up the method as follows: capillary temperature: 37 °C, sample tray temperature: 4 °C (*see* **Note 8**), capillary preconditioning: rinsing with 0.1 M NaOH for 1 min, water for 1 min, and BGE for 2 min, injection and separation procedure as summarized in Table 1 (*see* **Note 9**), detection at 195 nm, detection data rate 16 Hz, capillary postconditioning: rinsing with water for 1 min. A typical electropherogram obtained using this method is shown in Fig. 2.
5. Dilute 3.84 mM norketamine stock solution with IB to obtain working solutions for constructing the calibration curve (0, 40, 100, 200, 400, and 800 μM) (*see* **Note 10**).
6. Prepare the injection solutions by mixing 30 μL of norketamine working solution at the given concentration, prepared as described in **step 5** (use IB for blank), 20 μL of 3 mM NADPH, and 10 μL of IB in an Eppendorf vial and transfer 50 μL of the resulting solution into the glass vial insert. Mix 30 μL of norketamine working solution at the given concentration (use IB for blank) and 30 μL of CYP3A4 solution in an Eppendorf vial, and transfer 50 μL of the resulting solution into the glass vial insert. As a result, there will be six vials containing norketamine and NADPH mixtures and six vials containing CYP3A4 and norketamine, one pair per calibration point. Place the glass vials in the CE system.
7. Carry out the analyses with the incubation time set to 0 min and determine the peak areas of *S*-norketamine and *R*-norketamine.

Table 1
Overview of injection and separation setup

Step/time (min)	Event	Value	Duration	Vial	
				Inlet	Outlet
1	Injection—pressure	0.5 psi	3 s	Reagent mixture	Waste[a]
2	Insert inlet vial[b]		3 s	IB	
3	Injection—pressure	−0.5 psi	4 s	CYP3A4 solution	Waste[a]
4	Insert inlet vial[b]		3 s	IB	
5	Injection—pressure	0.5 psi	3 s	Reagent mixture	Waste[a]
6	Insert inlet vial[b]		3 s	IB	
7	Injection—pressure	−0.5 psi	4 s	CYP3A4 solution	Waste[a]
8	Insert inlet vial[b]		3 s	IB	
9	Injection—pressure	0.5 psi	3 s	Reagent mixture	Waste[a]
10	Insert inlet vial[b]		3 s	IB	
11	Injection—pressure	−0.5 psi	4 s	CYP3A4 solution	Waste[a]
12	Insert inlet vial[b]		3 s	IB	
13	Injection—pressure	0.5 psi	3 s	Reagent mixture	Waste[a]
14	Wait		10 min	IB[c]	BGE[c]
15/0.00	Separate—voltage	−20 kV	18 min	BGE[c]	BGE[c]
16/0.00	Apply pressure	0.2 psi	18 min	BGE	BGE
17/3.00	Autozero				

[a]Vial containing 1.5 mL of water (*see* **Note 7**)
[b]Dipping of the inlet capillary tip in the IB is used for its washing and thus minimizing the mass transfer between the vials of the enzyme and reactants
[c]Pairs of vials with equal liquid levels (*see* **Note 6**)
reagent mixture = substrate, NADPH, and inhibitor mixture solution

8. Plot the peak areas vs. known concentrations of norketamine standards and construct the calibration curve by linear regression analysis (*see* **Note 11**).
9. Dilute 1.68 mM racemic ketamine, 900 μM *S*-ketamine, and 780 μM *R*-ketamine working solutions with IB to obtain the following concentrations: 1680, 840, and 420 μM racemic ketamine; 900, 450, and 225 μM *S*-ketamine; and 780, 390, and 195 μM *R*-ketamine (*see* **Note 12**). Dilute 10 mM ketaconazole with IB to obtain 1.5; 3; 6; 9; 12; and 15 μM working solutions (*see* **Note 13**).
10. Mix 20 μL per ketamine, NADPH and inhibitor solutions (*see* **Note 13**) in an Eppendorf vial, use IB instead of inhibitor

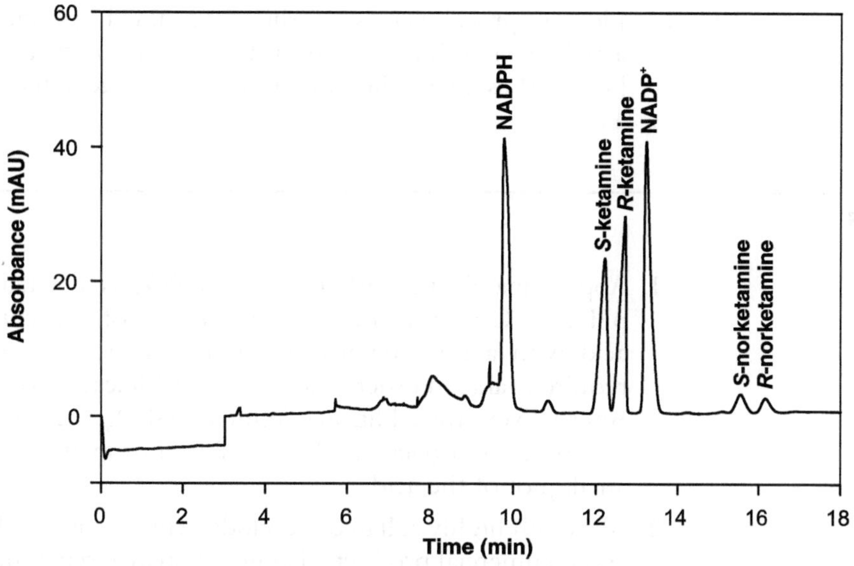

Fig. 2 Typical electropherogram obtained after in-capillary reaction with 200 nM CYP3A4, 400 μM racemic ketamine, and 1 mM NADPH. The reaction and separation conditions were as described in Subheading 3, **step 4**. The current observed during the separation was about −85 μA

solution to prepare the blank. Transfer 50 μL of the resulting solutions into the glass vial inserts. As a result, there will be nine sets of vials, one set per substrate (racemic ketamine and *S*- and *R*-ketamine) and its concentration prepared according to **step 9**, and each set comprised of seven vials (six inhibitor concentrations as described in **step 9** and a blank), i.e., 63 vials in total. Transfer 50 μL of CYP3A4 solution into the glass vial insert. Place the vials in the sample tray, which will be placed in the thermostated part of the CE system during analyses (*see* **Notes 8** and **14**).

11. Carry out the analyses and determine the concentration of *S*-norketamine and *R*-norketamine formed in each assay sample using the equation of the constructed calibration curve.
12. Calculate specific CYP3A4 activities as nmol product formed/min/nmol enzyme (*see* **Note 15**), plot the resulting values vs. the logarithm of inhibitor concentration and establish the IC_{50} value by nonlinear regression analysis according to the four-parameter logistic equation $Y = \text{Bottom} + (\text{Top} - \text{Bottom})/(1 + 10^{(\text{LogIC}_{50} - X) \times \text{HillSlope}})$, where Y represents the reaction rate, X is the logarithm of the inhibitor concentration, Bottom and Top are the lower and upper limits of the curve, corresponding to the norketamine formation rate with the highest and the lowest inhibitor concentration in the reaction mixture, and the HillSlope is the parameter describing the steepness of the fitted curve [19] (*see* **Note 16**).

13. Plot reciprocal values of the determined specific CYP3A4 activities vs. inhibitor concentration, construct a Dixon plot [20] and establish the inhibition type and K_i values (*see* **Note 17**).

4 Notes

1. Approximately 0.5 mL of 85% H_3PO_4 is needed to reach pH 2.5. Carefully add only tiny drops of the concentrated acid with a micropipette when the pH is approaching the required value in order to avoid over acidification of the buffer. Alternatively, the addition of diluted H_3PO_4 (e.g., 50% or 30%) may be more convenient due to a slower drop in pH induced by an aliquot of the acid.
2. Although highly sulfated γ-cyclodextrin solution is distributed as a commercial product, changes in analyte separation may be observed when a new bottle of chiral selector is used. Batch to batch differences and an uninvestigated ageing process of this product are the reasons for these changes. If optimization of the separation conditions is necessary, focus on adjusting the content of highly sulfated γ-cyclodextrin in the BGE, separation voltage, and concomitant pressure, as these parameters influence the migration times and resolution of the analytes.
3. The content of ethanol in the final incubation mixture formed inside the capillary by the presented method is below 1% v/v and has a negligible effect on CYP3A4 activity.
4. Prepare a fresh solution of NADPH for each assay. Note that NADPH is light-, temperature-, and pH-sensitive. For this reason, it is reasonable to weigh a certain amount of NADPH and then calculate the volume of the IB needed to obtain a 3 mM NADPH solution in order to keep the time interval during which solid NADPH is out of the freezer as short as possible.
5. The resulting incubation mixture formed inside the capillary by the presented method has a very limited volume of about 29 nL. Since recombinant enzymes from different suppliers and even batches may considerably differ in their concentration and activity, conduct preliminary experiments to ensure that there is a linear assay response with respect to the CYP3A4 concentration and the incubation time.
6. The BGE in the home vials, i.e., in the inlet and outlet vials used during separation, must be kept on the same level in order to prevent syphoning. The same is true for vials containing IB and BGE placed at the inlet and outlet, respectively, and the tips of the capillary during the incubation part of the analysis. If

very small peaks or nothing but baseline are observed in the electropherograms, it may be due to too high level of BGE in the outlet vial and leaking of the reaction mixture out of the capillary during incubation. On the other hand, if shorter migration times and lower peak resolution are observed, it may be due to a low level of BGE in the outlet vial causing the transport of the reaction mixture further into the capillary.

7. In our experience, injection into a vial filled with a liquid gives a somewhat higher repeatability than injection into an empty vial.
8. The Beckman Coulter Proteomelab PA 800 CE System is equipped with a thermostated sample tray, which is necessary for the prolonged storage of the CYP3A4 solution. The default minimal value is 4 °C. However, older systems may struggle to reach or maintain such a low temperature. Store the CYP3A4 solution in the sample tray when the temperature can be kept at ≤10 °C. In our experience, the enzyme should preserve its activity without significant changes under these conditions for more than 12 h. If this temperature requirement cannot be fulfilled or if a CE system without a thermostated sample tray is used, keep CYP3A4 on ice or in a fridge at 5 °C between analyses, because it loses its activity at room temperature. The same is true for temperature-sensitive NADPH; therefore handle the vial containing substrate, NADPH, and inhibitor solution in the same way.
9. The injection procedure summarized in Table 1 provides the seven-plug assembly schematically described in Fig. 1. Mathematical modeling of the in-capillary reactor showed a twofold dilution of the introduced plugs [16, 18].
10. Due to the rather low concentration of the commercially available norketamine stock solution, samples prepared in the presented way contain more than 1% methanol. However this does not cause any issues, because the incubation of the CYP3A4 reaction is not carried out during assay calibration.
11. The final concentration of each norketamine enantiomer can be determined as follows: one enantiomer represents half of the analyte content in the racemic solution, and the preparation of the samples as described in Subheading 3, **step 6,** leads to a further twofold dilution. The prepared 800 μM working solution of racemic norketamine thus provides the final 200 μM concentration of *S*- and *R*-norketamine in the calibration solution introduced into the capillary. Note that the twofold dilution provided by the alternative introduction of plugs of enzyme and reactants does not apply here, because norketamine standards are also added to the CYP3A4 solution plugs.

12. The final concentration of each ketamine enantiomer in the reaction mixture can be determined using a dilution factor of 12 for the racemate (racemate: factor 2, sample preparation: factor 3, mixing with CYP3A4 inside the capillary: factor 2, *see* **Note 9**) and a factor of 6 for single enantiomers (sample preparation: factor 3, mixing with CYP3A4 inside the capillary: factor 2). According to the recommendations of the Food and Drug Administration guidance, IC_{50} values should be determined in assays with different substrate concentrations to cover ranges above and below the values of the Michaelis-Menten constant (K_m). For this reason, the concentrations of ketamine suggested in Subheading 3, **step 9**, represent 0.5-fold, onefold, and twofold of the previously determined K_m values for racemic ketamine $K_m = 70$ μM, *S*-ketamine $K_m = 75$ μM, and *R*-ketamine $K_m = 65$ μM [17].
13. As was mentioned above, since diffusion-based mixing is generic, any other low-molecular-mass compound can be added to the reaction mixture instead of ketoconazole in order to test its putative inhibitory effect toward CYP3A4 activity. No additional optimization of the injection, mixing, or separation conditions is needed.
14. When the vials containing CYP and substrate, NADPH and inhibitor mixture solutions are kept on ice or in the fridge, it is necessary to place them in the CE system at the very same point in the analysis in order to ensure that the injection solutions have the same temperature in all the runs. Since viscosity and thus injected volume are temperature-dependent, such a protocol is crucial for maintaining the high method repeatability achieved in CE systems with thermostated sample trays, see [17].
15. The presented assay includes an incubation time of 10 min and a final CYP3A4 concentration in the reaction mixture of 200 nM.
16. The four-parameter logistic equation reflecting the non-unit value of Hill's coefficient is used due to the observation of a slight positive cooperativity in the binding of the substrate molecules in the previously published study [17]. If the Michaelis-Menten kinetics of the CYP3A4 reaction with ketamine is considered, the single-site competition model equation $Y = \text{Bottom} + (\text{Top} - \text{Bottom})/1 + 10^{X - \text{LogIC}_{50}}$ can be used for calculating IC_{50} values.
17. Although it is a common practice to determine the inhibition constants by the use of a Dixon plot, its applicability is limited to competitive and noncompetitive inhibitors. For this reason, experimental data from new putative CYP3A4 inhibitors should be analyzed by the concomitant use of Dixon and

Cornish-Bowden plots [21, 22] so that all inhibition types are covered. Alternatively, sets of samples containing an inhibitor concentration below and above the determined IC_{50} value and various concentrations of substrate can be analyzed and the inhibition type and K_i value consequently established by the use of a Lineweaver-Burk plot [19].

Acknowledgments

This work was supported by the Swiss National Science Foundation and by grant no. GA16-06106S from the Czech Science Foundation.

References

1. Agranat I, Wainschtein SR, Zusman EZ (2012) The predicated demise of racemic new molecular entities is an exaggeration. Nat Rev Drug Discov 11:972–973
2. McConathy J, Owens MJ (2003) Stereochemistry in drug action. Prim Care Companion J Clin Psychiatry 5:70–73
3. Nguyen LA, He H, Pham-Huy C (2006) Chiral drugs: an overview. Int J Biomed Sci 2:85–100
4. Brooks WH, Guida WC, Daniel KG (2011) The significance of chirality in drug design and development. Curr Top Med Chem 11:760–770
5. Glatz Z (2006) Determination of enzymatic activity by capillary electrophoresis. J Chromatogr B 841:23–37
6. Fan Y, Scriba GK (2010) Advances in-capillary electrophoretic enzyme assays. J Pharm Biomed Anal 53:1076–1090
7. Scriba GK, Abromeit H, Hense M, Fan Y (2013) Capillary electrophoretic enzyme assays. Methods Mol Biol 984:285–308
8. Ouimet CM, D'amico CI, Kennedy RT (2017) Advances in capillary electrophoresis and the implications for drug discovery. Expert Opin Drug Discov 12:213–224
9. Guengerich FP (2005) Human cytochrome P450 enzymes. In: Ortiz de Montellano PR (ed) Cytochrome P450: structure, mechanism, and biochemistry, 3rd edn. Kluwer Academic/Plenum Publishers, New York
10. Galetin A, Ito K, Hallifax D, Houston JB (2005) CYP3A4 substrate selection and substitution in the prediction of potential drug-drug interactions. J Pharmacol Exp Ther 314:180–190
11. Sinner B, Graf BM (2008) Ketamine. Handb Exp Pharmacol 182:313–333
12. Aroni F, Iacovidou N, Dontas I, Pourzitaki C, Xanthos T (2009) Pharmacological aspects and potential new clinical applications of ketamine: reevaluation of an old drug. J Clin Pharmacol 49:957–964
13. Machado-Vieira R, Salvadore G, DiazGranados N, Zarate CA Jr (2009) Ketamine and the next generation of antidepressants with a rapid onset of action. Pharmacol Ther 123:143–150
14. Řemínek R, Slezáčková L, Schejbal J, Glatz Z (2017) Development and comprehensive comparison of two on-line capillary electrophoretic methods for beta-secretase inhibitor screening. J Chromatogr A 1518:89–96
15. Okhonin V, Liu X, Krylov SN (2005) Transverse diffusion of laminar flow profiles to produce capillary nanoreactors. Anal Chem 77:5925–5929
16. Řemínek R, Zeisbergerová M, Langmajerová M, Glatz Z (2013) New capillary electrophoretic method for on-line screenings of drug metabolism mediated by cytochrome P450 enzymes. Electrophoresis 34:2705–2711
17. Řemínek R, Glatz Z, Thormann W (2015) Optimized on-line enantioselective capillary electrophoretic method for kinetic and inhibition studies of drug metabolism mediated by cytochrome P450 enzymes. Electrophoresis 36:1349–1357
18. Pelcová M, Řemínek R, Sandbaumhüter FA, Mosher RA, Glatz Z, Thormann W (2016) Simulation and experimental study of enzyme and reactant mixing in capillary electrophoresis

based on-line methods. J Chromatogr A 1471:192–200
19. Matoulsky H, Christopoulos A (2004) Fitting models to biological data using linear and nonlinear regression: a practical guide to curve fitting. Oxford University Press, New York
20. Dixon M (1953) The determination of enzyme inhibitor constants. Biochem J 55:170–171
21. Yoshino M, Murakami K (2009) A graphical method for determining inhibition constants. J Enzyme Inhib Med Chem 24:1288–1290
22. Cornish-Bowden A (1974) A simple graphical method for determining the inhibition constants of mixed, uncompetitive and noncompetitive inhibitors. Biochem J 137:143–144

Chapter 12

Enzymatic Bioautographic Methods

I. Ayelen Ramallo, Mario O. Salazar, and Ricardo L. E. Furlan

Abstract

Enzymatic bioautography enables the detection of enzyme inhibitors absorbed on a thin-layer chromatography plate. Therefore, it is an assay format that is particularly useful for the detection of inhibitors present in complex mixtures. The inhibition properties of compounds separated by thin-layer chromatography can be directly analyzed to produce an inhibition profile. Here, we describe the conditions to detect inhibitor of the enzymes xanthine oxidase and β-glucosidase immobilized on agar gel.

Key words Bioautographic method, Xanthine oxidase, β-Glucosidase, Immobilization in gel, Bio-guided isolation, Control assay

1 Introduction

Bioautography is the planar chromatographic analysis hyphenated with a biological detection method. Although the term bioautography was originally reserved for assays that included microorganisms, its use has now been extended to assays that use enzymes.

The main use of this type of assays is related to the biological evaluation of mixtures of molecules. By combining the separative capacity of the thin-layer chromatography (TLC), with a subsequent in situ determination of a bioactivity of the components of a mixture [1, 2] (Fig. 1), this assay format allows preassigning the biological activity, observed in a mixture, to one or a few of its components through the spots that they produce.

The compounds adsorbed in the stationary phase of a TLC plate diffuse into an aqueous medium (gelled or not) that usually contains enzyme and substrate. Sometimes other reagents are required to visualize enzyme activity (Fig. 2).

Here, we describe in detail two bioautographic assays based on the gel entrapment of xanthine oxidase (XO) and β-glucosidase (β-Glc), respectively. These enzymes are attractive targets in drug discovery; their inhibition is a validated strategy to treat different pathological processes.

Nikolaos E. Labrou (ed.), *Targeting Enzymes for Pharmaceutical Development: Methods and Protocols*, Methods in Molecular Biology, vol. 2089, https://doi.org/10.1007/978-1-0716-0163-1_12, © Springer Science+Business Media, LLC 2020

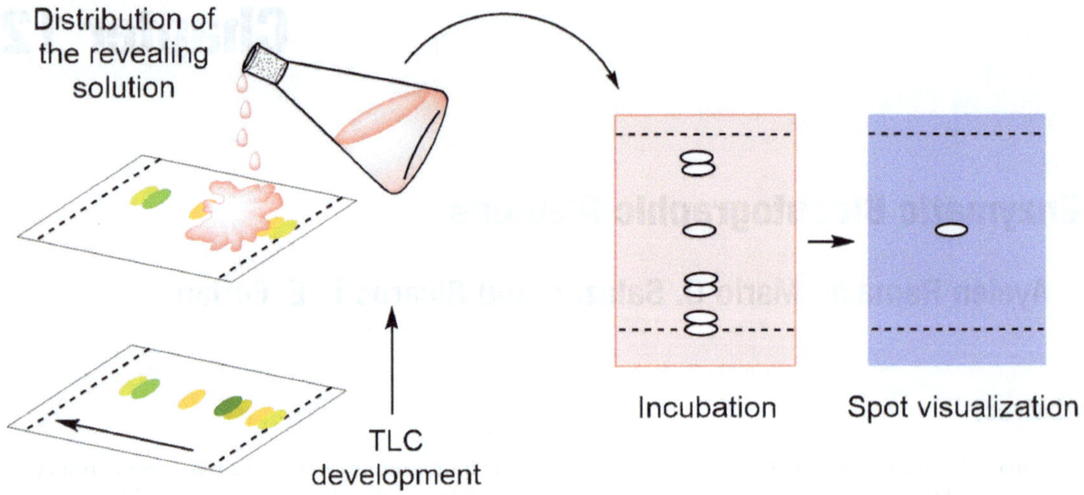

Fig. 1 General sequence of steps of the bioautographic methodology

Fig. 2 Representation of the cross-section of a bioautographic experiment

1.1 Xanthine Oxidase

Xanthine oxidase catalyzes the oxidation of hypoxanthine and xanthine to uric acid, producing the reactive oxygen species (ROS), superoxide radical ($O_2^{\cdot-}$), and hydrogen peroxide (Fig. 3) [3].

XO is implicated in gout, one of the most common inflammatory arthritides [4]. Patients with gout and hyperuricemia often have comorbid conditions, such as hypertension, chronic kidney disease, or an increased risk of myocardial infarction, stroke, and hypertension [5–8]. Importantly, the ROS generated by XO have been implicated in some pathophysiological processes, such as ischemia-reperfusion injury [9], chronic heart failure [10], and neurodegenerative diseases [11].

The XO autography described here is based on the detection of $O_2^{\cdot-}$ with nitro blue tetrazolium (NBT) [12]. In short, a staining solution containing a mixture of XO and NBT is distributed over a TLC plate. Then, the stained plate is submerged in xanthine substrate solution. The formed $O_2^{\cdot-}$ reacts with NBT (yellow color)

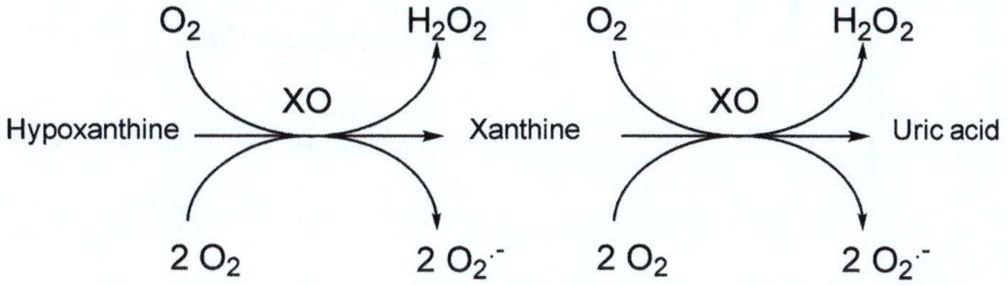

Fig. 3 The oxidation of hypoxanthine and xanthine to uric acid catalyzed by xanthine oxidase

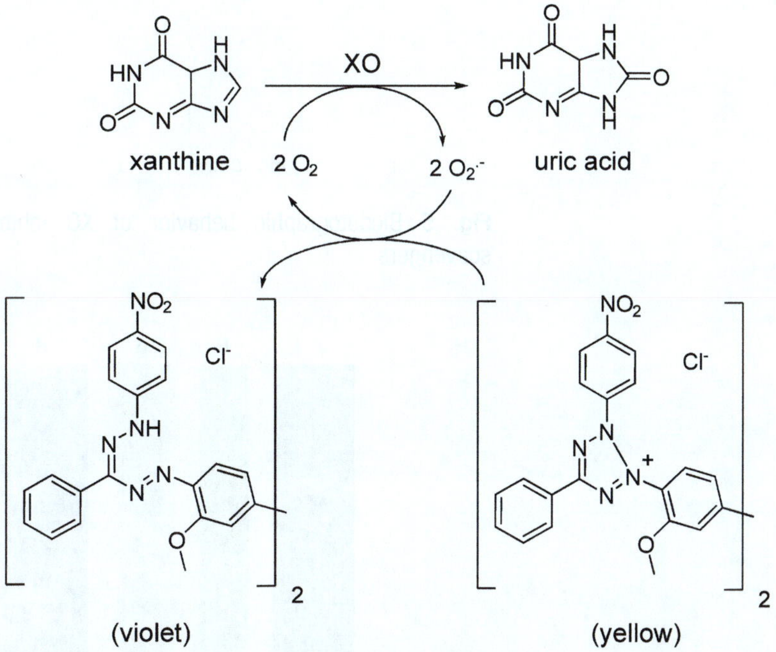

Fig. 4 NBT-based method for the detection of $O_2^{\cdot-}$ generated by XO

transforming it into its reduced form (violet color) (Fig. 4). A decreased amount of $O_2^{\cdot-}$, either by XO inhibition or by $O_2^{\cdot-}$ scavenging, will be detected as clear spots in a violet background. Therefore, a control assay is described here in order to differentiate pure XO inhibitors from superoxide radical scavengers (Fig. 5). In this assay, the $O_2^{\cdot-}$ is generated by a nonenzymatic approach through photoreaction of riboflavin [12], and only the $O_2^{\cdot-}$ scavengers will be observed as clear spots in a violet background. Using this methodology the inhibition of XO by *Foeniculum vulgare* Mill. brominated chemically engineered essential oil was observed. Bioautography guided fractionation of this mixture lead to identification of *(RS)*-2-bromo-1-(4-methoxyphenyl) propan-1-one [13] (Fig. 6).

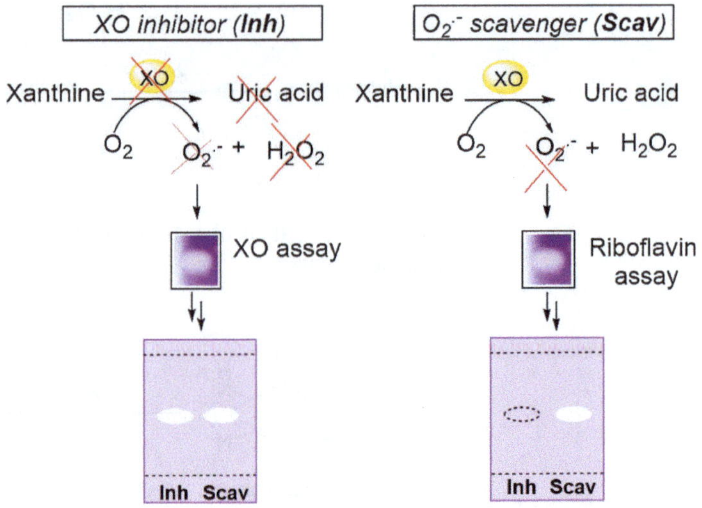

Fig. 5 Bioautographic behavior of XO inhibitor compounds versus $O_2^{\cdot-}$ scavengers

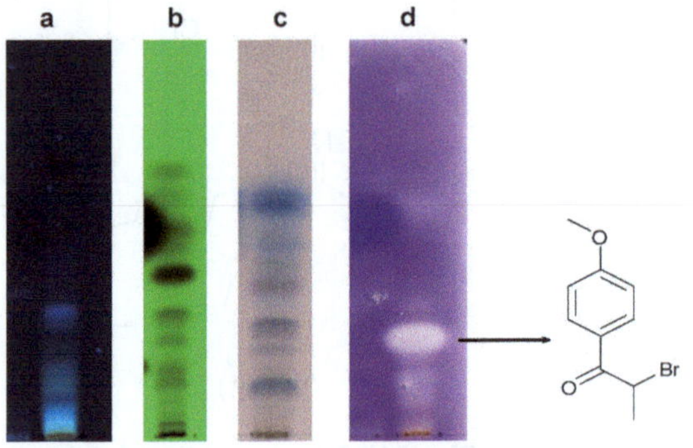

Fig. 6 TLC of *F. vulgare* chemically engineered essential oil revealed with: (**a**) UV 365 nm. (**b**) UV 254 nm. (**c**) Vanillin-sulfuric acid. (**d**) XO assay. Mobile phase hexane:ethyl acetate (90:10). Inhibition of XO by *(RS)*-2-bromo-1-(4-methoxyphenyl) propan-1-one

1.2 β-Glucosidase

β-Glucosidases (β-Glc) catalyze the hydrolysis of glycosidic bonds to release nonreducing terminal glucosyl residues from glycosides and oligosaccharides (Fig. 7) [14]. These enzymes are found universally in all domains of living organisms, playing a variety of functions. β-Glc inhibitors have several applications in agriculture and medicine [15, 16].

The β-Glc autography described here consists of the detection of esculetin, product of the enzymatic cleavage of esculin. In brief, the method consists of the distribution over a TLC plate of a staining solution containing β-Glc and $FeCl_3$. Then, the stained plate is submerged in the esculin solution [17].

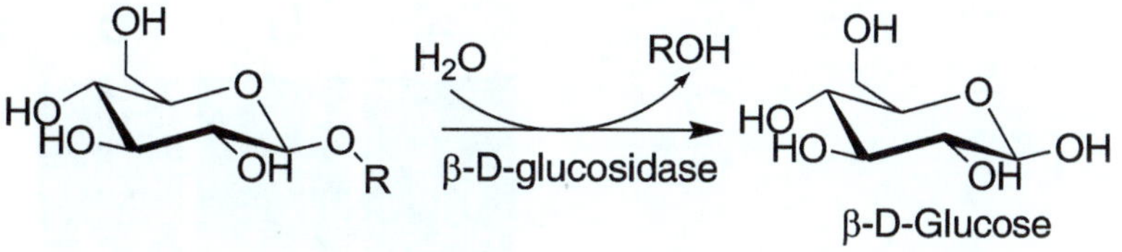

Fig. 7 The hydrolysis of glycosidic bonds catalyzed by β-D-glucosidase

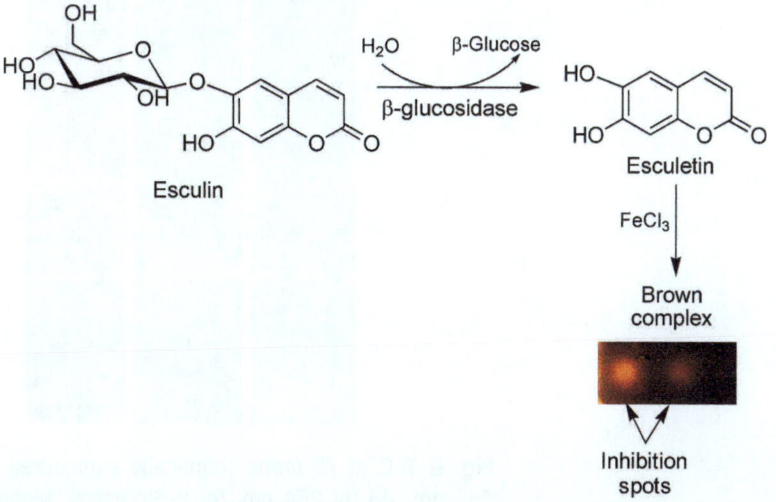

Fig. 8 Reaction of β-D-glucosidase with esculin and the subsequent formation of the brown complex between esculetin and Fe^{III+}

The enzyme cleaves the glycosidic bond of the esculin generating esculetin, which in the presence of Fe^{III+}, forms a brown complex. The reaction is illustrated in Fig. 8. The absence of esculetin by β-Glc inhibition will be observed as a clear spot in a brown background (Fig. 8). Using this methodology the β-Glc inhibitory activity by *Urtica urens* L. benzenesulfonyl chloride chemically engineered extract was detected. Bioautography guided fractionation of this mixture lead to the identification of N^{α},N^{τ}-dibenzenesulfonyl histamine [18] (Fig. 9).

2 Materials

Prepare all solutions using distilled water and analytical grade reagents. Prepare and store all reagents at 2–4 °C (unless otherwise specified). We do not add sodium azide to reagents.

The samples were separated on silica gel layers (*see* **Note 1**) with appropriate system solvent.

2.1 Xanthine Oxidase Assay

1. Aluminum-backed silica gel 60 F254 TLC layers (*see* **Note 2**).
2. Incubation chambers (*see* **Note 3**).

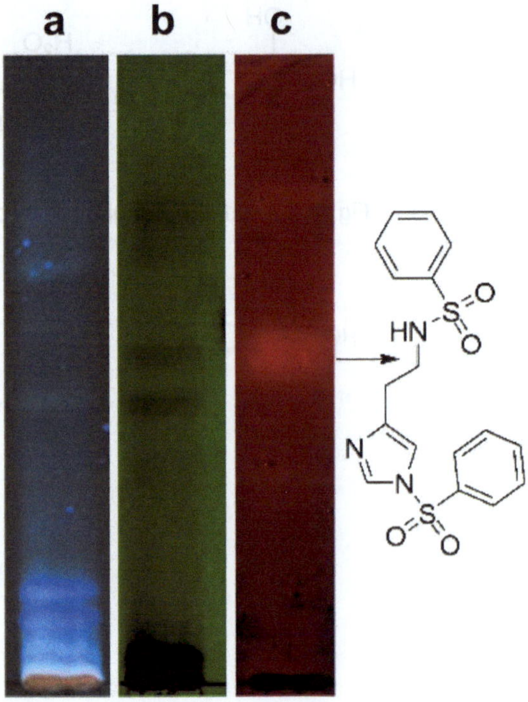

Fig. 9 TLC of *U. urens* chemically engineered extract revealed with: (**a**) UV 365 nm. (**b**) UV 254 nm. (**c**) β-Glc assay. Mobile phase hexane:ethyl acetate (50:50). Inhibition of β-Glc by N^{α},N^{τ}-di-benzenesulfonyl histamine

3. Agar powder.
4. Buffer phosphate 50 mM solution, pH 8.0: Dissolve 162.3 mg of K_2HPO_4 and 9.3 mg of KH_2PO_4 in 1 L distilled water. If necessary, set pH 8.0 with drops of HCl acid. Store the solution at 2–4 °C (see **Note 4**).
5. Xanthine 1.5 mM solution: Add 45.6 mg of xanthine to 100 mL of 50 mM phosphate buffer, pH 8.0, reflux the mixture up to complete solubilization, and store at room temperature (*see* **Note 5**).
6. XO 5 U/mL solution: Dissolve 50 U of XO in 10 mL of 50 mM phosphate buffer, pH 8.0, and store at 2–4 °C (*see* **Note 6**).
7. Nitro blue tetrazolium (NBT) 6.8 mM solution: Dissolve 55.8 mg of NBT in 10 mL of 50 mM phosphate buffer, pH 8.0, and store at 2–4 °C in darkness (*see* **Note 4**).
8. Ethylenediaminetetraacetic acid (EDTA) 10 mM solution: Dissolve 380.0 mg of EDTA in 10 mL of 50 mM phosphate buffer, pH 8.0, and store at 2–4 °C (*see* **Note 4**).
9. *N,N,N'N'*-Tetramethylethylenediamine (TEMED).

2.2 Riboflavin Assay

1. Aluminum-backed silica gel 60 F254 TLC layers (*see* **Note 2**).
2. Incubation chambers (*see* **Note 3**).
3. Agar powder.
4. Buffer phosphate 50 mM solution, pH 8.0: Dissolve 162.3 mg of K_2HPO_4 and 9.3 mg of KH_2PO_4 in 1 L distilled water. If necessary, set pH 8.0 with drops of HCl acid. Store the solution at 2–4 °C (*see* **Note 4**).
5. Nitro blue tetrazolium (NBT) 6.8 mM solution: Dissolve 55.8 mg of NBT in 10 mL of 50 mM phosphate buffer, pH 8.0, and store at 2–4 °C in darkness (*see* **Note 4**).
6. Ethylenediaminetetraacetic acid (EDTA) 10 mM solution: Dissolve 380.0 mg of EDTA in 10 mL of 50 mM phosphate buffer, pH 8.0, and store at 2–4 °C (*see* **Note 4**).
7. (−)-Riboflavin suspension: Dissolve 37.5 mg of riboflavin in 10 mL of 50 mM phosphate buffer, pH 8.0, and store at 2–4 °C in darkness (*see* **Notes 4** and 7).
8. *N,N,N'N'*-Tetramethylethylenediamine (TEMED).

2.3 β-Glucosidase Assay

1. Aluminum-backed silica gel 60 F254 TLC layers (*see* **Note 2**).
2. Incubation chambers (*see* **Note 3**).
3. Agar powder.
4. Acetate buffer 0.1 M solution, pH 5.0: Dissolve 15.0 g of sodium acetate trihydrate in 900 mL of distilled water. Mix and adjust pH with glacial acetic acid (*see* **Note 8**) and complete up to 1000 mL with distilled water. Store at 2–4 °C (*see* **Note 4**).
5. Ferric chloride hexahydrate 5% w/v solution: Dissolve 2.5 g of ferric chloride hexahydrate in 50 mL of 0.1 M sodium acetate buffer, pH 5.0 (*see* **Note 9**). Store at 2–4 °C (*see* **Note 10**).
6. β-Glucosidase 25 U/mL solution: Dissolve 25 U of β-glucosidase in 1 mL of 0.1 M acetate buffer, pH 5.0. Store at −20 °C.
7. β-Glucosidase 2.5 U/mL solution: Dilute 100 μL of 25 U/mL β-glucosidase in 900 μL of 0.1 M acetate buffer, pH 5.0. Store at −20 °C.
8. Esculin solution (0.2% w/v): Dissolve 0.2 g of esculin hydrate in 100 mL of 0.1 M acetate buffer, pH 5.0 (*see* **Note 11**). Store at 2–4 °C.

3 Methods

3.1 Xanthine Oxidase Assay

1. Mix 0.14 g of agar with 11.5 mL of phosphate buffer and dissolve by heating in a microwave oven or on a heating plate (*see* **Note 12**).
2. Keeping the agar solution at 40 °C in a water bath, add 1.5 mL of EDTA solution (*see* **Note 13**) and mix carefully (*see* **Note 14**) avoiding the formation of bubbles.
3. Keeping the agar solution at 40 °C in a water bath, add 600 μL NBT solution, 200 μL XO solution (*see* **Note 15**), and 28 μL TEMED (*see* **Note 14**). Mix carefully avoiding the formation of bubbles and distribute evenly over the TLC layer (*see* **Notes 16** and **17**).
4. Once the gel formation is achieved, put the plate in the incubation chamber and cover it with the xanthine solution. Incubate in a stove at 37 °C for 20 min. A dark purple matrix is generated due to the reduction of NBT by the superoxide radical anion formed during the oxidation of xanthine by XO. Enzyme inhibitors and radical scavengers will be visualized as clear spots (*see* **Note 18**). In order to differentiate true inhibitors from superoxide scavengers, it is necessary to carry out the control assay (Subheading 3.2) in which radical anion superoxide is formed in a nonenzymatic way through photoreaction of riboflavin.

3.2 Riboflavin Assay

1. Mix 0.14 g of agar with 11.5 mL of phosphate buffer and dissolve it by heating in a microwave oven or on a heating plate (*see* **Note 12**).
2. Keeping the agar solution at 40 °C in a water bath, add 1.5 mL of EDTA solution (*see* **Note 13**) and mix carefully (*see* **Note 14**) avoiding the formation of bubbles.
3. Keeping the agar solution at 40 °C in a water bath, add 600 μL NBT solution, 50 μL riboflavin suspension, and 28 μL TEMED (*see* **Note 14**). Mix carefully avoiding the formation of bubbles and distribute evenly over the TLC layer (*see* **Notes 16** and **17**). Keep the TLC in the dark during gel formation.
4. Once gel formation is achieved in the dark, put the plate under a fluorescent lamp (at a distance of around 10 cm) for 5 min approximately. A dark purple matrix is generated due to the reduction of NBT by the superoxide radical anion formed during the photoreaction of riboflavin. Radical scavengers will be visualized as clear spots (*see* **Note 18**).

3.3 β-Glucosidase Assay

1. Mix 0.105 g of agar with 15 mL of acetate buffer and dissolve it by heating in a microwave oven or on a heating plate (*see* **Note 12**).
2. Incorporate 1.8 mL of ferric chloride solution to the heated agar solution and mix carefully (see **Note 14**) avoiding the formation of bubbles and keep it at 50 °C.
3. Keeping the agar solution at 40 °C in a water bath, add 50 μL of β-glucosidase solution (*see* **Note 15**). Mix carefully avoiding the formation of bubbles and distribute evenly over the TLC layer (*see* **Note 17**).
4. Once the gel formation is achieved, put the plate in the incubation chamber for 120 min.
5. After this incubation period, cover the TLC plate with the esculin solution. Place it in a stove and incubate at 37 °C for 120 min. The enzymatic assay consists of the hydrolysis of esculin so that it releases esculetin, which reacts with the pale yellow ferric chloride salt to produce a dark brown complex (Fig. 7). Enzyme inhibitors will be visualized as clear spots. Since false negatives could be produced by direct reaction of certain compounds with ferric chloride, a parallel experiment can be carried out in which the TLC layer was not immersed in the esculin solution.

4 Notes

1. The method is compatible with different TLC stationary phases (Cellulose, Diol, and RP18 Silica gel), but note that the detection limit could be different according to the polarity of the studied inhibitor compounds.
2. The protocols are set for 10 × 10 cm TLC size, for other sizes, the quantities and volumes need to be proportionally adjusted, except for xanthine substrate solution that depends only on the size of the incubation chamber.
3. The incubation chamber is a plastic or glass lidded container, such as a Petri dish.
4. The solution can be stored at 2–4 °C for 6 months, or frozen for a longer period. However, the solution must be discarded if suspended particles are observed. Impurities or fungi development will affect the assay quality.
5. Xanthine solution should be freshly prepared before use. Heating up to the boiling temperature can be applied to speed up xanthine dissolution in the buffer solution.
6. We use xanthine oxidase from bovine milk ammonium sulfate suspension. Attempts to use xanthine oxidase lyophilized powder were not successful; the desired color on the TLC plate could not be seen on this assay conditions.

7. Shake the riboflavin suspension immediately before use to guarantee a homogenous aliquot.
8. Approximately 2 mL of glacial acetic acid is necessary for pH adjustment.
9. It is strongly discouraged to heat the buffer solution to improve the solubility of ferric salt.
10. Do not use the ferric chloride solution if it presents a dark brown color.
11. Sonication can be used to speed up esculin dissolution in the acetate buffer. Heating is strongly discouraged because substrate degradation occurs.
12. The agar will be considered completely solubilized when a text could be read watching through its container. Microwave oven solubilization is faster than solubilization in a heating plate.
13. If the agar is melt in a hot plate, the EDTA solution can be added before the melting process. However, if a microwave oven is used for agar melting, the EDTA solution must be added after the melting process; otherwise, no change in color will be observed.
14. In order to avoid bubble formation, mixing should be carried out gently by hand, avoiding vigorous shaking.
15. Temperatures higher than 50 °C can produce partial or total inactivation of the enzyme.
16. The color of the staining solution is pale yellow. If the color is grayish or pale purple, probably NBT was reduced by some impurities present in the materials (i.e., detergent). If so, do not use it because it can diminish the assay sensitivity.
17. The homogeneous distribution of the staining solution over the TLC plate is crucial for the assay quality. In order to achieve that and to avoid overflow, the staining solution should be distributed starting from the center of the TLC plate. A pipet tip can be used, if necessary, to distribute the staining solution near the edges or to collapse bubbles that could spoil the gel. Notice that the gel is carefully distributed over the TLC layer without any limiting walls on the sides. Superficial tension is the only force that maintains the revealing solution on the plate, making important to perform the experiment over a flat surface without any slope, and not to move the TLC layer until complete gelation. This process should be carried out fast enough in order to avoid gel formation before the solution is evenly distributed on the whole plate.
18. The presence of a purple color more intense than the matrix indicates the presence of compounds that increase the production of superoxide.

References

1. Cieśla ŁM, Waksmundzka-Hajnos M, Wojtunik KA, Hajnos M (2015) Thin-layer chromatography coupled with biological detection to screen natural mixtures for potential drug leads. Phytochem Lett 11:445–454
2. Bräm S, Wolfram E (2017) Recent advances in effect-directed enzyme assays based on thin-layer chromatography. Phytochem Anal 28:74–86
3. Pacher P, Nivorozhkin A, Szabó C (2006) Therapeutic effects of xanthine oxidase inhibitors: renaissance half a century after the discovery of allopurinol. Pharmacol Rev 58:87–114
4. Choi HK, Mount DB, Reginato AM (2005) Pathogenesis of gout. Ann Intern Med 143:499–516
5. Kim SC, Schneeweiss S, Choudhry N, Liu J, Glynn RJ, Solomon DH (2015) Effects of xanthine oxidase inhibitors on cardiovascular disease in patients with gout: a cohort study. Am J Med 128:653.e7–e653.e16
6. Kim SY, Guevara JP, Kim KM, Choi HK, Heitjan DF, Albert DA (2009) Hyperuricemia and risk of stroke: a systematic review and meta-analysis. Arthritis Rheum 61:885–892
7. Grayson PC, Kim SY, LaValley M, Choi HK (2010) Hyperuricemia and incident hypertension: a systematic review and meta-analysis. Arthritis Care Res (Hoboken) 63:102–110
8. Kim SY, Guevara JP, Kim KM, Choi HK, Heitjan DF, Albert DA (2010) Hyperuricemia and coronary heart disease: a systematic review and meta-analysis. Arthritis Care Res (Hoboken) 62:170–180
9. Linas SL, Whittenburg D, Repine JE (1990) Role of xanthine oxidase in ischemia/reperfusion injury. Am J Phys 258:F711–F716
10. Doehner W, Jankowska EA, Springer J, Lainscak M, Anker SD (2016) Uric acid and xanthine oxidase in heart failure—emerging data and therapeutic implications. Int J Cardiol 213:15–19
11. Kim GH, Kim JE, Rhie SJ, Yoon S (2015) The role of oxidative stress in neurodegenerative diseases. Exp Neurobiol 24:325–340
12. Ramallo IA, Zacchino SA, Furlan RLE (2006) A rapid tlc autographic method for the detection of xanthine oxidase inhibitors and superoxide scavengers. Phytochem Anal 17:15–19
13. Garcia P, Ramallo IA, Salazar MO, Furlan RLE (2016) Chemical diversification of essential oils, evaluation of complex mixtures and identification of a xanthine oxidase inhibitor. RSC Adv 6:57245–57252
14. Ketudat Cairns JR (2010) β-Glucosidases. Cell Mol Life Sci 67:3389–3405
15. Asano N (2003) Glycosidase inhibitors: update and perspectives on practical use. Glycobiology 13:93R–104R
16. Borges de Melo E, da Silveira Gomes A, Carvalho I (2006) α- and β-Glucosidase inhibitors: chemical structure and biological activity. Tetrahedron 62:10277–10302
17. Salazar MO, Furlan RLE (2007) A rapid tlc autographic method for the detection of glucosidase inhibitors. Phytochem Anal 18:209–212
18. Salazar MO, Osella MI, Ramallo IA, Furlan RLE (2018) N^{α}-arylsulfonyl histamines as selective β-glucosidase inhibitors. RSC Adv 8:36209–36218

Chapter 13

High-Throughput Assessment of Metabolism-Induced Toxicity of Compounds on a 384-Pillar Plate

Soo-Yeon Kang, Kyeong-Nam Yu, Pranav Joshi, and Moo-Yeal Lee

Abstract

A variety of oxidative and conjugative enzymes are involved in the metabolism of compounds including drugs, which can be converted into toxic metabolites by Phase I drug-metabolizing enzymes (DMEs), such as the cytochromes P450 (CYP450s), and/or detoxified by Phase II DMEs, such as UDP-glucuronosyltransferases (UGTs), sulfotransferases (SULTs), and glutathione *S*-transferases (GSTs). Traditionally, primary hepatocytes containing a complete set of DMEs have been widely used as a gold standard to assess metabolism-induced compound toxicity. However, primary hepatocytes are expensive, have high donor variability in expression levels of DMEs, and rapidly lose liver-specific functions when the cells are maintained under standard in vitro cell culture conditions over time. To address this issue and rapidly profile metabolism-induced drug toxicity, we have developed a 384-pillar plate, which is complementary to conventional 384-well plates. In this chapter, we provide step-by-step procedures for three-dimensional (3D) cell printing on the 384-pillar plate coupled with DMEs and compounds in the 384-well plate for high-throughput assessment of metabolism-induced toxicity.

Key words Metabolism-induced toxicity, Drug-metabolizing enzymes, 384-Pillar plate, 3D cell culture, High-throughput screening (HTS)

1 Introduction

The drug discovery process is an investment-intensive, high-risk endeavor that results in low yield of effective and safe drugs [1, 2]. Approximately 30% of drug candidates are failed in clinical trials due to adverse drug reactions (ADRs), including drug-induced liver injury (DILI). The liver contains a variety of drug-metabolizing enzymes (DMEs) including (but not limited to) cytochromes P450 (CYP450s), UDP-glucuronosyltransferases (UGTs), sulfotransferases (SULTs), and glutathione *S*-transferases (GSTs), which are involved in the initial clearance of drugs from the body and generate drug metabolites, some of which are unstable and toxic [3]. Thus, assessing compound-induced toxicity in pre-clinical evaluations is utmost importance in drug discovery to weed

Nikolaos E. Labrou (ed.), *Targeting Enzymes for Pharmaceutical Development: Methods and Protocols*, Methods in Molecular Biology, vol. 2089, https://doi.org/10.1007/978-1-0716-0163-1_13,

out potentially toxic drug candidates in the liver. However, incorporating physiological levels of drug metabolism into high-throughput screening (HTS) of compounds is still challenging due to intrinsic limitations of primary hepatocytes, as emphasized by the US EPA and the NIH in "The Transform Tox Testing Challenge: Innovating for Metabolism" [4].

In response to the need of high-throughput, metabolism-induced toxicity assessment, we have developed several microarray chip platforms, including the metabolizing enzyme toxicology assay chip (MetaChip), the data analysis toxicology assay chip (DataChip), and the transfected enzyme and metabolism chip (TeamChip) [5–7]. However, these microarray chip platforms were too small in size (25 mm × 75 mm) for robotic systems commonly adopted for HTS assays and difficult to retrofit existing HTS assays. Thus, we have developed a new 384-pillar plate that can be coupled with conventional 384-well plates for 3D cell cultures and high-throughput, high-content imaging (HCI) assays (Fig. 1) [8]. Human cell types in biomimetic hydrogels can be printed on the 384-pillar plate rapidly using a microarray spotter for organotypic 3D cell cultures, which can be combined with DMEs and test compounds in the 384-well plate for metabolism-induced toxicity assays. The 384-pillar plate sandwiched with the 384-well plate is compatible with existing HTS equipment such as microtiter plate readers and automated fluorescent microscopes, which allows for rapid absorbance, fluorescence, and luminescence measurements in situ. Any combinations of DMEs can be used in the 384-pillar/well plates to emulate metabolic reactions in the human liver and evaluate augmented toxicity and detoxification by drug metabolism. Thus, this approach could potentially decipher toxicodynamic and toxicokinetic traits of drugs and provide insight into complicated toxicology pathways and related ADRs in early stages of drug discovery.

2 Materials

2.1 Culture of Human Embryonic Kidney (HEK) 293 Cells in T-75 Flasks

1. Human embryonic kidney (HEK) 293 cell line (ATCC or the American Type Culture Collection).
2. Complete Dulbecco's modified eagle media (DMEM): DMEM is supplemented with 10% fetal bovine serum (FBS) and 1% penicillin/streptomycin (P/S). Store at 4 °C until the expiration date.
3. Phosphate-buffered saline (PBS; 1×).
4. Trypsin/EDTA (0.25%). Store at 4 °C.
5. Moxi Z mini automated cell counter (MXZ001, ORFLO Technologies) (*see* **Note 1**).

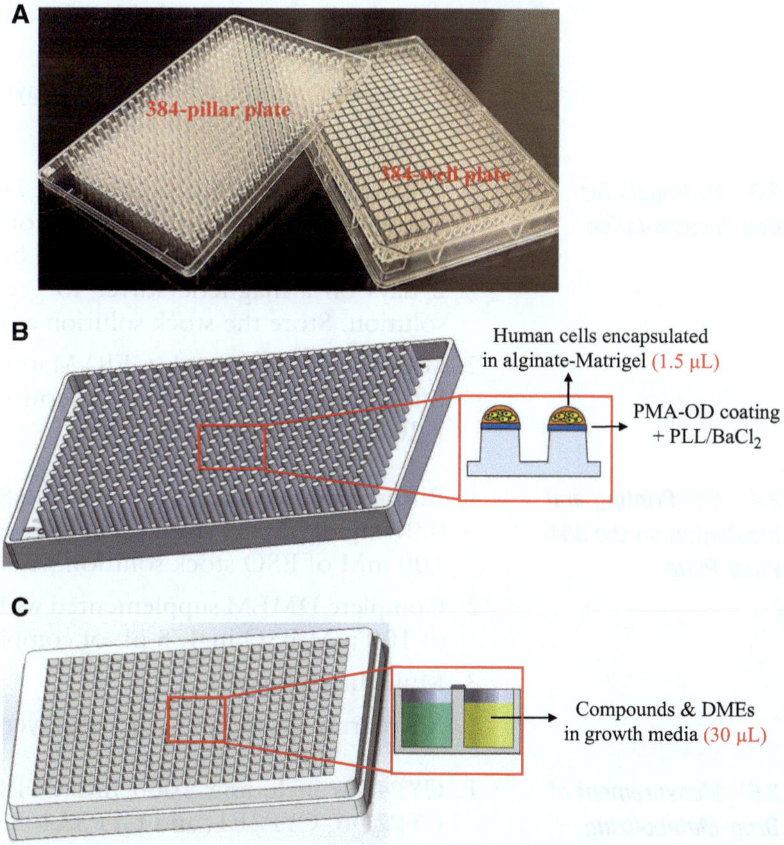

Fig. 1 The 384-pillar plate platform for metabolism-induced toxicity assays: (**a**) Picture of the 384-pillar plate and the 384-well plate, (**b**) Schematic of the 384-pillar plate with human cells encapsulated in biomimetic hydrogels. (**c**) Schematic of the 384-well plate containing drug-metabolizing enzymes (DMEs) and test compounds

6. Moxi Z cassette.
7. CO_2 incubator. Set to 5% CO_2 and 37 °C.
8. Centrifuge.

2.2 Surface Coating and Gelation

1. Cell culture plate (384-pillar plate) (MBD Korea, Republic of Korea).
2. Poly(maleic anhydride *alt*-1-octadecene) (PMA-OD) working solution (Sigma-Aldrich): Dissolve 200 mg PMA-OD in 20 mL of pure ethanol to make 1% (w/v) PMA-OD stock solution. Dilute 1% (w/v) PMA-OD further in ethanol to prepare 0.01% (w/v) PMA-OD working solution.
3. Poly-L-lysine (PLL) (Sigma-Aldrich) and barium chloride ($BaCl_2$) (Sigma-Aldrich) solution: Mix 333 μL of 0.01% (w/v) PLL, 166 μL of 100 mM $BaCl_2$, and 501 μL of sterile

deionized water to make 1 mL (*see* **Note 2**). Store at 4 °C until use.

4. Microarray spotter (S+ MicroArrayer, Advanced Technology Inc. (ATI), Republic of Korea).

2.3 Hydrogels for Cell Encapsulation

1. Alginate solution (Sigma-Aldrich): Dissolve 0.3 g low-viscosity alginic acid sodium salt in 10 mL of sterile deionized water in 20 mL of a sterile glass sample vial by continuously stirring for 2 days on a magnetic stirrer to get 3% (w/v) alginate stock solution. Store the stock solution at 4 °C.
2. Growth factor reduced (GFR) Matrigel (Thermo Fisher Scientific): 8 mg/mL stock from Thermo Fisher Scientific. Store at −20 °C (*see* **Note 3**).

2.4 Cell Printing and Incubation on the 384-Pillar Plate

1. Buthionine sulfoximine (BSO) (Sigma-Aldrich): Dissolve 0.01 mg of BSO in sterile deionized distilled water and prepare 100 mM of BSO stock solution (*see* **Note 4**).
2. Complete DMEM supplemented with 50 μM BSO: Add 25 μL of 100 mM BSO in 475 μL of complete DMEM.
3. Multichannel pipette.
4. Microarray spotter (S+ MicroArrayer).

2.5 Measurement of Drug-Metabolizing Enzyme (DME) Activity

1. CYP450s including 1000 nM CYP1A2, CYP2B6, CYP2C9, CYP2D6, CYP2E1, and CYP3A4 (Thermo Fisher).
2. β-Nicotinamide adenine dinucleotide phosphate sodium salt ($NADP^+$) (Sigma-Aldrich): Prepare 10 mM $NADP^+$ in 100 mM potassium phosphate buffer (pH 8.0).
3. Vivid® NADP-regeneration system (Thermo Fisher): Add 333 mM glucose-6-phosphate and 30 U/mL glucose-6-phosphate dehydrogenase in 100 mM potassium phosphate buffer (pH 8.0) to prepare 100× NADP-regeneration system.
4. Mixture of CYP450, $NADP^+$, and a Vivid® NADP-regeneration system: Mix 200 μL of 1000 nM CYP450 isoform, 100 μL of 10 mM $NADP^+$, and 100 μL of the NADP-regeneration system.
5. Fluorogenic substrate stock solutions (Thermo Fisher): Add acetonitrile in 0.1 mg Vivid® BOMCC and 0.1 mg Vivid® EOMCC to prepare 2 mM BOMCC stock solution (for CYP2B6, CYP2C9, and CYP3A4) and 2 mM EOMCC stock solution (for CYP1A2, CYP2D6, and CYP2E1). Further dilute the fluorogenic substrates 200-fold in complete DMEM to get a final concentration of 10 μM.
6. Five mg/mL UGT1A4 (Corning).
7. Working solution of 4-Methylumbelliferone (4-MU) (Sigma-Aldrich): Mix 100 mM 4-MU stock solution in methanol with

complete DMEM to prepare 100 μM of 4-MU working solution.

8. UGT Reaction Mix A (UDPGA) solution (Corning).
9. Microtiter plate reader (Synergy H1, BioTek Instrumentation, Inc.).

2.6 Compound and DME Treatment

1. Compound stock solutions (Sigma-Aldrich): Dissolve model compounds including benzo[*a*]pyrene, aflatoxin B1, cyclophosphamide, 2-naphthlyamine, acrylamide, doxorubicin, 6-aminochrysene, 8-methoxypsoralen, and 4-nitrophenol in dimethyl sulfoxide (DMSO) to prepare 50 mM stock solutions.
2. Complete DMEM supplemented with 50 μM BSO and 0.5% DMSO (*see* **Note 5**).
3. Baculosome® Plus (control) (Thermo Fisher).
4. CYP450s including 1000 nM CYP1A2, CYP2B6, CYP2C9, CYP2D6, CYP2E1, and CYP3A4 (Thermo Fisher).
5. Five mg/mL UGT1A4 (Corning).
6. $NADP^+$ in 100 mM potassium phosphate buffer (pH 8.0).
7. NADP-regeneration system (Thermo Fisher).
8. UDPGA: Dilute 25 mM UDPGA in 50 μM BSO-supplemented complete DMEM to get a final concentration of 0.75 mM UDPGA.

2.7 Measurement of Cell Viability

1. Saline solution: Dissolve 8.1 g of NaCl and 2.9 g of $CaCl_2{\cdot}H_2O$ in 1 L of sterile deionized water to prepare a saline solution containing 140 mM NaCl and 20 mM $CaCl_2$.
2. Calcein AM stock (Sigma-Aldrich): Dissolve 1 mg of calcein AM in 1 mL DMSO to prepare a stock solution of 1.6 mM calcein AM. After further dilution in DMSO, store aliquots of 1 mM calcein AM in a −20 °C freezer protected from light until use.
3. CellTiter-Glo® luminescent cell viability kit (Promega) (*see* **Note 6**).
4. Orbital shaker (Fisherbrand™ platform rockers) (Fisher Scientific).
5. Automated fluorescence microscope (S+ scanner) (ATI, Republic of Korea): Green fluorescent cell images were obtained at ×4 magnification with the Olympus UPLFLN ×4 (numerical aperture (NA) 0.13, f-number 26.5, and depth of field (DOF) ~32.3 μm) (Olympus, Japan) and a green filter (XF404 from Omega Optical). S+ scanner has four filter channels for detecting multicolor, blue, green, and red fluorescent dyes, individually or simultaneously.

2.8 Image Processing and Data Analysis

1. ImageJ (NIH).
2. GraphPad Prism 4 (GraphPad Software).

3 Methods

3.1 Functionalization of the 384-Pillar Plate (Timing: 1 day)

1. Coat a 384-pillar plate with 0.01% (w/v) PMA-OD by stamping the 384-pillar plate in a lid of the 384-well plate containing 20 mL of 0.01% (w/v) PMA-OD. Check the small droplets of PMA-OD formed on the 384-pillar plate after immersion.
2. Place the 384-pillar plate facing up in a sterile bioassay plate and dry for 3–4 h at room temperature (*see* **Note 7**).
3. Print 2 μL of the 0.0033% PLL and 16.66 mM $BaCl_2$ mixture on the 384-pillar plate coated with PMA-OD using a S+ Micro-Arrayer and let it dry at room temperature overnight (Fig. 2).

3.2 Preparation of Cell Suspension (Timing: 30 min)

1. Culture human cells such as HEK 293 cells in a T75 flask with 12–15 mL complete DMEM at 37 °C in a 5% CO_2 incubator. Grow HEK 293 cells until approximately 80% confluency with growth medium change every 2 days.
2. Once the cells are 80% confluent, carefully remove the old medium from the T75 flask by aspirating the medium and rinse the cells with 5 mL of 1× DPBS at pH 7.4 (*see* **Note 8**).
3. Add 2 mL of 0.25% trypsin/EDTA into the T75 flask and incubate for 3–5 min at 37 °C in a 5% CO_2 incubator to detach the cells.
4. Add 7 mL of complete DMEM into the T75 flask to inhibit the trypsin activity. Aspirate and dispense the growth medium several times to break apart big cell clumps and create uniform cell suspension.
5. Transfer the cell suspension into a 15-mL conical tube.
6. Centrifuge the cell suspension at 500 × *g* for 3 min at room temperature.
7. Gently aspirate out the supernatant without disturbing the cell pellet.
8. Add 1 mL of complete DMEM and resuspend the cell pellet thoroughly.
9. Take 2 μL of cell suspension out and mix it with 98 μL of complete DMEM at 50-fold dilution for cell counting.
10. Insert a Moxi Z cassette into Moxi Z mini automated cell counter to measure cell density.
11. Take 75 μL of the diluted cell suspension and load it in the Moxi Z cassette to measure cell density (number of cells/mL).

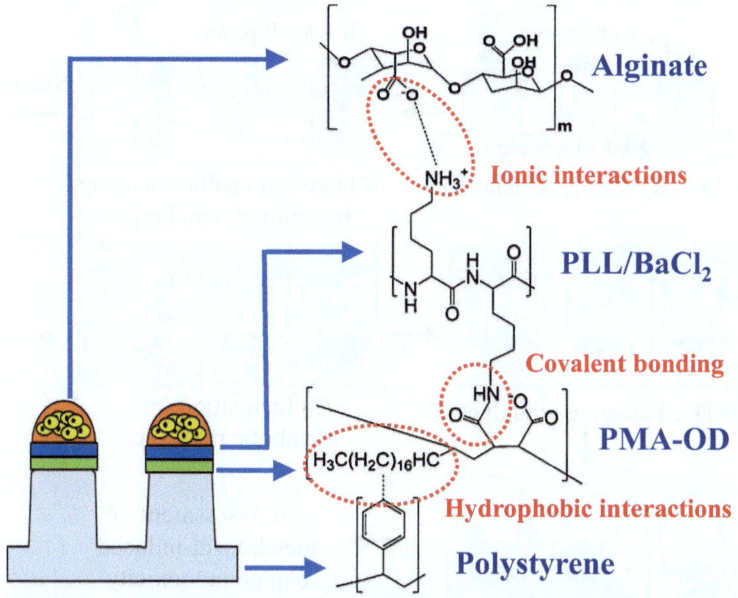

Fig. 2 Surface chemistry of the 384-pillar plate for cell encapsulation in alginate

12. Dilute the 1 mL cell suspension with complete DMEM to obtain a final concentration of 1.07 × 10^6 cells/mL.

3.3 Cell Printing and Encapsulation on the 384-Pillar Plate for 3D Cell Culture (Timing: 1 day)

1. Mix 625 μL of HEK 293 cell suspension (1.07 × 10^6 cells/mL) with 250 μL of 3% (w/v) alginate and 125 μL of 8 mg/mL GFR Matrigel to obtain a final cell concentration of 0.67 × 10^6 HEK 293 cells/mL in a mixture of 0.75% (w/v) alginate and 1 mg/mL GFR Matrigel (*see* **Note 9**).
2. Place the 384-pillar plate with dried PLL and $BaCl_2$ spots on the chilling chip deck at 4–7 °C. Turn on the chiller at least 30 min before cell printing to prevent water evaporation on the 384-pillar plate.
3. Print 1.5 μL of the HEK 293 cell suspension (0.67 × 10^6 cells/mL) in the mixture of alginate and GFR Matrigel (final 1000 cells/pillar) on the PLL-$BaCl_2$ coated 384-pillar plate (Fig. 3).
4. After 4 min of gelation on the chilling chip deck, sandwich the 384-pillar plate containing cells with the 384-well plate containing 50 μL of complete DMEM and incubate for 4–6 h at 37 °C in the 5% CO_2 incubator (*see* **Note 10**).
5. To replace the 384-well plate with old growth medium, remove the 384-pillar plate with cells and sandwich it onto a 384-well plate containing fresh complete DMEM supplemented with 50 μM BSO (*see* **Note 4**).

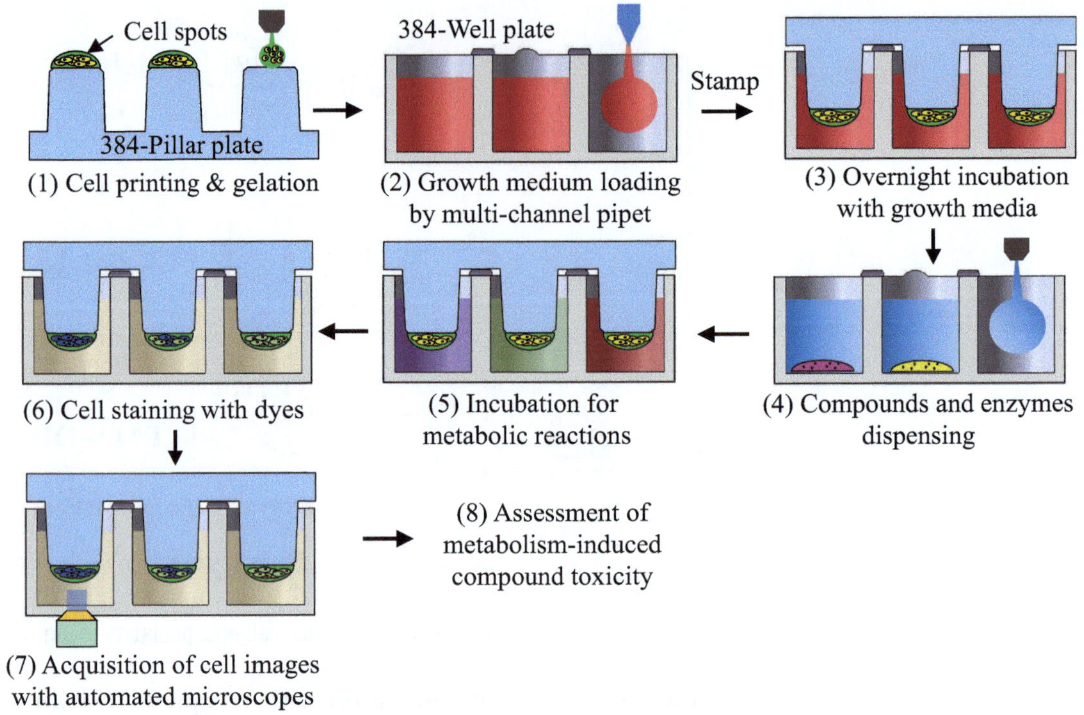

Fig. 3 Simplified schematic of experimental procedures for metabolism-induced toxicity assays with human cells in hydrogels on the 384-pillar plate and compounds and DMEs in the 384-well plate

6. Incubate the sandwiched 384-pillar/well plate at 37 °C in the 5% CO_2 incubator overnight before compound and enzyme treatment.

3.4 Measurement of Enzyme Activity with Fluorogenic Substrates (Timing: 2–3 h)

3.4.1 CYP450 Activity Assay

1. Prepare a mixture of 1000 nM CYP450 isoform, 100 mM $NADP^+$, and Vivid® NADP-regeneration system (100×) at 2:1:1 ratio.
2. Add 47.5 μL of 10 μM fluorogenic substrates in complete DMEM into each 384-well. Use BOMCC for CYP2B6, CYP2C9, and CYP3A4 and EOMCC for CYP1A2, CYP2D6, and CYP2E1.
3. Dispense 2.5 μL of the CYP450-NADP-regeneration system mixture in the 384-well plate using the microarray spotter.
4. Immediately record the changes in fluorescence intensity using the microtiter plate reader at an excitation wavelength of 405 nm and an emission wavelength of 460 nm.

3.4.2 UGT Activity Assay

1. Add 30 μL of 100 μM 4-MU in each 384-well.
2. Dispense 1 μL of UDPGA in the 384-well plate followed by 2.5 μL of 5 mg/mL UGT1A4 using the microarray spotter.

3. Immediately record decrease in fluorescence intensity using the microtiter plate reader at an excitation wavelength of 372 nm and an emission wavelength of 445 nm.

3.5 Preparation of Test Compounds in a 384-Well Plate (Timing: 2–3 h)

1. Warm up 50 mM of compound stock solutions in DMSO at room temperature.
2. Prepare the highest dosage of compounds (250 μM) by mixing 7.5 μL of the compound stock solution with 1493 μL of complete DMEM supplemented with 50 μM BSO.
3. For twofold serial dilution, add 750 μL of complete DMEM supplemented with 50 μM BSO in eight 1.5 mL Eppendorf tubes and 750 μL of complete DMEM supplemented with 0.5% DMSO and 50 μM BSO (DMSO-alone control) in one Eppendorf tube.
4. Take 750 μL of the highest dosage (in **step 2**) and transfer it to the next Eppendorf tube containing 750 μL of complete DMEM supplemented with 50 μM BSO. Mix the solution well by aspirating and dispensing at least 5 times.
5. Repeat **step 4** sequentially until the last Eppendorf tube with the DMSO-alone control remains as a solvent-only control. The final dilution factors in DMEM are 1:200, 1:400, 1:800, 1:1600, 1:3200, 1:6400, 1:12,800, and 1:25,600. Perform the twofold serial dilution for each test compound.
6. Dispense 30 μL of the diluted compound solutions at eight dosages and one control in a new 384-well plate using a multichannel pipette.
7. Place the 384-well plate containing the diluted compound solutions on the chilling chip deck at 4–7 °C prior to DME printing.

3.6 Printing DMEs in the 384-Well Plate Containing Test Compounds (Timing: 2 h)

1. Turn on the chiller connected to the 96-well plate deck and maintain the temperature at 4–7 °C before DME printing.
2. For Phase I enzymes, thaw CYP 450s, $NADP^+$, and Vivid® NADP-regeneration system on ice (*see* **Note 11**).
3. Thaw Baculosome® Plus on ice as no DME control.
4. Prepare a CYP450 solution in a 96-well plate by mixing 125 μL of 1000 nM CYP450 isoform, 62.5 μL of 10 mM $NADP^+$, and 62.5 μL of Vivid® NADP-regeneration system (100×). Keep the plate on ice before use.
5. For Phase II enzyme, thaw 25 mM UDPGA and UGT1A4 on ice.
6. Dilute 25 mM UDPGA in complete DMEM supplemented with BSO to prepare a final concentration of 0.75 mM UDPGA and print 1 μL of 0.75 mM UDPGA in the

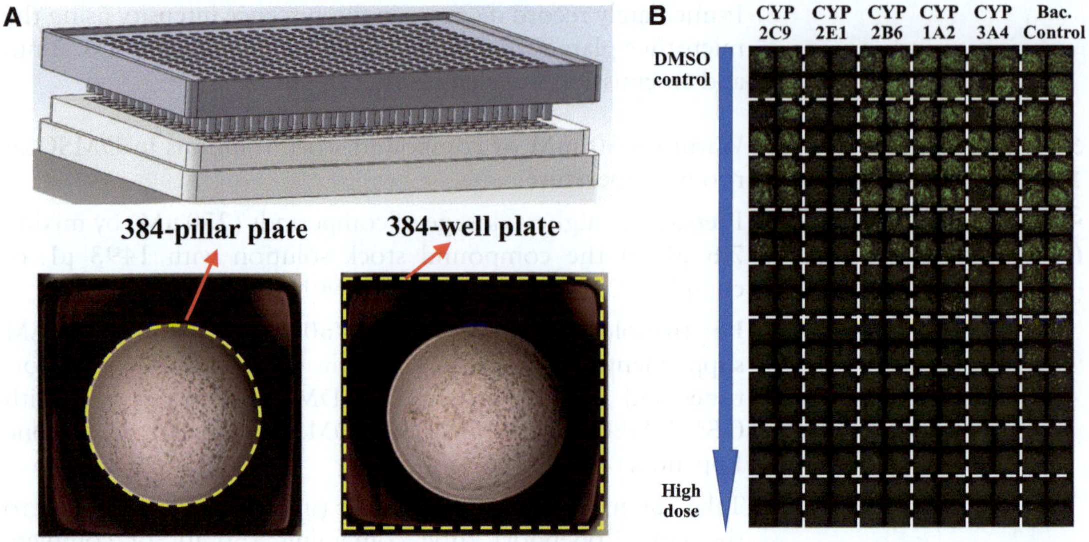

Fig. 4 Metabolism-induced toxicity assays with HEK 293 cells on the 384-pillar plate and drug-metabolizing enzymes (DMEs) and acetaminophen in the 384-well plate: (**a**) Microscopic images of the 384-pillar plate with HEK 293 cells sandwiched with the 384-well plate containing acetaminophen and CYP450s. The yellow dotted square represents the 384-well and the yellow dotted circle indicates the 384-pillar. (**b**) Scanned images of HEK 293 cells on the 384-pillar plate exposed to acetaminophen and CYP450s. HEK 293 cells were encapsulated in alginate-Matrigel on the 384-pillar plate, cultured in 3D, exposed to acetaminophen and CYP450s in the 384-well plate for 24 h, stained with calcein AM, and scanned with the S+ scanner. HEK 293 cells were treated with eight concentrations of acetaminophen and one DMSO control coupled with five CYP450s and one baculosome control

UGT1A4 region in the 384-well plate containing the diluted compound solutions before printing of UGT1A4.

7. Print 2.5 μL of one Baculosome® Plus control, four CYP450 isoforms, and UGT1A4 in six regions in the 384-well plate containing the diluted compound solutions (four replicates per compound dosage) (Fig. 4).
8. Separate the 384-pillar plate containing HEK293 cells from the 384-well plate containing growth medium and sandwich the 384-pillar plate with the 384-well plate containing six DMEs and one diluted compound.
9. Incubate the sandwiched plates for 24 h at 37 °C in the 5% CO_2 incubator prior to cell staining with calcein AM.

3.7 Staining HEK 293 Cells with Calcein AM for the Measurement of Cell Membrane Integrity (Timing: 2 h)

1. Prepare a working solution of calcein AM by adding 6 μL of 1 mM calcein AM stock in 12 mL of the saline solution to get a working concentration of 0.5 μM calcein AM (*see* **Note 12**).
2. Prepare a staining plate by adding 40 μL of 0.5 μM calcein AM in each 384-well and keep it in the dark at room temperature.

3. Prepare two washing plates containing 60 μL of the saline solution in each 384-well.
4. Separate the 384-pillar plate from the 384-well plate containing six DMEs and one compound and sandwich the 384-pillar plate with the washing plate once for 10 min to rinse the cells before staining.
5. Separate the 384-pillar plate from the washing plate, sandwich it with the staining plate, and then incubate the sandwiched plates for 60 min in the dark for cell staining.
6. After cell staining, rinse the 384-pillar plate with a fresh washing plate for 10 min to remove excess calcein AM on the cell spots.

3.8 Acquisition of Fluorescent Images with S+ Scanner (Timing: 1–2 h)

1. Load the sandwiched plates into S+ scanner and open an appropriate plate file from the list of files. (*see* **Note 13**).
2. Select the green filter channel as the green fluorescent dye is used for staining the cells on the 384-pillar plate.
3. Set the exposure time for the green filter based on histogram and brightness of the fluorescent images to obtain desired fluorescence images while avoiding photobleaching of fluorescence.
4. Scan the sandwiched 384-pillar plate to obtain green fluorescent images of cell spots.
5. Save the image files in a chosen folder.

3.9 Measurement of Intracellular ATP Levels Using CellTiter-Glo® Luminescent Cell Viability Assay Kit (Timing: 30 min)

1. Prepare a 384-well plate containing 40 μL of CellTiter-Glo® reagent solution.
2. Separate the 384-pillar plate from the washing plate and sandwich it with the 384-well plate containing CellTiter-Glo® reagent solution.
3. Shake the sandwiched plates on the orbital shaker for 2 min to induce cell lysis.
4. Incubate the sandwiched plates at room temperature for 10 min to stabilize the signal.
5. Record changes in luminescence using the microtiter plate reader at 560 nm emission wavelength.

3.10 Image Processing and Data Analysis (Timing: 3–4 h)

1. Open one representative image in ImageJ software.
2. Convert acquired image to gray scale (16-bit) using Type function in Image tab.
3. Apply a desired threshold by selecting a threshold algorithm in Threshold sub-function in Adjust function in Image tab to label cellular regions.

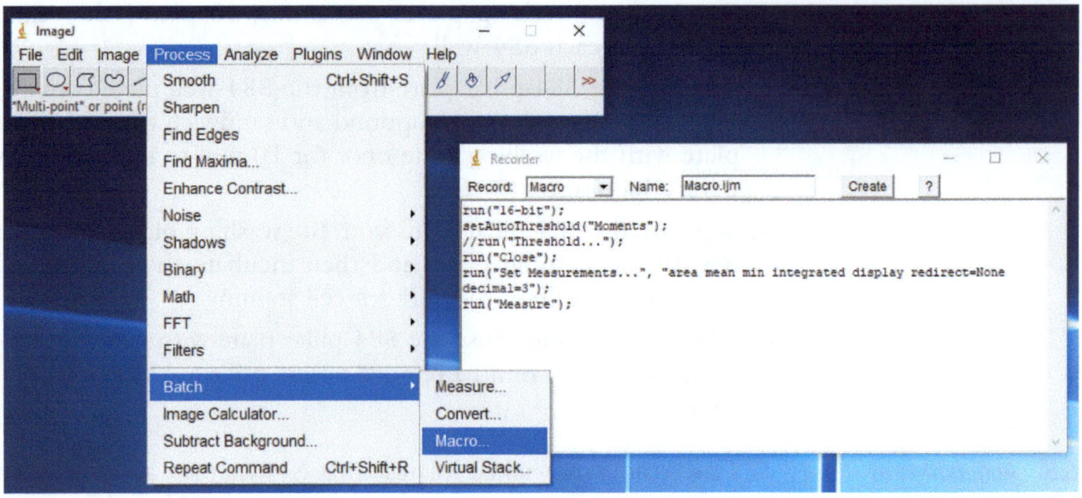

Fig. 5 Image processing by ImageJ software

4. Select parameters such as display label and integrated density using Set Measurements function in Analyze tab.
5. Quantify the fluorescence intensity from the image using Measure part in Analyze tab. A result window will pop up with the fluorescent intensities (Fig. 5).
6. By applying batch processing macro, intensities from hundreds of images can be extracted by recording the above single image steps saved as a plugin.
7. Save the data after batch processing (Excel file).
8. Normalize fluorescent intensities from the cells exposed to DMEs and compound with no compound controls to calculate 100% live cell using Prism 4 (Fig. 6).
9. Convert the test compound concentration to their respective logarithms.
10. Obtain dose–response curves and IC_{50} values using the following equation (Fig. 7):

$$Y = \text{Bottom} + \left[\frac{\text{Top} - \text{Bottom}}{1 + 10^{(\text{LogIC}_{50} - X) \times H}}\right]$$

where IC_{50} is the concentration of a compound in which 50% of the cellular response inhibited, H is the slope of the dose–response curve, X is the logarithm concentration of the compound, and Y is the cellular response, starting from the top

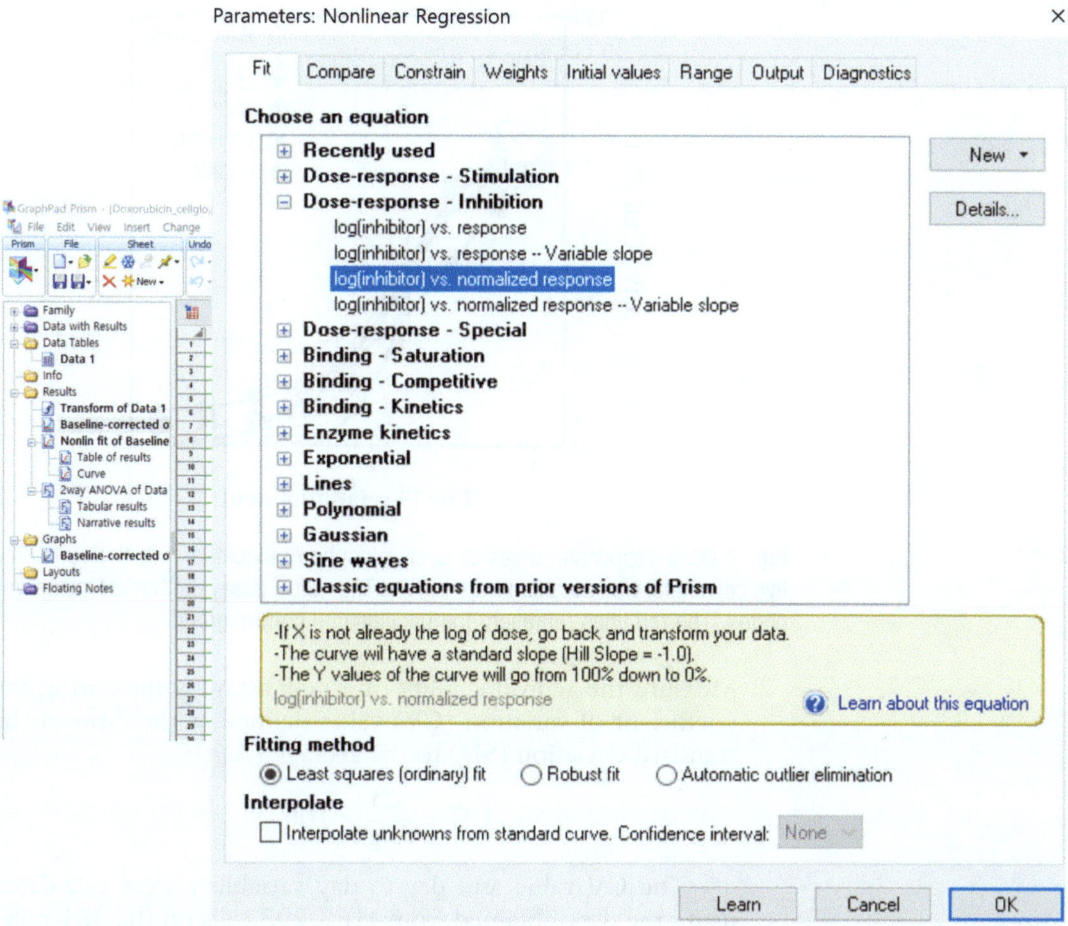

Fig. 6 Screenshot of GraphPad Prism to calculate IC_{50} values of compounds at different enzyme conditions

plateau (Top) of the sigmoidal curve to the bottom plateau (Bottom).

3.11 Assay Validation (Timing: 1–2 h)

1. Assess the robustness of a new assay by calculating Z factors with the equation below:

$$Z' = \frac{(\mathrm{Avg}_{\mathrm{Max}} - 3\mathrm{SD}_{\mathrm{Max}}) - (\mathrm{Avg}_{\mathrm{Min}} + 3\mathrm{SD}_{\mathrm{Min}})}{\mathrm{Avg}_{\mathrm{Max}} - \mathrm{Avg}_{\mathrm{Min}}}$$

where $\mathrm{Avg}_{\mathrm{Max}}$ is an average of maximum luminescence intensity from fully viable HEK293 cells on the 384-pillar plate, $\mathrm{SD}_{\mathrm{Max}}$ is a standard deviation of maximum luminescence intensity, $\mathrm{Avg}_{\mathrm{Min}}$ is an average of minimum luminescence intensity from dead cells affected by the highest dose of a compound, $\mathrm{SD}_{\mathrm{Min}}$ is a standard deviation of minimum luminescence intensity (*see* **Note 14**).

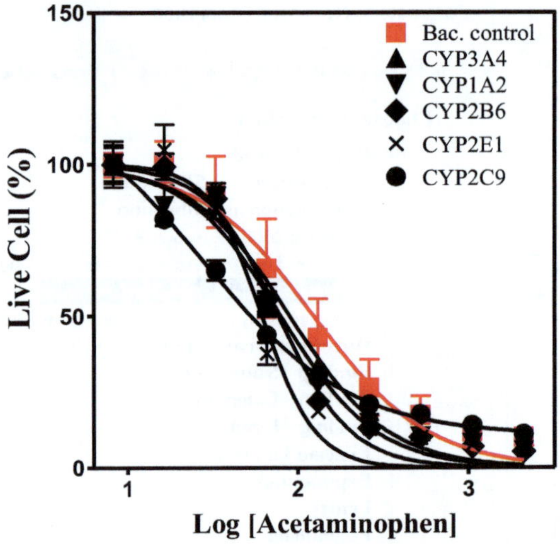

Fig. 7 Dose–response curves of acetaminophen incubated with CYP450s. The live cell signal was obtained from the CellTiter-Glo® assay on the 384-pillar/well plates. The red lines represent the baculosome control group

2. Measure the reproducibility of a new assay by measuring the coefficient of variation (CV) value defined as the ratio of the standard deviation (SD) to the average (Avg):

$$CV = \frac{SD}{Avg} \times 100$$

The CV value and day-to-day variability were calculated using the data obtained from HEK 293 cells on the 384-pillar plate exposed to no compound and no enzyme and stained with the CellTiter-Glo® luminescent cell viability kit (*see* **Note 15**).

3.12 Statistical Analysis (Timing: 1–2 h)

One-way analysis of variance (ANOVA) was performed to figure out statistically significant difference in IC_{50} values between no enzyme control and multiple DME conditions. The symbols indicate as follows: $*$ for $p < 0.05$, $**$ for $p < 0.01$, and $***$ for $p < 0.001$. No indication means $p > 0.05$.

1. Input at least three different sets of IC_{50} values from individual DME and compound conditions in GraphPad Prism 4.
2. Select "Analyze these data" under "Analyze" and select "One-way ANOVA (and nonparametric)" under "Column analyses" (Fig. 8).
3. Choose Test name as "One-way analysis of variance" under "Choose test" and select Test name "Bonferroni: Compare all pairs of columns" under "Post test" and select 0.05 (95%

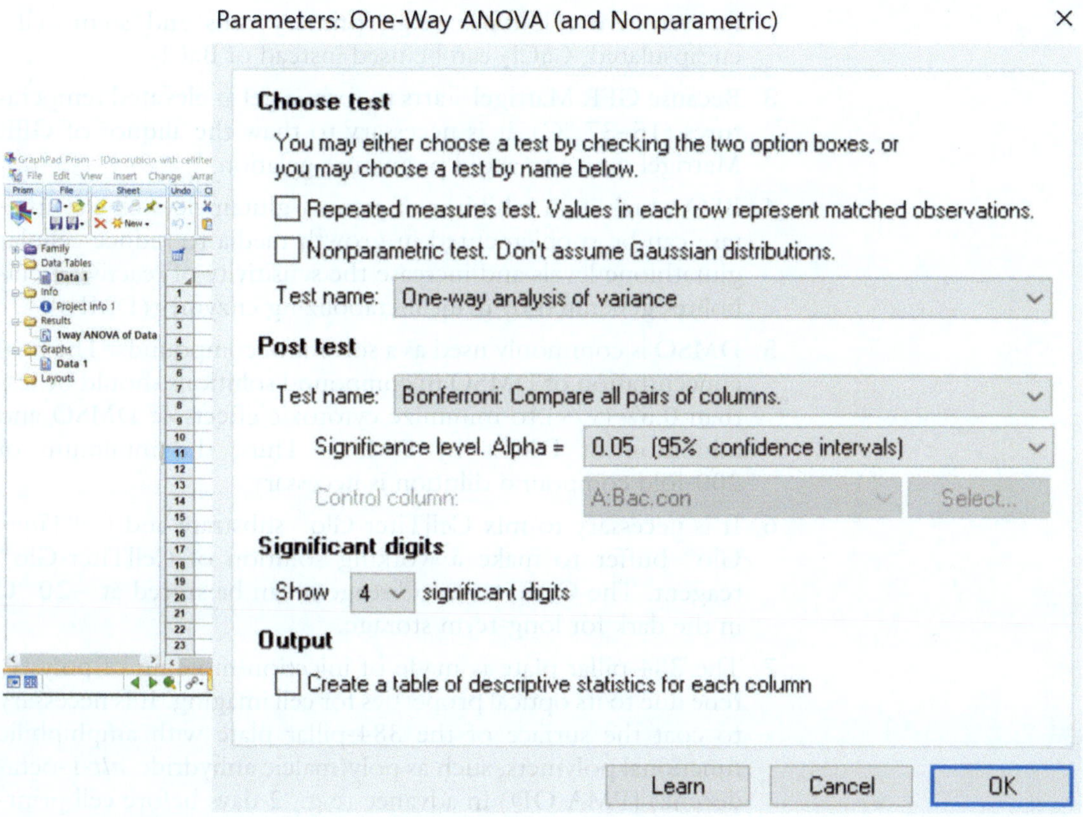

Fig. 8 Screenshot of GraphPad Prism to Perform ANOVA Analysis

confidence intervals) and select "4" under "Significant digits" and press "OK."

4. Enter the "Data 1" under "Graphs" for checking the results. "Data 1" showed the table with mark the * if any enzyme has significance.

4 Notes

1. The cell density is measured as the number of cells per mL of growth medium using the Moxi cell counter. For accurate cell counting, it is important to maintain good cell suspension in a proper cell density. Typically, 50–500-fold diluted cell suspension is used; thus, actual cell density is 50–500-fold higher than the cell density measured because of the dilution factor.

2. In general, $BaCl_2$ is used when relatively robust cell lines are encapsulated in alginate and high concentrations of phosphate ions are included in growth media, which could degrade alginate gel over time. In case the basal toxicity of $BaCl_2$ is an issue

for sensitive cell types (e.g., primary cells and stem cells) encapsulated, $CaCl_2$ can be used instead of $BaCl_2$.

3. Because GFR Matrigel starts to form a gel at elevated temperatures (15–37 °C), it is necessary to thaw the aliquot of GFR Matrigel on ice to avoid premature gelation.
4. BSO which is an inhibitor of gamma-glutamylcysteine synthetase can be supplemented in growth media to reduce cellular glutathione levels and increase the sensitivity of reactive metabolites generated by drug-metabolizing enzymes (DMEs).
5. DMSO is commonly used as a solvent of compounds. The final concentration of DMSO in compound solutions should be less than 0.5% (v/v) to minimize cytotoxic effects of DMSO and inhibition of DMEs by DMSO. Thus, the minimum of 200-fold compound dilution is necessary.
6. It is necessary to mix CellTiter-Glo® substrate and CellTiter-Glo® buffer to make a working solution of CellTiter-Glo® reagent. The CellTiter-Glo® reagent can be stored at −20 °C in the dark for long-term storage.
7. The 384-pillar plate is made of injection molding of polystyrene due to its optical properties for cell imaging. It is necessary to coat the surface of the 384-pillar plate with amphiphilic functional polymers, such as poly(maleic anhydride-*alt*-1-octadecene) (PMA-OD) in advance (e.g., 2 days before cell printing) to enhance cell spot attachment. PMA-OD provides hydrophobic affinity to polystyrene on the 384-pillar plate surface and reactivity to hydrophilic polymers, such as poly-L-lysine (PLL). Maleic anhydride groups on PMA-OD can react with amine groups on PLL. Hydrophilic alginate spots with human cells can be robustly attached on the 384-pillar plate via PLL and $BaCl_2$ chemistry. PLL ionically binds to negatively charged alginate spots while $BaCl_2$ is added to promote cross-linking of alginate.
8. PBS should be dispensed gently from the side of the T-flask to prevent nonspecific cell detachment from the T-flask surface.
9. GFR Matrigel is mixed with alginate to provide biomimetic cellular microenvironments for cell culture. Cells in growth medium, alginate, and GFR Matrigel should be mixed gently to create uniform cell suspension while minimizing air bubble entrapment.
10. To remove excess $BaCl_2$ from the cell spots on the 384-pillar plate and minimize cytotoxic effects from $BaCl_2$, we can sandwich the 384-pillar plate with cell spots twice with the 384-well plate with fresh growth media.
11. DMEs and the NADP-regeneration system can be denatured at room temperature rapidly. Thus, these samples must be kept

on ice prior to dilution in growth medium. DMEs are often contaminated with microbials. Thus, it is necessary to check their sterility prior to use. If necessary, add antibiotics in growth medium to avoid microbial contamination. In addition, DMEs contain cryoprotectants which can be cytotoxic. Thus, it is necessary to dilute DMEs at least 7–20 folds. The typical concentration of CYP450s is 1000 nM.

12. This volume (12 mL) is sufficient for staining one 384-pillar plate. Depending on the number of 384-pillar plates needed to be stained, readers can adjust the amount of the staining solution.
13. The files contain information for XYZ coordinates of the pillar plate used and the magnification of lens used for cell imaging. In general, ×4 magnification is selected for high-throughput acquisition of cell images from the 384-pillar plate.
14. An assay with the *Z′* factor between 0.5 and 1 is considered as highly robust.
15. The CV value below 25% is generally acceptable for high-throughput assays.

Acknowledgments

This work was partially supported by the US Environmental Protection Agency (US EPA Transform Tox Testing Challenge), Medical & Bio Device (MBD) Korea, the Cleveland State University (Faculty Innovation Fund), and the National Institutes of Health (NIEHS R01ES025779).

References

1. May JE, Xu J, Morse HR et al (2009) Toxicity testing: the search for an in vitro alternative to animal testing. Br J Biomed Sci 66:160–165
2. Paul SM, Mytelka DS, Dunwiddie CT et al (2010) How to improve R&D productivity: the pharmaceutical industry's grand challenge. Nat Rev Drug Discov 9:203–214
3. Lee M-Y, Clark DS, Dordick JS (2006) Human P450 microarrays for in vitro toxicity analysis: toward complete automation of human toxicology screening. J Assoc Lab Autom 11:374–380
4. EPA. Transform Tox Testing Challenge. https://www.challenge.gov/assets/document-library/Transform-Tox-Testing-Challenge-Stage-2-Update1.pdf
5. Lee M-Y, Park CB, Dordick JS et al (2005) Metabolizing enzyme toxicology assay chip (MetaChip) for high-throughput microscale toxicity analyses. Proc Natl Acad Sci 102:983–987
6. Lee M-Y, Kumar RA, Sukumaran SM et al (2008) Three-dimensional cellular microarray for high-throughput toxicology assays. Proc Natl Acad Sci 105:59–63
7. Kwon SJ, Lee DW, Shah DA et al (2014) High-throughput and combinatorial gene expression on a chip for metabolism-induced toxicology screening. Nat Commun 5:1–12
8. Yu KN, Kang SY, Hong S et al (2018) High-throughput metabolism-induced toxicity assays demonstrated on a 384-pillar plate. Arch Toxicol 92:2501–2516

Chapter 14

Droplet-Based Microfluidics Methods for Detecting Enzyme Inhibitors

Abraham Ochoa, Frida Trejo, and Luis F. Olguín

Abstract

Sub-nanoliter droplets produced in microfluidic devices have gained an enormous importance for performing all kinds of biochemical assays. One of the main reasons is that the amounts of reagents employed can be reduced in approximately five orders of magnitude compared to conventional microplate assays. In this chapter, we describe how to carry out the design, fabrication, and operation of a microfluidic device that allows performing enzyme kinetics and enzyme inhibition assays in droplets. This procedure can be used effectively to screen a small size library of compounds. Then, we describe how to use this droplet microfluidic setup to screen for potential inhibitor compounds eluted from a coupled high-performance liquid chromatography (HPLC) system that separates crude natural extracts.

Key words Microdroplets, Microfluidic device, Droplet microfluidics, Droplet-based assays, Enzyme kinetics, Enzyme inhibition, Enzymatic assays, High-throughput screening, Drug discovery, Natural products

1 Introduction

Microfluidic droplets are monodisperse microscopic liquid compartments (usually water), with typical volumes of nano to femtoliter, separated by an immiscible phase (mineral, silicon, or fluorinated oil) [1]. They are produced by intersecting streams of water and oil in microfluidic channels of tenths to hundred microns width and T-junction or flow-focusing geometries. Microfluidic droplets can harbor all sorts of biomolecules (enzymes, nucleic acids, etc.) to set up almost any type of biochemical assay with the advantage of reducing in approximately five orders of magnitude the amounts of reagents used in conventional microplate assays. Another outstanding characteristic of the droplets produced with microfluidic devices is that they are monodisperse (standard deviation of less than 3% in their volume) [2, 3]. Therefore, it is easy to compare and quantify the extent of the reactions occurring inside them. As the droplets can be continuously produced at rates from a

Nikolaos E. Labrou (ed.), *Targeting Enzymes for Pharmaceutical Development: Methods and Protocols*, Methods in Molecular Biology, vol. 2089, https://doi.org/10.1007/978-1-0716-0163-1_14,

few hertz to some megahertz [4], they have been used as microreactors for ultra-high-throughput screening [5].

To perform enzymatic assays in microdroplets, a continuous stream of an enzyme solution is combined with its substrate just before being intersected by a stream of immiscible oil to form the aqueous droplets. Inside these droplets, the enzyme catalyzes the transformation of substrate to product while traveling through a long microfluidic channel. After certain distance (equivalent to a determined reaction time), the concentration of the accumulated product is interrogated. Commonly, the product concentration is measured at many different points in the droplets pathway, allowing to determine the initial rate [6, 7] or the whole course of reaction [8, 9]. For this purpose, analytical methods [10] such as absorbance [7], electrochemistry [11], and mass spectrometry [12] have been used for enzyme kinetics. However, fluorescence microscopy [6] and laser-induced fluorescence [13] are among the most popular for their sensitivities, fast acquisition, and relatively low cost compared to others (e.g., mass spectrometry) [14].

In its simplest form, detection of enzyme inhibitors in microfluidic droplets consists of incorporating a potential inhibitor inside the droplets together with the enzyme and substrate. Then, by comparing the initial rate or an endpoint measurement with the inhibitor-free reaction, the inhibition percentage can be calculated. As the droplets are filled continuously, the amount of inhibitor inside the droplets can be varied systematically with the aid of automated syringe pumps [11, 15, 16]. An alternative is to make a continuous dilution of the tested molecule by injecting certain amount in a continuous flow (Aris-Taylor dispersion). By this method, enough points at different inhibitor concentrations can be measured to obtain the half-maximal inhibitory concentration (IC_{50}) [17–19]. These methods are relatively straightforward to set up if the inhibitor activity of a few compounds must be tested. However, when a medium or big size compound library has to be screened, the procedure is not trivial anymore, as it is unpractical to switch between thousands of syringes. One solution has been to "sip" each compound from a 96-well plate format and introduce it into a set of droplets preserving the identity of each molecule [12, 20]. The main disadvantages are that the compounds have to be first arranged in the plates and that complex automation is needed. Another emerging strategy has been to create droplets filled with microbeads carrying one type of compound each. After encapsulation with an enzymatic assay, the compounds are released by UV-light dosage and the droplets showing inhibitory effects are identified by fluorescence [21]. However, it has not been shown how the compounds will be identified once many different beads are used. Finally, another reported approach to encapsulate different compounds into the droplets is to use liquid chromatography or capillary electrophoresis to separate compounds from a complex mixture and encapsulate them as they come out of the chromatographic column [22, 23]. Using this procedure, neuraminidase activity inhibition assays in droplets were performed with the different HPLC-eluted peaks from a crude plant extract [24]. To unravel the full potential of droplet-based ultra-high-throughput assay for enzyme inhibitors, more research and clever setups are needed.

Here, we describe a procedure to design and fabricate a microfluidic chip for enzyme inhibition assay based on microfluidic droplets and using fluorescent image analysis for product quantification, presenting economic alternatives for instrumentation and reagents in the notes. We exemplify the use of the device with an inhibition assay employing a known inhibitor and a potential natural crude extract. We also describe the setting up of an HPLC-coupled microdroplet inhibition assay to screen for potential inhibitors from natural products separated in the HPLC.

2 Materials

2.1 Design and Fabrication of the Microfluidic Device

Transparency Photomask

1. Personal computer with software for drawing (Adobe Illustrator, AutoCAD, Inkscape, CorelDRAW, etc.).

Master Mold Fabrication

1. Spin Coater (WS-650, Laurell Technologies Corp. or any other spin coater).
2. Two hot plates topped with aluminum foil (temperature range: ambient-250 °C).
3. High-intensity UV lamp, 100 W (B100AP, Analytik Jena US LLC).
4. Silicon wafer (diameter: 76 mm, single side polished, thickness: 381 μm, orientation: <100>, Silicon Valley Microelectronics, Inc.).
5. SU-8 3035 photoresist (MicroChem Corp.).
6. Developer: 1-Methoxy-2-propyl acetate (propylene glycol monomethyl ether acetate) and isopropanol.
7. Two square flat clear glass plates, 14 × 14 × 0.3 cm ($L \times W \times H$).
8. Clean room facilities or Chemistry laboratory with compressed air and fume hood in which all the lights have been covered with orange or yellow cellophane wrap to block UV and near-UV light to avoid SU-8 photoresist polymerization. Ensure good cleaning to avoid excess of dust.

Polydimethylsiloxane (PDMS) Replica Fabrication

1. Polystyrene Petri dishes (90 mm × 15 mm).
2. Polydimethylsiloxane (Sylgard 184 silicone elastomer kit; Dow Corning Corp.).
3. Disposable smooth plastic cups and spoons.
4. Digital balance, vacuum desiccator, and scalpel or razor blade.
5. Laboratory oven (temperature range: ambient—75 °C).
6. Biopsy punch, 0.75 mm diameter (Electron Microscopy Sciences).
7. Glass cover slides, 75 mm × 50 mm.
8. Corona Treater (BD-20AC, Electro Technic Products) or any oxygen plasma oven.
9. Scotch Magic tape (3 M).

2.2 Microfluidic Device Operation

1. PTFE tubing (0.56 mm ID × 1.07 mm OD, Cole-Parmer).
2. Disposable 1.0 mL syringes with 22G needle.
3. Syringe pumps (NE-300 or NE-1002X New Era).
4. Fluorinated oil HFE-7500 (3 M, Novec) (*see* **Note 1**).
5. PFPE–PEG–PFPE surfactant. It can be synthetized according to references [25–27] or bought from RAN Biotechnologies, USA (FluoroSurfactant) or Dolomite, UK (Pico-Surf 1).
6. Enzyme, substrate, and product in buffer solutions.

2.3 Fluorescent Image-Based Enzymatic Assays in the Microfluidic Device

1. Fluorescent Inverted Microscope preferably with a camera port and with ×4, ×10, and ×20 objectives.
2. Fluorescence illumination (e.g., PhotoFluor LM-75, 89 North).
3. Fluorescence filter cubes. In this protocol we used DAPI-50LP-A (Semrock Inc.) to capture 4-methylumbelliferone (4-MU) fluorescence.
4. High-speed camera (Phantom Miro M110, Vision Research) or conventional small digital camera (e.g., Canon Power Shot SX260 HS).
5. Universal digital camera adapter for mounting the conventional digital camera to the microscope eyepiece (ocular) (Celestron LLC).
6. Multifunction data acquisition (DAQ) device (USB-6351, National Instruments) (optional).
7. High-performance liquid chromatography (HPLC) instrument (optional).

3 Methods

3.1 Design and Fabrication of the Microfluidic Device

This section describes the design and fabrication of the microfluidic device shown in Fig. 1a with rectangular channels of 150-μm-width and 100-μm-height. The design contains two zones. The first one is the droplet formation zone with a flow-focusing geometry formed by the intersection of the three inlets for aqueous phase and one inlet for oil phase (Fig. 1c). The second region is the droplet

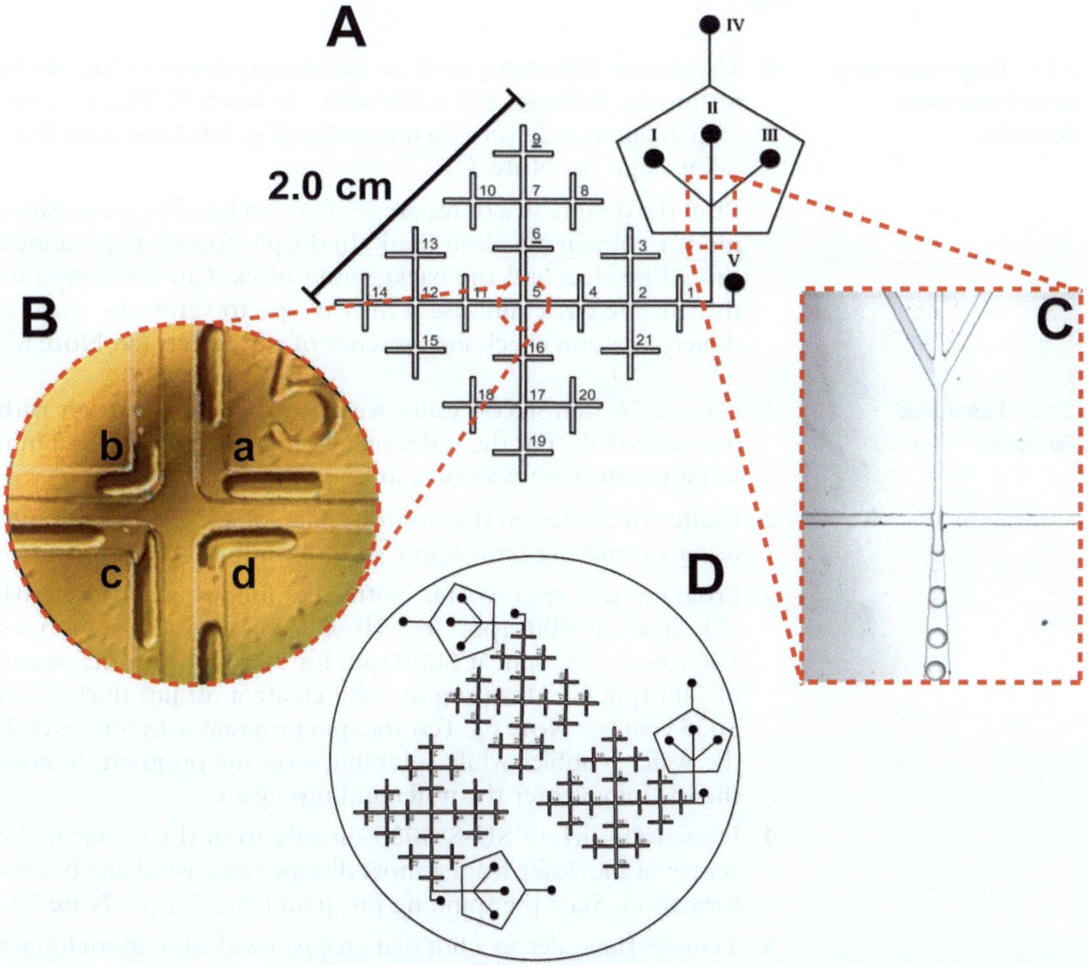

Fig. 1 (**a**) Microchip design used to carry out the enzymatic assays. A fractal-shape channel arranged in 21 different nodes composes the detection area traced within a square of 2 × 2 cm. At each node, four different droplets can be simultaneously analyzed. *I*, *II*, and *III* are the inlets for the aqueous phase components. *IV* is the inlet for the oil phase and *V* is the outlet. (**b**) Bright-field image at node five in the center of the microfluidic device. Here, four droplets with proportionally spaced residence times (15, 40, 65, and 90% of the total) can be observed. (**c**) The flow-focusing design allows the droplet formation. (**d**) Arrangement of a photomask in which three replicas of the microfluidic device were drawn within a circle of 65 mm diameter

detection zone, consisting of a 28-cm-long single channel following a fractal shape inspired from a previous work [28]. With these channel dimensions, the droplet residence time is 35.2 s when the total flow is 430 μL/h (250 μL/h for oil flow and 60 μL/h for each aqueous phase component) (*see* **Note 2**). The fractal shape design allows to monitor the droplets at 21 different nodes marked in Fig. 1a. On each node four droplets with different resident times can be simultaneously monitored in a single microscope image. At the node five, in the center of the fractal zone (Fig. 1b), it is possible to monitor droplets of 15, 40, 65, and 90% of the total residence time.

3.1.1 Design and Printing of the Transparency Photomask

1. Use Adobe Illustrator to draw the design shown in Fig. 1a by setting the line stroke at 0.150 mm (*see* **Note 3**). Three microchip designs can fit on a 76 mm wafer (Fig. 1d) leaving a 0.5 cm safety edge (*see* **Note 4**).
2. Print the design on a transparency film (acetate film) in negative mode to obtain the photomask. In the photomask the channels should be clear and the background black. Cut the design by the outside circle and use a microscope to verify the channel dimensions and check the presence of any defects (*see* **Note 5**).

3.1.2 Master Mold Fabrication

1. Clean a 76-mm silicon wafer with isopropanol and dry it with pressurized air. Put the wafer on a hotplate at 200 °C for 5 min to evaporate any traces of water.
2. Center the wafer on the rotating stage of a leveled spin coater using a centering tool. Apply vacuum to hold it in place.
3. Program the spin coater with the following instructions: (1) Spin at 500 rpm for 10 s using an acceleration of 100 rpm/s. (2) Spin at 2000 rpm for 30 s with an acceleration of 300 rpm/s. This program will create a 50 μm thick SU-8 3035 coat (*see* **Note 6**). Test the spin program with the wafer. If the wafer wobbles while spinning, stop the program, remove the vacuum, center the wafer and test again.
4. Dispense 3 mL of SU-8 3035 directly from the bottle at the center of the wafer from a short distance and avoid any bubble formation. Start the spinning program (Fig. 2a) (*see* **Note 7**).
5. Transfer the wafer to a hot plate topped with aluminum foil set to 105 °C and bake it for 15 min (*see* **Note 8**) (Fig. 2b).
6. To obtain channels of 100-μm-height, repeat **steps 4** and **5** over the same wafer. This will add a second 50-μm-resin layer over the previous one, giving a final 100-μm-thick layer (*see* **Note 9**).
7. Place the photomask obtained in **step 2** of Subheading 3.1.1 on top of the coated wafer and put them between two glass plates, so that the photomask and wafer remain flat and in full

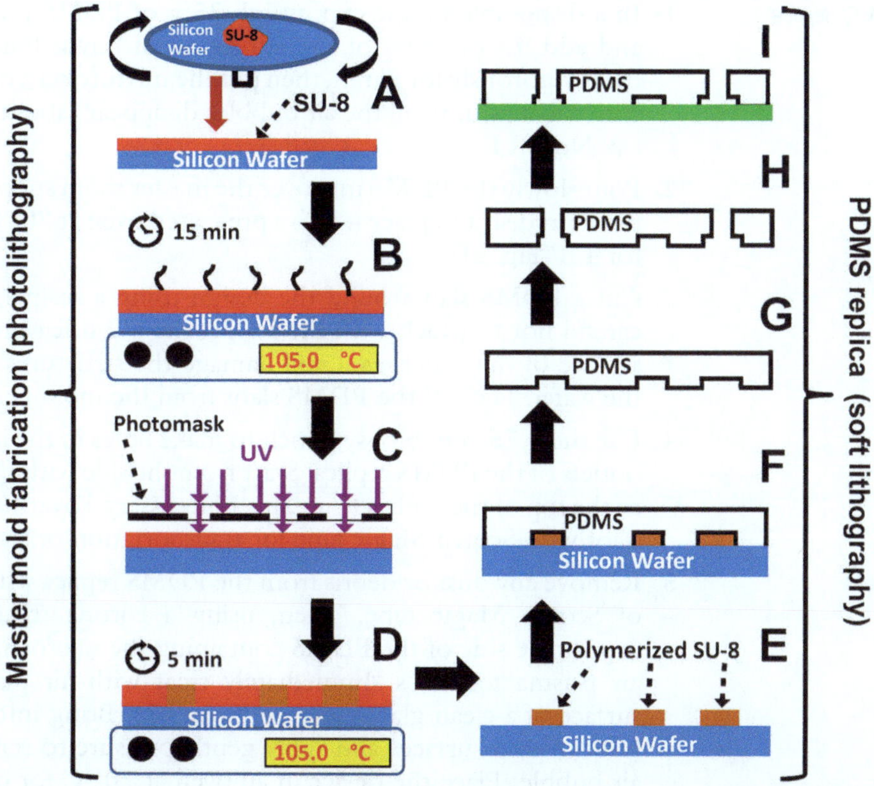

Fig. 2 Microfluidic device fabrication scheme. On the left, the master mold fabrication via photolithography is described, and on the right, the process of obtaining PDMS-based devices via soft lithography is explained. (**a**) Spin coat photoresist. (**b**) Soft bake. (**c**) Expose photoresist through photomask. (**d**) Postexposure bake. (**e**) Develop to obtain master mold. (**f**) Pour PDMS over the master mold and cure. (**g**) Peel off the cured PDMS slab. (**h**) Punch tubing inlets in the PDMS replica. (**i**) Seal the PDMS replica to a glass slide

contact. Set them under the UV lamp for 45 s (Fig. 2c) (*see* **Note 10**).

8. Heat the wafer for 1 min on a hotplate topped with aluminum foil at 75 °C, then transfer the wafer to a second hotplate topped with aluminum foil at 105 °C for 5 min to finish the polymerization reaction of SU-8 (Fig. 2d) (*see* **Note 8**).
9. Immerse the wafer in a crystallization dish with the developer 1-methoxy-2-propyl acetate and shake constantly for about 10 min. Take the wafer out of the developer and rinse it with isopropyl alcohol. If a white film is observed during this process, immerse the wafer in the dish with fresh developer for 5 min or until no white streaks are observed. Dry it with clean pressurized air (Fig. 2e).
10. Hard-bake the wafer on a hotplate at 200 °C for 15 min. Remove it and let it reach ambient temperature. Put it in a clean plastic Petri dish.

3.1.3 (PDMS) Replica Fabrication

1. In a disposable plastic cup, weigh 25 g of PDMS prepolymer and add 9% on mass of the curing agent (cross-linker). Mix them vigorously for 3 min, then put the mixture on a desiccator with vacuum until all the air bubbles disappear (about 30 min) (*see* **Note 11**).
2. Pour slowly the PDMS mix over the master mold supported on the Petri dish and place it into a preheated oven at 70 °C to cure for 4 h (Fig. 2f).
3. Cut a PDMS slab around the design using a scalpel. Be very careful not to touch with the scalpel the SU-8 features or the surface of the wafer, as it will damage the structures or break the wafer. Peel off the PDMS slabs from the mold (Fig. 2g).
4. Use the 0.75-mm biopsy punch to make holes in the inlets and outlets of the PDMS replica. Start from the side with the design to the top of the replica (Fig. 2h). If necessary, cover the PDMS slabs with Scotch Magic tape for transportation or storage.
5. Remove any dust or debris from the PDMS replica with a piece of Scotch Magic tape. Then, using a corona treater [29], expose the side of the PDMS containing the microchannels to air plasma for 15 s. Immediately treat with air plasma the surface of a clean glass slide (*see* **Note 12**). Bring into contact both treated surfaces and apply gentle pressure to remove any air bubble. Place the device in an oven at 70 °C for overnight (Fig. 2i).

3.2 Operation of the Microfluidic Device

The following procedure describes how to generate water in oil microfluidic droplets with the device shown in Fig. 1a. The aqueous phases are buffer and 4-methylumbelliferone (4-MU) solutions, as these reagents will be used to build a calibration curve of 4-MU (Subheading 3.7), which is a fluorescent product of many enzymatic hydrolytic reactions.

1. Fill a disposable 1.0 mL syringe with a solution of 20 μM of 4-MU or the fluorescent product of the enzymatic reaction in 50 mM Tris–HCl buffer pH 8.0 (*see* **Note 13**).
2. Fill two disposable 1.0 mL syringes with 50 mM Tris-HCl buffer.
3. Fill a 1.0 mL disposable syringe with fluorinated oil HFE-7500 with 0.4% w/w of PFPE–PEG–PFPE surfactant (*see* **Note 14**).
4. Cut PTFE tubing of appropriate length and connect it to each of the syringe needles with the help of tweezers.
5. Load the syringes on the syringe pumps and purge them to expel the air in the tubing (*see* **Note 15**).
6. Connect the free end of each tubing to the microfluidic chip inlets according to Fig. 1a starting with the fluorinated oil at

the inlet IV, following with the buffer at the inlets I and III, and finishing with the product 4-MU at the inlet II. As the inlets are slightly narrower than the tubing, it will be hold tightly (*see* **Note 16**).

7. Place the microfluidic chip in the inverted microscope and pump the fluorinated oil into the chip at a flow of 250 μL/h. With the help of the microscope, observe that the fluorinated oil travels through all channels. It is preferable that fluorinated oil makes contact with the channels before aqueous solutions.
8. Pump each of the aqueous phases at 60 μL/h and wait until all the oil is expelled from their microchannels and the droplet formation is stable (*see* **Note 17**). Use a 4× objective and the fast camera or digital camera to monitor the entire droplet flow path to check for any obstructed channels or droplet breakup (*see* **Note 18**).

3.3 Determining the Average Speed and the Residence Time of the Droplets in the Microfluidic Device from Video Images

It is important to know the droplet velocity through the microchannel, as this will determine for how long the reaction has evolved at each position of the fractal design. With this information it is possible to decide the droplet enzyme concentration and the nodes at which the reaction product will be analyzed. As mentioned in **Note 2**, it is possible to estimate the droplet speed from the cross-sectional area of the microchannel and the total flow. But because these two variables could have systematic errors, and droplet speed is sensitive to many factors [30], it is highly recommended to experimentally determine the droplet speed.

1. Use the bright field and the 4× objective to observe the flowing droplets and record 15-s videos at different positions in the microfluidic device (from the droplet formation zone to the node of interest). If a high-speed video camera is used, configure the settings to record 50 frames per second (change it if other flows are used) and continue to **step 2**. If a conventional digital camera is used, the video mode will normally be set up at 24 frames per second and continue to **step 5** (*see* **Note 19**).
2. Open the high-speed video with the Phantom Camera Control (PCC) software and make a distance calibration: *Play > Measurements > Calibration > Calibrate.* Set the first calibration point at one edge of the microchannel and the second at the opposite wall. Enter the experimental channel width to calibrate the software (150 μm for the design in Fig. 1).
3. Determine the speed of one droplet:

   Play > Measurements > Instant measurements > Active > Distance, Angle and Speed: 2 points

   Set the first point at the front edge of one droplet, advance some frames and set the second point at the front edge of the same droplet. The distance, speed, and acceleration will appear in the Instant Measurements Section. Repeat the measurement with ten different droplets at the consecutive positions of the microchannel.

4. To determine the droplet residence time at the different fragments of the channel, divide the distance traveled by a droplet by the droplet speed. Add the droplet residence times of all the fragments to calculate the total residence time.
5. For videos taken with a conventional digital camera, Tracker-Video Analysis and Modeling Tool (https://physlets.org/tracker/) is free software useful to determine the speed of droplets.
6. If necessary, adjust the flow rates of oil and water to modify the residence time of the droplets in the microfluidic chip.

3.4 Capture of Droplet Fluorescent Videos

1. Turn on the fluorescence lamp and choose the corresponding fluorescence filter cube in the microscope. In this case DAPI cube for 4-MU.
2. Focus the ×20 objective at the node five (Figs. 1b and 3a) of the fractal zone and capture the fluorescence droplet intensity by any of the following methods (*see* **Note 20**).

3.4.1 High-Speed Image Capture

Using a high-speed camera it is possible to capture droplet images faster than the droplet frequency and record some fluorescence images of each droplet in the experiment, giving precise information about the frequency and shape of the droplets. However, this method might need all the RAM memory of the camera to record only a few minutes of the experiment depending on the image size.

1. Set the camera sample rate at 50 frames per second and the exposure time at 20,000 μs (*see* **Note 21**).
2. Record a 30-s video at the selected node (*see* **Note 22**).

3.4.2 Intermittent High-Speed Image Capture

Sometimes it is necessary to record flowing droplets for a long period of time in which the contents of the droplets gradually change (e.g., Subheading 3.10). In these cases the camera in-built memory will not be enough to store all the corresponding droplets. An alternative is to capture intermittently a small fraction of droplets through the entire event. To do this, the fast camera can be externally controlled to impose a low sampling rate, but still using short exposure times.

1. Use the F-SYNC connector (BNC) of the camera to synchronize it with a multifunction data acquisition device (DAQ) coupled to a computer.
2. Use NI Signal Express software to control the DAQ and to create an analog square signal at a frequency of 5 pulses per

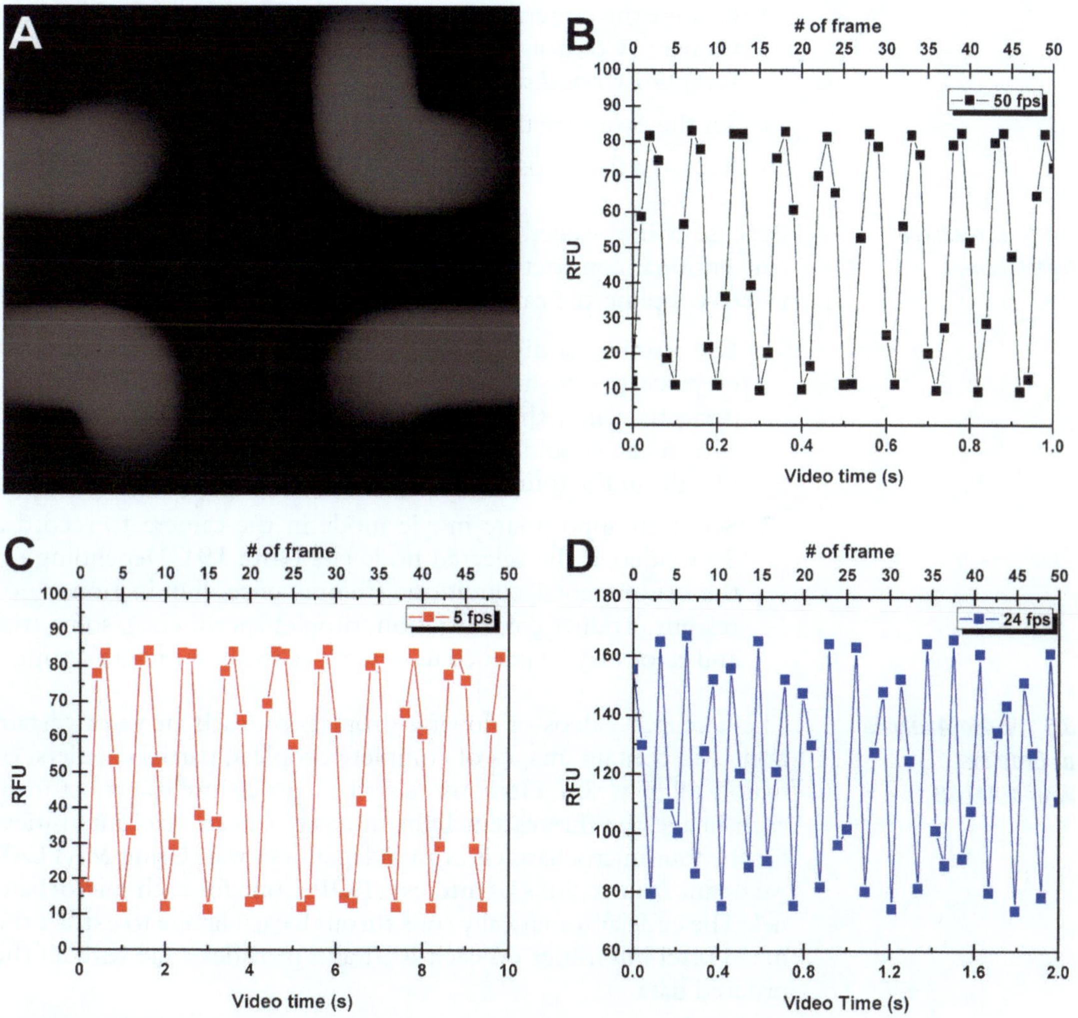

Fig. 3 (**a**) Fluorescence image at node five in the center of the microfluidic device observed with ×20 microscope objective. (**b**) Typical recording of the fluorescence signal in relative fluorescence units (RFU) for curve "a" at node five. Each oscillation represents one droplet passing through the analyzed area during 1-s video at 50 frames per second (fps). The maximum value in each oscillation corresponds to a measurement in the center of a droplet, while the minimum corresponds to a measurement made in the oil (between two droplets). (**c**) Typical recording of the fluorescence signal for videos taken at 5 fps. In this case, each oscillation does not represent one droplet, but an envelope signal of 1 s of a video taken at 50 fps. Note that video time scales are proportionally different on each graph. (**d**) Typical recording of the fluorescence signal for videos taken at 24 fps with a conventional digital camera. Each oscillation represents one droplet. However, at this capture rate it is not possible to make a measurement at the droplet center in every case. Therefore, there is more noise in the obtained traces with respect to the videos taken at 50 fps

second (or the desired image capture frequency), 2.5 V of amplitude and 2.5 V of offset (higher voltage could damage the camera).

3. Change the camera settings to External control in the Phantom Camera Control (PCC) software: *Live* > *Advanced settings* > *External control.*
4. Set the exposure time at 20,000 μs in the PCC.
5. Record a fixed position video for 2 min (*see* **Note 23**).

3.4.3 Conventional Digital Camera

Because a high-speed camera is expensive, we have also used a conventional compact digital camera to record fluorescent image videos coupling the camera to the ocular port.

1. Use a universal digital camera adapter to fix the camera to an eyepiece of the microscope. Move the *x*, *y*, and *z* stages to perfectly center the camera lens with the microscope eyepiece. The image should be centered on the screen and the intensity of light in the four channels should look the same.
2. Select an appropriate movie mode in the camera to record a 30-s video at the selected node (*see* **Note 19**). Depending on the experimental conditions (fluorescence illumination, fluorescent product concentration, droplet speed, etc.), some trial and error may be needed to determine the best camera settings.

3.5 Video Analysis of Droplets with MATLAB

The acquired videos of flowing droplets on each curve of certain node will contain images of complete droplets, partial droplets, or mainly oil (Fig. 3a). Here, we describe a procedure to extract from each image the fluorescent light intensity (mean pixel intensities) for the four microchannels in an independent way. Using MATLAB we define four regions of interest (ROIs), one for each microchannel. The code automatically runs through each image to extract the mean pixel intensities on each ROI and provides a file with all the ordered data.

1. Transform the videos to a format with the extension *.mov.* Appropriate free software to do this is "*aTube Catcher.*"
2. Write in MATLAB the following code:

```
a=roipoly();
b=roipoly();
c=roipoly();
d=roipoly();
[This line is intentionally left blank]
for frames=Initial frame : final frame
disp(frames)
tempImage=double(Video name(:,:,1,frames)); a2=tempImage
(a==1);
a3(frames)=mean(a2);
b2=tempImage(b==1);
b3(frames)=mean(b2);
c2=tempImage(c==1);
```

```
13- c3(frames)=mean(c2);
14- d2=tempImage(d==1);
15- d3(frames)=mean(d2);
16- end
17- %%
18- a3Data=[(1:frames)',a3'];
19- b3Data=[(1:frames)',b3'];
20- c3Data=[(1:frames)',c3'];
21- d3Data=[(1:frames)',d3'];
```

3. Import a selected video of droplets into MATLAB by dragging the file into the MATLAB command window.
4. Rename the file and change the line 8 of the code to include the video file name.
5. To obtain the total number of frames in the video, write the command:

```
>> size(Video name)
```

   The program will return three numbers related to the frames characteristics and a fourth number corresponding to the total number of frames in the video.
6. Modify line 6 of the code to set the starting and finishing frames to analyze.
7. In the MATLAB command window write:

```
>> imagesc(Video name(:,:,:, # of desire frame))
```

   The program will open the specified frame in the last command. This image will be used to select the regions of interest to be analyzed and any of the frames is fine for this purpose.
8. Run the code in MATLAB. Using the computer mousse, define the first ROI by drawing a small square inside the curve "a" of the fractal image opened in the last step. Close the square with a double click. Draw one square for each of the other three channels in the fractal image to define the rest of the ROIs (*see* **Note 24**).
9. MATLAB will start the analysis and when finished it will show at the Workspace the variables a3Data, b3Data, c3Data, and d3Data corresponding to each ROI previously drawn in the same order that were made. Open the file and copy the data to be further analyzed in other software (Excel, Origin, etc.). Expected graphs of these results are shown in Fig. 3b–d depending on the capture method employed in Subheading 3.4.

3.6 Analyzing MATLAB Data with Origin

After the video analysis with MATLAB, a broad range of fluorescence intensity measurements are obtained depending on the image captured in each frame of the video (remember images will contain complete droplets, partial droplets, or mainly oil). An ideal measurement is obtained when the ROI coincides with the droplet center (Fig. 4a). Variations in these measurements are due to small differences in the droplet content and intrinsic noise of the fluorescence technique. A second case of measurement occurs when the ROI coincides with the edge or a fraction of a droplet (Fig. 4b, c). In the last case, an underestimated fluorescence intensity value will be obtained. The opposite extreme to the ideal measurement is acquired when the ROI coincides with the space between the two droplets (Fig. 4d). The analysis of the MATLAB results from a typical 750-frame video taken at 50 fps is shown as a histogram in Fig. 4e. The classes with higher fluorescence intensities are integrated by frames in which the measurement was made at the droplet center or with a slight deviation. The lower intensity fluorescence classes correspond to droplet fractions or oil measurements. To consider only the values in which adequate measurements were made, a threshold is established. By observing the fluorescent intensity vs. time plots (Fig. 3b–d) and checking manually some frames, it is possible to select the appropriate threshold. In this way, histograms such as the one shown in Fig. 4f are obtained. What follows is a procedure for sorting these data.

1. Import the MATLAB data for each analyzed video to Origin.
2. Plot a histogram for each of the four curves analyzed at node 5. Set the bin size to two units and adjust the axis. Examine the histogram, the fluorescent intensity vs. time plots and inspect some video frames to establish a threshold value that allows retaining those measurements made in microdroplet centers (high fluorescence intensities) and neglecting measurements made in the oil or droplet fractions (low fluorescence intensities).
3. Sort the data in the worksheet in descending order with respect to the fluorescence intensity column and eliminate all the data below of the established threshold.
4. Average the signals for each data set and plot them as a function of the independent variable (concentration, residence time, etc.).

3.7 Constructing a Calibration Curve of the Fluorescent Product

1. Using the configuration described in Subheading 3.2, get different 4-MU concentrations in the droplets by making proportional changes in the flows rates of the aqueous stream maintaining the total flow at 180 μL/h. For example, set the flow of the 4-MU solution (20 μM in the syringe) at 90 μL/h and each of the two buffer solutions at 45 μL/h. With these

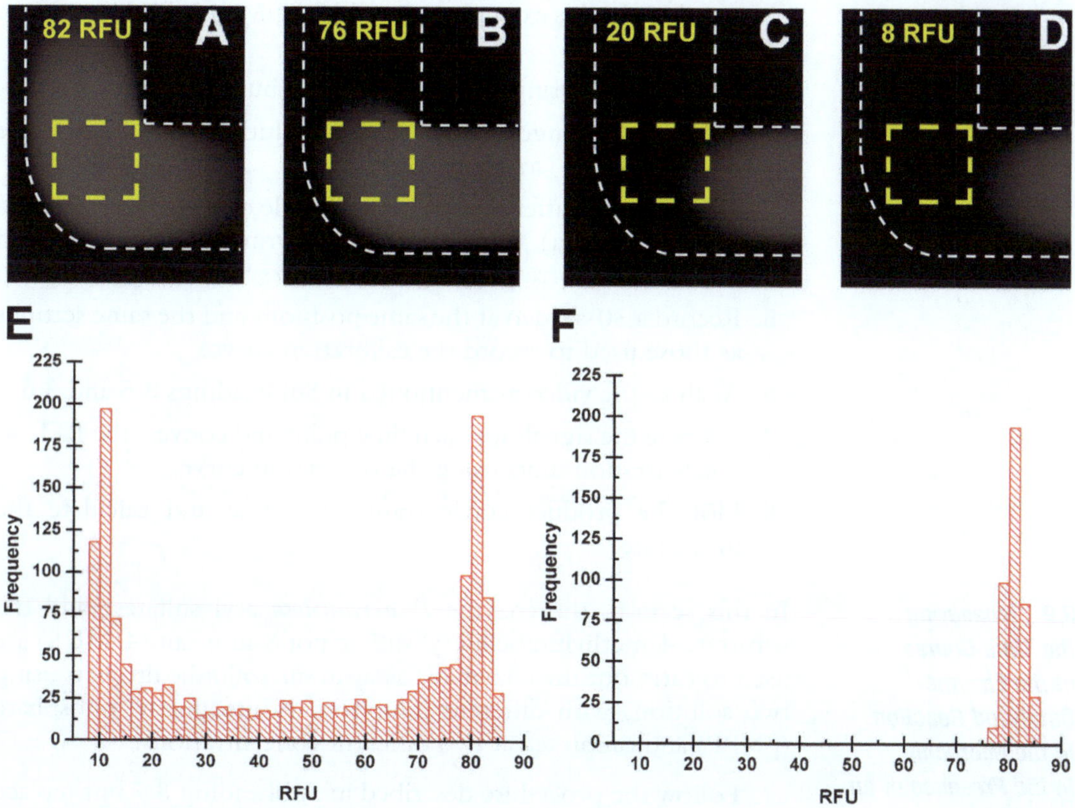

Fig. 4 Examples of obtained fluorescence images of flowing microdroplets with 10 μM of 4-MU. Their corresponding fluorescence intensities after MATLAB analysis of the ROIs are shown. (**a**) Measurement at the droplet center. (**b**, **c**) Measurement of partial droplets. (**d**) Measurement of oil between two droplets. (**e**) Histogram of the fluorescence intensities obtained after the analysis of a 15-s video taken at 50 fps (750 frames in total). (**f**) Histogram of fluorescence intensities after selecting only the frames in which measurements were made in the center of droplets

rates the 4-MU concentration in droplets will be 10 μM (*see* **Note 25**).

2. For each fluorescent product concentration in the droplets acquire the corresponding video (Subheading 3.4) and analyze it (Subheadings 3.5 and 3.6) to obtain the corresponding average droplet fluorescence of each concentration.
3. Construct a calibration curve of average droplet fluorescence vs. product concentration (Fig. 5a, b).

3.8 Measuring the Time Course or Initial Rate of the Enzyme-Catalyzed Reaction in the Droplets

In this section, the enzyme *P. aeruginosa* aryl sulfatase and the substrate 4-methylumbelliferyl sulfate potassium salt (4-MUS) are used to carry out the enzymatic assay in microfluidic droplets.

1. Fill a syringe with a 3× stock solution of substrate in the reaction buffer. The substrate will be diluted threefold with the other two aqueous components when they form the

droplets. In this example we used 19.8 μM of 4-MUS (*see* **Note 26**).

2. Fill a second syringe with the reaction buffer.
3. Fill a third syringe with a 3× stock solution of enzyme in the reaction buffer (*see* **Note 27**).
4. Connect the buffer syringe in the middle of the aqueous stream (inlet II, Fig. 1a). Connect the other syringes to the device and produce the corresponding droplets (*see* Subheading 3.2).
5. Record a 30-s video at the same positions and the same settings as those used to record the calibration curve.
6. Analyze the video as mentioned in Subheadings 3.5 and 3.6.
7. Average the signals for each time point and convert the RFU to concentration units using the calibration curve.
8. Plot the product concentration vs. time and calculate the initial rate.

3.9 Measuring the Time Course of the Enzyme-Catalyzed Reaction in the Microchip in the Presence of an Inhibitor

In this section, the enzyme *P. aeruginosa* aryl sulfatase and the substrate 4-methylumbelliferyl sulfate potassium salt (4-MUS) are used to carry out the enzymatic assay in microfluidic droplets using two solutions with different inhibitors: p-nitrophenyl phosphate (pNPP) and onion tea at two different concentrations.

1. Follow the procedure described in Subheading 3.7 but replace the syringe containing the reaction buffer (inlet II Fig. 1a) with a 3× stock solution of the substance to be measured as inhibitor in the reaction buffer (*see* **Note 28**).
2. Follow the same procedure described in Subheading 3.8 for the determination of the initial rate.
3. Compare the initial rate with/without the inhibitor to determine the inhibition percentage [31].
4. Expected results for an enzyme-catalyzed reaction in the absence or presence of an inhibitor are shown in Fig. 5c, d.

3.10 Coupling HPLC Peak Separation with Inhibition Measurements in Microfluidic Devices

In this section, we describe an HPLC-droplet enzymatic assay to identify the potential inhibitors from a crude natural extract separated in the HPLC (Fig. 6a). First, the exit of the HPLC is coupled to the inlet II of the device shown in Fig. 6b. However, the flow coming from the HPLC is much higher than the flows from the enzyme and substrate; therefore, we have designed a purging channel connected to a syringe pump with withdrawal capabilities to remove the excess of fluid.

1. Modify the design of the microfluidic chip shown in Fig. 1a to include an extra port for draining the excess of liquid coming from the HPLC before the droplet formation zone [24]

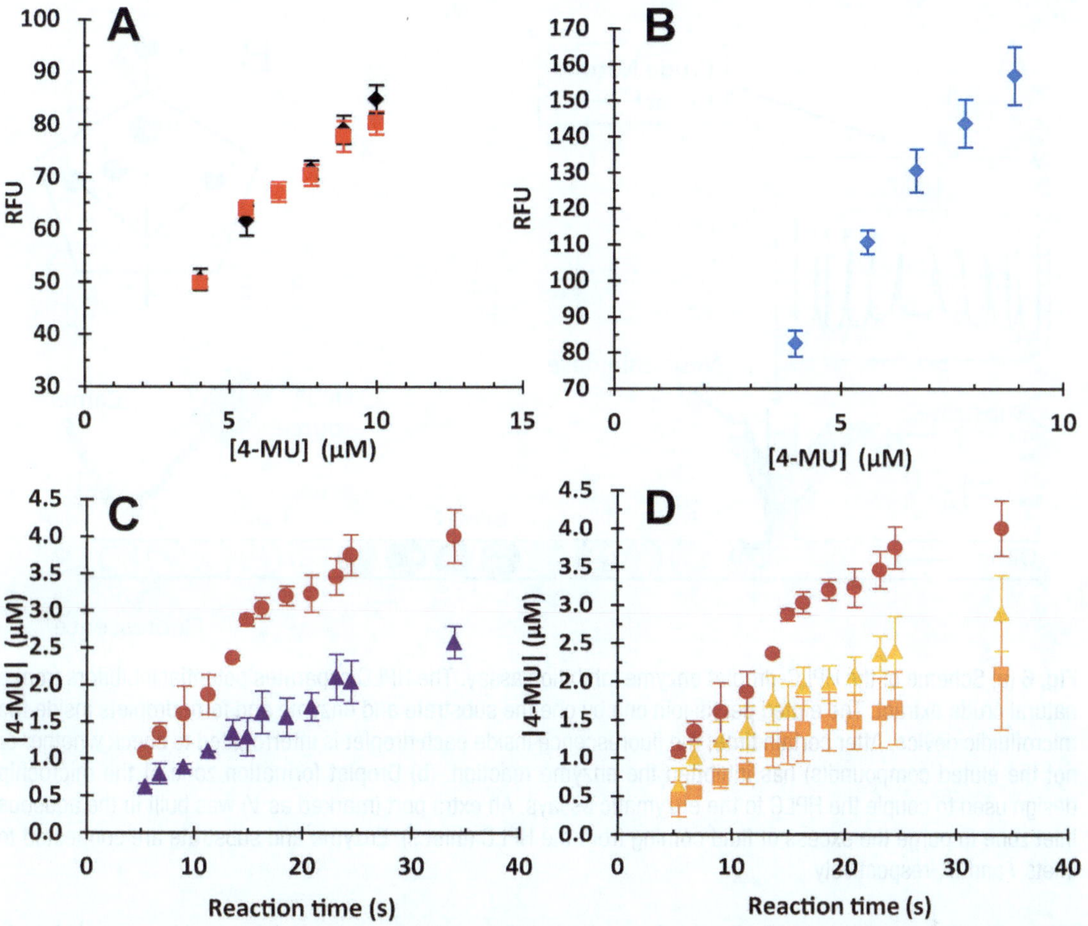

Fig. 5 (**a**) Calibration curve of RFU as function of 4-MU concentration inside the droplets. The curve was constructed by altering the flow rate in the buffer and 4-MU syringes. By changing the 4-MU concentration in the syringe, lower concentrations can be measured. Black diamonds represent the curve obtained with videos taken at 50 fps and 20,000 μs of exposure time with the Phantom camera. Red squares are the calibration curve obtained with videos taken at 5 fps and 20,000 μs of exposure time with the Phantom camera. (**b**) Calibration curve obtained with a Cannon camera at 24 fps and by altering the flow rate in the buffer and 4-MU syringes. (**c**) Time courses obtained in the droplet microfluidic device for the catalyzed hydrolysis of 4-MUS (6.6 μM) by *P. aeruginosa* aryl sulfatase enzyme (29 nM) (circles) and the same reaction but in the presence of pNPP (1.98 mM) as inhibitor (triangles). To obtain these results the reaction progress was measured at nodes 5, 7, and 12. (**d**) Time course obtained in the droplet microfluidic device for the catalyzed hydrolysis of 4-MUS (6.6 μM) by *P. aeruginosa* aryl sulfatase enzyme (29 nM) (circles). Also shown are the time courses in the presence of onion tea as inhibitor in concentration 1× and 2× (triangles and squares, respectively). The reaction progress was measured at nodes 5, 7, and 12

(Fig. 6b). Then, fabricate the PDMS-based microfluidic device (*see* **Note 29**).

2. Take 1 mL of water with a disposable 20 mL syringe. Cut PTFE tubing of appropriate length and connect it to the syringe needle with the help of tweezers.

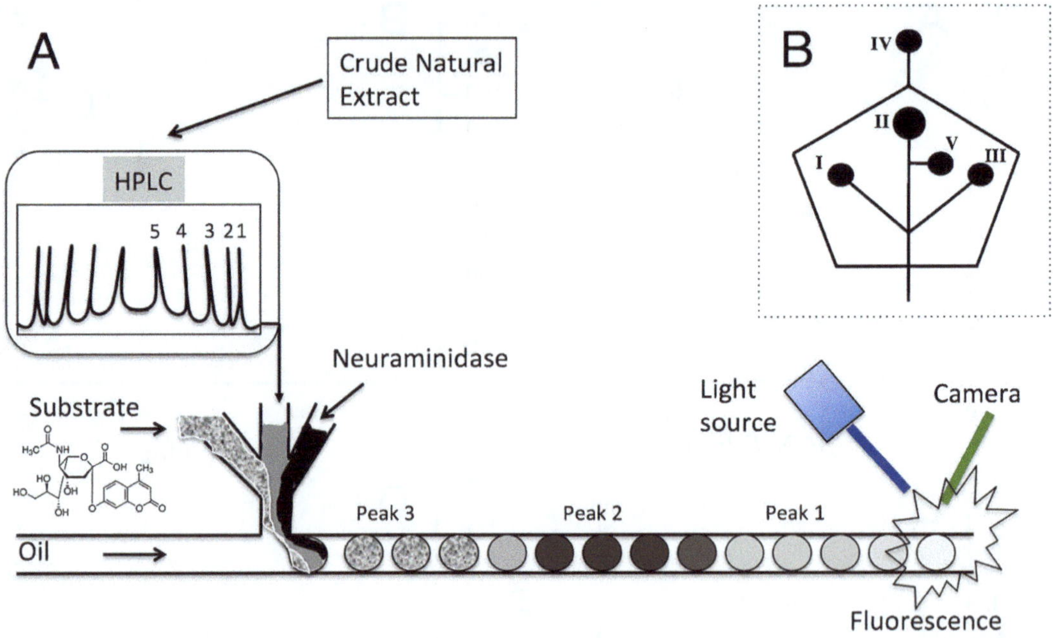

Fig. 6 (**a**) Scheme of the HPLC-droplet enzyme inhibition assay. The HPLC separates potential inhibitors from a natural crude extract. The eluted peaks join one by one the substrate and enzyme and form droplets inside the microfluidic device. After certain time, the fluorescence inside each droplet is interrogated to check whether or not the eluted compound(s) has inhibited the enzyme reaction. (**b**) Droplet formation zone of the microchip design used to couple the HPLC to the enzymatic assays. An extra port (marked as *V*) was built in the aqueous inlet zone to purge the excess of fluid coming from the HPLC (inlet *II*). Enzyme and substrate are connected to inlets *I* and *III*, respectively

3. Load the syringe on a syringe pump and purge it to expel the air in the tubing. Then set the withdraw mode in this pump but do not start it yet.
4. Load the syringes with the enzyme and substrate (*see* Subheading 3.8) into the syringe pumps and purge them to expel the air in the tubing.
5. Connect the free end of each tubing to the microfluidic chip inlets according to Fig. 6b.
6. Connect the end of tubing coming from the HPLC detector to the microchip according to Fig. 6b.
7. Pump the fluorinated oil into the chip at a flow of 250 μL/h.
8. Pump the substrate and enzyme aqueous phases at 60 μL/h and wait until droplet formation is stable.
9. Set the HPLC with an isocratic flow (0.1 mL/min) of 85% water and 15% methanol and an appropriate wavelength for detection (*see* **Note 30**).
10. Start to run the HPLC and immediately start the pump in withdrawal mode at a flow rate of 5940 μL/h. If it is necessary

to change the HPLC flow, adjust the pump flow in withdrawal mode such that the total HPLC flow upon joining with the other assays components is 60 μL/h.

11. Wait until droplet formation is stable and then make a blank injection of 20 μL.
12. Start recording a video of the flowing droplets at node five of the microchip using the *Intermittent high-speed image capture* (*see* Subheading 3.4.2). The video length should be at least equal to the elution program time. This experiment will create a base line for comparing the results obtained with the injection of a crude extract sample.
13. Analyze the video with MATLAB (Subheading 3.5) and identify and select the resulting fluorescence measurements at the center of the droplets (local maxima or peaks) using appropriate software (*e.g.* Wolfram Mathematica or MATLAB). Create a list of the selected fluorescence values with their corresponding times for each curve of the node five.
14. Repeat **steps 11–13** with a 20-μL injection of the crude natural extract.
15. Plot as a function of time the selected fluorescence values obtained in **step 13** from the blank injection. Use one graph for the data of each curve from node five (Fig. 7, blue line). On each graph, plot the fluorescence data corresponding to the crude extract injection (Fig. 7, red line) and the HPLC chromatogram of the crude extract (Fig. 7, black dashed line). Search for peaks with possible inhibitory activity by comparing both fluorescence signals. A decrease of the signal from droplets containing the crude extract with respect to the one with only the blank injection, strongly suggests enzymatic inhibition (Fig. 7).

4 Notes

1. HFE 7500 is the preferred choice of oil to perform droplet-based assay as it minimize the partition of organic molecules into the oil and reduces the exchange of compounds between droplets. In the presence of the biocompatible PFPE–PEG–PFPE surfactant, it permits to have very stable droplets. A second less favored option would be to use mineral oil with Span 80, but caution should be taken when the calibration curves and enzymatic assays are made as it is necessary to take into account the partition of organic molecules into the oil.
2. The droplet residence time can be estimated with the equation: $Q = vA$, where Q is the total flow (oil + water), A is the

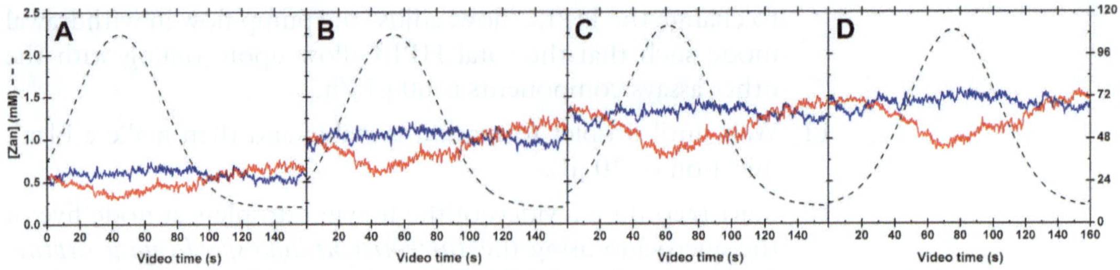

Fig. 7 Droplet neuraminidase enzyme inhibition assays with the inhibitor zanamivir injected and separated in a coupled HPLC. The capacity of zanamivir to inhibit *C. perfringens* neuraminidase was tested in microdroplets formed in a fractal-shape microchip coupled to the exit of an HPLC. A 20-μL injection of a 20 mg/mL solution of zanamivir was eluted from an HPLC column. As the liquid came out of the HPLC, it was introduced in the microfluidic device and mixed with the neuraminidase and its substrate 4-MUNANA in droplets. The concentration of zanamivir in the flow coming out of the HPLC followed the increasing and then decreasing concentration gradient observed in the chromatogram (dashed line in plots *A–D*, left axis). Therefore, the formed droplets contained a concentration gradient as well. The red lines in graphs *A–D* correspond to the maximum fluorescence signals of the enzyme reaction in droplets together with zanamivir following the HPLC concentration gradient. The measurements were made in curves "a" to "d," respectively, at node five of the microfluidic device through 160 s of the same video. The blue lines correspond to the maximum fluorescence signals of the enzyme reaction in droplets without zanamivir (blank injection). It can be observed that the red trace showed a smaller fluorescence than the blue line, indicating enzyme inhibition. Also, the fluorescence of the droplets containing zanamivir continuously decrease as droplets with higher inhibitor concentration reached the detection zone, and then, when the zanamivir content decreased again, the droplet fluorescence recovered to the baseline level of the reaction without inhibitor. Reprinted with permission from ref. 24 Copyright (2017) American Chemical Society

cross section channel, and v is the droplet speed. The latest can be expressed as $v = \frac{d}{t}$, where t is the droplet residence time and d is the distance traveled by a droplet in the microchip. Therefore, $t = \frac{d \cdot A}{Q}$ and consequently, if the droplet residence time in the microchip needs to be modified, it is possible to make changes in the total flow or in the channel dimensions.

3. The microfluidic channels can be designed with any computer-aided design (CAD) programs like AutoCAD or Inkscape. We have found that Adobe Illustrator is easier to use without any expert skills.
4. During the spin coating process (Subheading 3.1.2) the resist forms an uneven edge in the wafer, therefore it is recommended not to draw any design at 0.5 cm near the edge.
5. Using a commercial high-resolution printer it is possible to obtain small features up to 10–20 μm. Check availability with your local printing companies. The photomask can be also printed in a low-resolution printer with features up to 100–200 μm. However, some low-resolution printers can change considerably the channel dimensions. It is possible to make corrections by adjusting the dimensions in the computer

design according to the error of the printer, which can be calculated by characterizing the masks with a microscope.

6. The thickness of the SU-8 layer depends on the photoresist viscosity and the rotational speed. There are many SU-8 presentations with different viscosities, see http://www.microchem.com/Prod-SU8.htm. Choose the one that better matches the required thickness (height) of the microchannels. We use the SU-8 3000 series as they show better adherence to the silicon wafer than the SU-8 2000 series.
7. Storing the SU-8 photoresist at +4 °C will extend its shelf life. However, as the photoresist viscosity is sensitive to temperature, the bottle should be brought to room temperature at least 16 h before using it. The bottle should never be opened if it is cold, as this will condense water in the photoresist.
8. The recommended temperature by the SU-8 manufacturer is 95 °C, but we set the temperature ten degrees above as the aluminum foil lowers the temperature ~10 °C. The aluminum foil will keep your hot plate clean. This heating step eliminates the resin solvents so at the end of the baking time the resin must not flow.
9. Using SU-8 3050 a 100-μm-thick layer can be obtained in a single coating step. However, if only one SU-8 resin is available in the laboratory, it is possible to coat again over a deposited resin. We have successfully coated SU-8 at least three times to obtain features of up to 150-μm-height.
10. Set the distance between the wafer and the light bulb at 30 cm. Be sure to place the wafer in the center of the UV irradiance to ensure receiving the maximum light intensity. If more than one design is in the same photomask, it is better to expose only one design at a time and cover the others with a black paper.
11. Some people prefer to pour the mixture directly over the master mold supported on the Petri dish and introduce it to the desiccator with vacuum until all the air bubbles disappear. In this case it is advisable to rest the mixture ~15 min before applying vacuum to avoid spillage.
12. Some authors recommend cleaning the glass slides with piranha solution (mix of H_2SO_4 and H_2O_2). However, we have successfully used glass slides cleaned with ethanol or acetone followed by deionized water, drying with pressurized air and 1 h inside an oven at 70 °C.
13. The buffer and pH should be chosen accordingly with the specific enzyme assay.
14. Many articles report the usage of PFPE–PEG–PFPE surfactant at a concentration of 2.0% w/w, especially when droplets are going to be manipulated outside the microfluidic channel.

However, if droplets will be confined in microchannels, lower surfactant concentration of up to 0.1% w/w could be used with good droplets stability. PFPE–PEG–PFPE surfactant concentration of 2.0% w/w or more could promote molecular exchange between droplets by micellar activity [32].

15. Air bubbles in the tubing seriously affect the droplet formation, so purge the syringes until all bubbles come out of the tubing. Also, ensure the syringe pump is programed with the proper inner diameter of the syringe barrel.
16. A 45° cut might help to introduce the tubing. The PDMS replicas should be prepared with a thickness of at least 0.5 cm. If they are made thinner, the tubing will be prone to fall down.
17. Commonly, no more than 10 min is necessary to stabilize the droplet formation. This can be verified by observing the zone where the three aqueous channels intersect (Fig. 1c) using the fluorescence mode of the microscope. The flow of the 4-MU solution should be exactly in the middle channel and look free of strong vibrations. If this is not the case, check for any leakage in the device inlets. Many times, reintroducing the tubing into the microchip can eliminate these leakages.
18. Sometimes small PDMS debris or dust can partially block the channels and disrupt the droplets. They might be removed by pressing with tweezers the PDMS surface above the debris. Another option is to purge the line by releasing a high amount of oil and water using the purge button of the syringe pumps and then wait for droplet stabilization. If this does not work, replace the device.
19. Most digital cameras will record videos at 24 or 30 frames per second depending in the resolution and configuration. This information is important for image processing and it is possible to calculate it by dividing the total number of frames in a video (obtained with MATLAB as described in Section 3.5) by the total video time length.
20. We recommend the node five because four droplets with proportionally spaced residence times (15, 40, 65, and 90% of the total) can be observed in a single image. It is wise to measure the fluorescence at some other nodes to check for homogeneity in the device.
21. Reducing the exposure time permits to increase the frames per second. However, the camera detector will collect less fluorescent light, so, depending on the fluorophore concentration and other experimental conditions, verify the limits of detection.

22. Using the recommended flows in Subheading 3.2, the droplet frequency is ~8 droplets per second. Therefore, a 30 s video will record ~240 droplets.
23. Remember that the sampling rate of this method is low, so videos of longer duration are needed to have a good amount of droplet images.
24. The function *roipoly* in MATLAB allows drawing any closed figure in the analyzed images and the program returns the mean pixel intensity inside that figure.
25. For reliable results the ratio between the flows should not exceed 4:1.
26. It is recommended that the substrate concentration inside the droplets equals the K_m value of the enzyme for the best chance to identify compounds with competitive and uncompetitive inhibition behavior [31].
27. Use an enzyme concentration inside the droplets that would give ~15% of the reaction progress in the curve "d" of the node five at the center of microchip (249 mm from the droplet formation zone). With these conditions, in one static shot video, it is possible to calculate the initial rate with four points from measurements at curves "a," "b," "c," and "d." For certain experiments in which the complete progress curve of the enzyme-catalyzed reaction is needed, estimate the enzyme concentration so that the extent of reaction in the curve "d" of the node five is ~95–100%. Then, measure the product concentration at some other nodes. Fit the data to the integrated rate law of Michaelis-Menten equation to obtain the K_m and V_{max} values [33].
28. To test different inhibitor concentrations without changing the syringes content, fill two syringes as follows: Syringe one, a mixture of substrate at 1.5× its droplet concentration and inhibitor at 3× its droplet concentration, dissolved in the reaction buffer. Syringe two, substrate at 1.5× its droplet concentration in the reaction buffer. Then, change the flow rate of these two syringes and maintain the flow rate of the enzyme solution. This will produce droplets of variable inhibitor concentration while keeping constant the enzyme and substrate concentrations. Typically, a flow ratio of 4:1 between syringes one and two is the maximum difference before problems are caused as the liquid of the lower flow might not enter the chip at the set flow rate.
29. Use an adequate biopsy punch to make the hole in the HPLC inlet of the PDMS replica according to the external diameter of the HPLC exit tubing. If an UHPLC with a flow rate <100 μL/h is used this purging port might be avoided.

30. We successfully carried out the separation of an ethanolic natural crude extract under these conditions in a C-18 analytical column (Raptor ARC-18, 2.7 μm, 50 × 2.1 mm, Restek) [24]. Other isocratic or gradient elution can be used, but at more than 50% methanol the droplet formation is compromised.

Acknowledgments

The authors thank CONACyT (284240) and PAIP-UNAM (5000-9023) for financial support.

References

1. Shang L, Cheng Y, Zhao Y (2017) Emerging droplet microfluidics. Chem Rev 117:7964–8040. https://doi.org/10.1021/acs.chemrev.6b00848
2. Nie Z, Seo MS, Xu S et al (2008) Emulsification in a microfluidic flow-focusing device: effect of the viscosities of the liquids. Microfluid Nanofluid 5:585–594. https://doi.org/10.1007/s10404-008-0271-y
3. Umbanhowar PB, Prasad V, Weitz DA (2000) Monodisperse emulsion generation via drop break off in a coflowing stream. Langmuir 16:347–351. https://doi.org/10.1021/la990101e
4. Shim JU, Ranasinghe RT, Smith CA et al (2013) Ultrarapid generation of femtoliter microfluidic droplets for single-molecule-counting immunoassays. ACS Nano 7:5955–5964. https://doi.org/10.1021/nn401661d
5. Price AK, Paegel BM (2016) Discovery in droplets. Anal Chem 88:339–353. https://doi.org/10.1021/acs.analchem.5b04139
6. Damean N, Olguin LF, Hollfelder F et al (2009) Simultaneous measurement of reactions in microdroplets filled by concentration gradients. Lab Chip 9:1707–1713. https://doi.org/10.1039/b821021g
7. Gielen F, van Vliet L, Koprowski BT et al (2013) A fully unsupervised compartment-on-demand platform for precise nanoliter assays of time-dependent steady-state enzyme kinetics and inhibition. Anal Chem 85:4761–4769. https://doi.org/10.1021/ac400480z
8. Sjostrom SL, Joensson HN, Svahn HA (2013) Multiplex analysis of enzyme kinetics and inhibition by droplet microfluidics using picoinjectors. Lab Chip 13:1754–1761. https://doi.org/10.1039/c3lc41398e
9. Hess D, Rane A, Demello AJ, Stavrakis S (2015) High-throughput, quantitative enzyme kinetic analysis in microdroplets using stroboscopic epifluorescence imaging. Anal Chem 87:4965–4972. https://doi.org/10.1021/acs.analchem.5b00766
10. Zhu Y, Fang Q (2013) Analytical detection techniques for droplet microfluidics—a review. Anal Chim Acta 787:24–35. https://doi.org/10.1016/j.aca.2013.04.064
11. Han Z, Li W, Huang Y, Zheng B (2009) Measuring rapid enzymatic kinetics by electrochemical method in droplet-based microfluidic devices with pneumatic valves. Anal Chem 81:5840–5845. https://doi.org/10.1021/ac900811y
12. Sun S, Kennedy RRT (2014) Droplet electrospray ionization mass spectrometry for high throughput screening for enzyme inhibitors. Anal Chem 86:9309–9314. https://doi.org/10.1021/ac502542z
13. Vazquez B, Qureshi N, Oropeza-Ramos L, Olguin LF (2014) Effect of velocity on microdroplet fluorescence quantified by laser-induced fluorescence. Lab Chip 14:3550–3555. https://doi.org/10.1039/c4lc00654b
14. Dressler OJ, Casadevall i, Solvas X, deMello AJ (2017) Chemical and biological dynamics using droplet-based microfluidics. Annu Rev Anal Chem 10:1–24. https://doi.org/10.1146/annurev-anchem-061516-045219
15. Mao Z, Guo F, Xie Y et al (2015) Label-free measurements of reaction kinetics using a droplet-based optofluidic device. J Lab Autom 20:17–24. https://doi.org/10.1177/2211068214549625

16. Baccouche A, Okumura S, Sieskind R et al (2017) Massively parallel and multiparameter titration of biochemical assays with droplet microfluidics. Nat Protoc 12:1912–1932. https://doi.org/10.1038/nprot.2017.092
17. Cai L-F, Zhu Y, Du G-S, Fang Q (2012) Droplet-based microfluidic flow injection system with large-scale concentration gradient by a single nanoliter-scale injection for enzyme inhibition assay. Anal Chem 84:446–452. https://doi.org/10.1021/ac2029198
18. Miller OJ, El A, Mangeat T et al (2012) High-resolution dose–response screening using droplet-based microfluidics. Proc Natl Acad Sci 109:378–383. https://doi.org/10.1073/pnas.1113324109
19. Gu S, Lu Y, Ding Y et al (2013) Droplet-based microfluidics for dose-response assay of enzyme inhibitors by electrochemical method. Anal Chim Acta 796:68–74. https://doi.org/10.1016/j.aca.2013.08.016
20. Wang X-L, Zhu Y, Fang Q (2014) Coupling liquid chromatography/mass spectrometry detection with microfluidic droplet array for label-free enzyme inhibition assay. Analyst 139:191–197. https://doi.org/10.1039/c3an01917a
21. Price AK, MacConnell AB, Paegel BM (2016) HνSABR: photochemical dose-response bead screening in droplets. Anal Chem 88:2904–2911. https://doi.org/10.1021/acs.analchem.5b04811
22. Theberge AB, Whyte G, Huck WTS (2010) Generation of picoliter droplets with defined contents and concentration gradients from the separation of chemical mixtures. Anal Chem 82:3449–3453. https://doi.org/10.1021/ac1005316
23. Keyon ASA, Guijt RM, Bolch CJ, Breadmore MC (2014) Droplet microfluidics for postcolumn reactions in capillary electrophoresis. Anal Chem 86:11811–11818. https://doi.org/10.1021/ac5033963
24. Ochoa A, Álvarez-Bohórquez E, Castillero E, Olguin LF (2017) Detection of enzyme inhibitors in crude natural extracts using droplet-based microfluidics coupled to HPLC. Anal Chem 89:4889–4896. https://doi.org/10.1021/acs.analchem.6b04988
25. Holtze C, Rowat AC, Agresti JJ et al (2008) Biocompatible surfactants for water-in-fluorocarbon emulsions. Lab Chip 8:1632–1639. https://doi.org/10.1039/b806706f
26. Zhu S, Edmonds WF, Hillmyer MA, Lodge TP (2005) Synthesis and self-assembly of highly incompatible polybutadiene- poly(hexafluoropropylene oxide) diblock copolymers. J Polym Sci B Polym Phys 43:3685–3694. https://doi.org/10.1002/polb.20621
27. Fischlechner M, Schaerli Y, Mohamed MF et al (2014) Evolution of enzyme catalysts caged in biomimetic gel-shell beads. Nat Chem 6:791–796. https://doi.org/10.1038/nchem.1996
28. Hirata K, Ichii T, Suzuki H et al (2012) Fractal-shaped microchannel design for a kinetic analysis of biochemical reaction in a delay line. Microfluid Nanofluid 13:273–278. https://doi.org/10.1007/s10404-012-0958-y
29. Haubert K, Drier T, Beebe D (2006) PDMS bonding by means of a portable, low-cost corona system. Lab Chip 6:1548–1549. https://doi.org/10.1039/b610567j
30. Ferraro D, Serra M, Filippi D et al (2019) Controlling the distance of highly confined droplets in a capillary by interfacial tension for merging on-demand. Lab Chip 9:136–146. https://doi.org/10.1039/C8LC01182F
31. Copeland RA (2005) Evaluation of enzyme inhibitors in drug discovery. Wiley, Hoboken
32. Skhiri Y, Gruner P, Semin B et al (2012) Dynamics of molecular transport by surfactants in emulsions. Soft Matter 8:10618–10627. https://doi.org/10.1039/c2sm25934f
33. Exnowitz F, Meyer B, Hackl T (2012) NMR for direct determination of Km and Vmax of enzyme reactions based on the Lambert W function-analysis of progress curves. Biochim Biophys Acta Proteins Proteomics 1824:443–449. https://doi.org/10.1016/j.bbapap.2011.10.011

Chapter 15

Ligand Fishing: An Approach for the Discovery of Inhibitors from Complex Biological Mixtures

Evangelia G. Chronopoulou, Christina Varotsou, Nikolaos Georgakis, Georgios Premetis, Elisavet Ioannou, and Nikolaos E. Labrou

Abstract

Ligand fishing is a convenient bioanalytical screening method that is based on the affinity selection of a ligand from a complex biological sample by an immobilized receptor. It is a versatile affinity-based screening approach and it has found application in multiple interacting pairs such as enzyme-inhibitor/activator, antigen-antibody, receptor-ligand, and protein-protein. Important parameters that affect the successful operation of the method are the high specificity and strong binding affinity of the interacting pair (e.g., enzyme-ligand complex) and the elution of the bound ligand from the complex. This chapter provides protocols for the synthesis of affinity adsorbent and its application in off-line ligand-fishing procedure for a 6His-tagged glutathione transferase (GST).

Key words Metal affinity chromatography, Screening active compounds, Bioanalytical screening technique, GST enzyme natural inhibitors

1 Introduction

Ligand fishing is an efficient approach for the discovery of ligands toward an immobilized enzyme or protein from complex biological mixtures (Table 1) based on the principle of affinity chromatography [1–4]. The method can be performed in an online and off-line mode, depending on whether the separation and analysis are performed at the same or different time, respectively [5]. Several affinity-based techniques have been used for ligand fishing such as centrifugation, ultrafiltration, equilibrium dialysis, microdialysis, magnetic beads, and affinity chromatography [2, 5–8]. His-tag epitope proteins meet very high affinity with the immobilized metal ion, making them a stable "bait" for selecting bioactive compounds, proteins, or peptides [9].

Ligand fishing using immobilized enzyme have gain significant attention as an easy and effective screening method [10–12] for trapping enzyme inhibitors or activators. Human GSTs are targets

Nikolaos E. Labrou (ed.), *Targeting Enzymes for Pharmaceutical Development: Methods and Protocols*, Methods in Molecular Biology, vol. 2089, https://doi.org/10.1007/978-1-0716-0163-1_15,

Table 1
Advantages of ligand-fishing approach

Advantages
Screening and identifying bioactive compounds or biomolecules in one experiment
Identify protein function
Cost-effectiveness
Can be performed in vitro and in vivo
Easily applicable in a lot of pipelines such as in drug discovery
Specificity, efficiency, and less sample preparation

for the design of anticancer drugs since they participate directly or indirectly in the development of cancer cell drug resistance phenomenon [1, 13]. Therefore, GST inhibitors are useful to sensitize drug-resistant tumors.

This chapter describes protocols for the synthesis of immobilized metal affinity adsorbent and its application as matrix for ligand fishing of a 6His-tagged GST (Figs. 1 and 2).

2 Materials

2.1 Synthesis of Metal Affinity Absorbent

1. Agarose-based support (e.g., Sepharose CL 6B, Sigma-Aldrich).
2. NaOH solution (1 M).
3. Epichlorohydrin (purity 99%).
4. Iminodiacetic acid.
5. Bicarbonate buffer ($NaHCO_3$) (0.1 M, pH 9).
6. Aqueous ethanol solution (20%, v/v).
7. Vacuum pump and Gooch crucible.

2.2 Dialysis of 6His-tagged Purified GST

1. 6His-tagged GST, previously purified by affinity chromatography [1]. The 6His epitope can be located at the C-terminal or N-terminal. Instead of GST, other 6His-tagged protein or enzyme can be used.
2. Dialysis buffer: sodium phosphate buffer (50 mM, pH 8), containing 300 mM NaCl and 10 mM imidazole.
3. Dialysis cellulose membrane tubes, with typical molecular weight cutoff of 14 kDa and plastic clips.

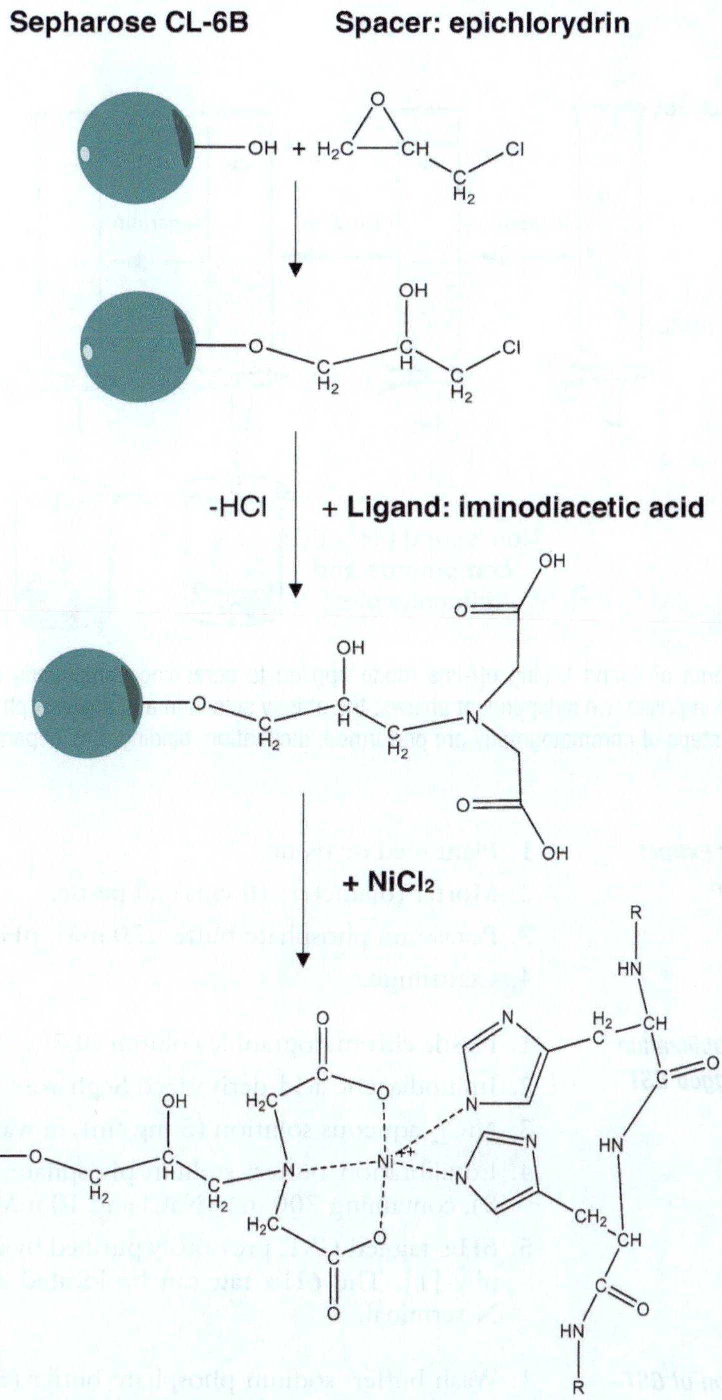

Fig. 1 Synthesis of immobilized metal affinity adsorbent. Epichlorohydrin is the spacer for the binding of iminodiacetic acid, a common chelating ligand for metal immobilization

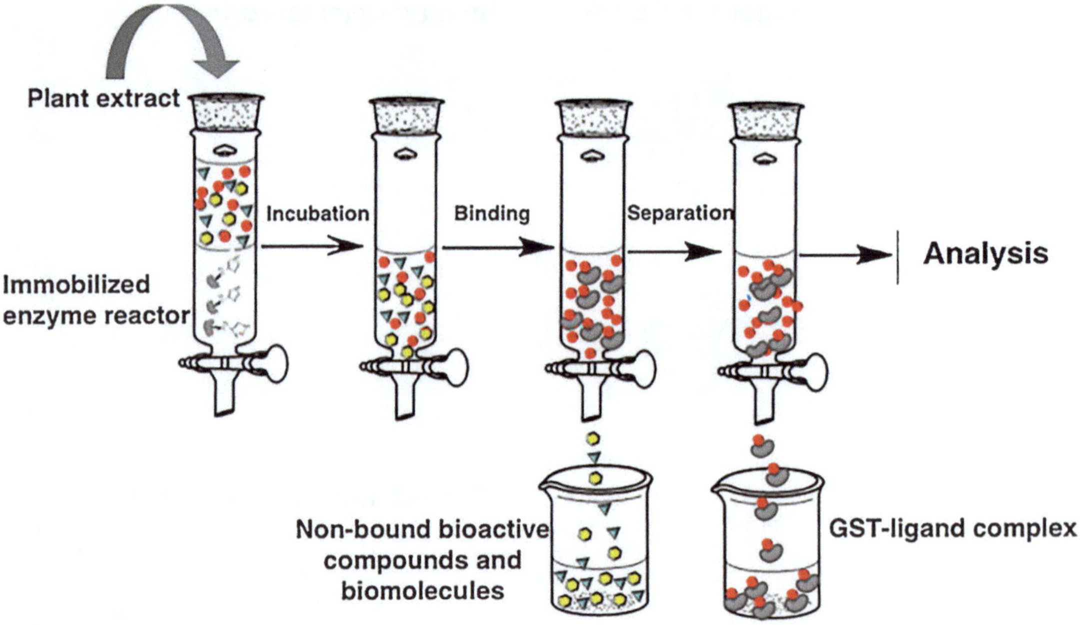

Fig. 2 Scheme of ligand fishing off-line mode applied to screening compounds from plant extracts. The procedure comprises two independent phases: the affinity selection and analysis phase. In the former phase, three main steps of chromatography are performed: incubation, binding, and separation

2.3 Plant Extract Preparation

1. Plant seed or tissue.
2. Mortar (diameter: 10 cm) and pestle.
3. Potassium phosphate buffer (20 mM, pH 7).
4. Centrifuge.

2.4 Immobilization of 6His-tagged GST

1. Plastic chromatographic column (0.5 × 5 cm).
2. Iminodiacetic acid-derivatized Sepharose CL 6B.
3. $NiCl_2$ aqueous solution (5 mg/mL in water).
4. Equilibration buffer: sodium phosphate buffer (50 mM, pH 8), containing 300 mM NaCl and 10 mM imidazole.
5. 6His-tagged GST, previously purified by affinity chromatography [1]. The 6His tag can be located at the C-terminal or N-terminal.

2.5 Elution of GST–Ligand Complex

1. Wash buffer: sodium phosphate buffer (50 mM, pH 8), containing 300 mM NaCl.
2. Elution buffers: sodium phosphate buffer (50 mM, pH 8), containing 300 mM NaCl and different concentrations of imidazole (50, 100, 150, 200, 250, and 300 mM, pH 8).

2.6 Regeneration and Storage of Metal-Ligand Adsorbents

1. EDTA aqueous solution (0.1 M).
2. Aqueous ethanol solution (20%, v/v).

2.7 Assay for Measuring GST Activity

1. Potassium phosphate buffer (0.1 M, pH 6.5).
2. Glutathione (GSH, 75 mM). Weigh 46.09 mg of GSH and dissolve in 2 mL double-distilled water. Store at 4 °C.
3. 1-Chloro-2,4-dinitrobenzene (CDNB, 30 mM). Weigh 12.15 mg of CDNB and dissolve in 2 mL ethanol. Store at 4 °C.

3 Methods

Analytical grade chemicals and double-distilled water were used to prepare the buffers for the synthesis of metal affinity absorbent, enzyme immobilization, and ligand binding. All buffers were stored at 4 °C.

3.1 Synthesis of Metal Affinity Absorbent

The iminodiacetic acid is coupled to Sepharose via epichlorohydrin (Fig. 1) [14].

1. Wash Sepharose (10 g) with 100 mL ddH_2O and filter with Gooch crucible in vacuum pump.
2. Weigh the dried Sepharose and add the appropriate volume of 1 M NaOH [the mass/volume ratio (w/v) should be equal to 1].
3. Leave the suspension under shaking overnight at room temperature.
4. Add 1 mL epichlorohydrin. Agitate the mixture for 6 h at room temperature.
5. Wash the epichlorohydrin-derivatized Sepharose with water (100 mL) for removing unreacted epichlorohydrin. Filter the Sepharose with Gooch crucible in vacuum pump. Keep the dried Sepharose in a glass vial.
6. Add iminodiacetic acid (1 g of iminodiacetic acid diluted in 3 mL of 0.1 M $NaHCO_3$ buffer, pH 9). Leave the suspension shaking at room temperature for 42 h (*see* **Note 1**).
7. After completion of the reaction, wash the Sepharose with ddH_2O (100 mL) for removing unreacted iminodiacetic acid and filter with Gooch crucible in vacuum pump.
8. Store the adsorbent in a glass vial in aqueous ethanol solution (20%, v/v) at 4 °C.

3.2 Dialysis of 6His-tagged Purified GST

1. Transfer the 6His-tagged GST in dialysis membrane.
2. Leave the dialysis membrane under stirring in dialysis buffer for at least 4 h at 4 °C.

3.3 Preparation of Plant Extract

1. Weigh 1 g of plant tissue and transfer it to the mortar with 3 mL of 20 mM phosphate buffer, pH 7 [the ratio of plant tissue and buffer should be equal to 1/3 (w/v)]. Crush the plant tissue in the mortar with the pestle. Squeeze the homogenate through cheesecloth, and collect the extract in a beaker.
2. Clarify the extract by centrifugation (14,000 × *g*, 15 min).
3. Collect the supernatant and clarify by filtration through a cellulose filter (0.45 μm pore size).

3.4 Immobilization of 6His-tagged GST

1. Filter the dialyzed enzyme sample through 0.4 μm pore-sized filter or centrifuge to remove any insoluble material.
2. Wash the iminodiacetic acid-Sepharose with 10 bed vol. of ddH_2O and then add 5 bed vol. of $NiCl_2$ aqueous solution (5 mg/mL) (*see* **Note 2**).
3. Continue by washing the adsorbent with 10 bed vol. of water and equilibration buffer (*see* **Note 3**).
4. Load 0.5–5 mL of the purified 6His-tagged GST (*see* **Note 4**) to the column at a linear flow rate of 10–20 cm/mL. Collect the unbound fraction (F1).
5. Assay for enzyme activity and for total protein the F1 fraction.

3.5 Elution of GST-Ligand Complex with Imidazole

1. Apply the plant extract (1–3 bed vol. plant extract) to the affinity adsorbent. Collect the unbound fraction (F2).
2. Wash the column with 10 bed vol. of equilibration buffer and 6 bed vol. of wash buffer. Collect in one fraction (F3).
3. Elute the bound enzyme with 3 bed vol. of elution buffer (*see* **Note 5**). Collect the eluted fractions in separate tubes (F_{50}, F_{100}, F_{150}, F_{200}, F_{250}, and F_{350}).
4. Assay the eluted fractions for enzyme activity and for total protein. Verify the eluted enzyme through enzymatic assay or SDS-PAGE gel.
5. Extract the bound ligands by suitable organic solvent [4]. Proceed to HPLC-MS to identify bound ligands [5].

3.6 Regeneration and Storage of Metal-Ligand Adsorbents

Metal affinity adsorbents may be effectively regenerated by applying 10 bed vol. of EDTA (0.1 M). After regeneration, wash the column with 10 bed vol. of water and add 1 bed vol. of 20% (v/v) aqueous ethanol solution for the storage at 4 °C.

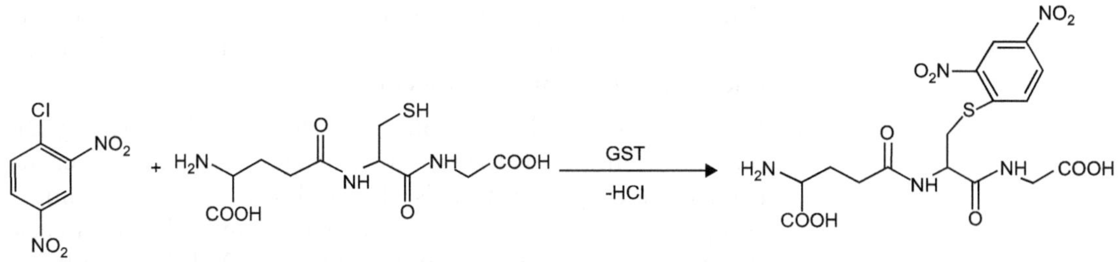

Fig. 3 The conjugation reaction between 1-chloro-2,4-dinitrobenzene (CDNB) and glutathione (GSH), catalyzed by GST

3.7 Continuous Assay for Measuring GST Activity

Determination of enzyme activities can be carried out by measuring the increase in absorbance at 340 nm due to the conjugation of CDNB with GSH (Fig. 3).

1. Add the buffer in reference and sample cuvette and then add the substrates (GSH and CDNB) in final concentrations 2.5 and 1 mM, respectively.
2. Add enzyme sample and mix by inversion two times. Monitor the absorbance at 340 nm continuously for 2 min (*see* **Note 6**).
3. Calculate the enzymatic activity (U/mL) of the loaded enzyme and F1 fraction in order to define the amount of enzyme (units) that was immobilized.
4. Measure the GST activity in every fraction (*see* **Note 7**).

4 Notes

1. IDA is a small size, hydrophilic tridentate chelator that can be complexed with divalent metals (Cu^{2+}, Ni^{2+}, Zn^{2+}, and Co^{2+}), keeping metal coordination sites available to be occupied by suitable electron-donor groups from protein [15–17]. Similarly, other chelating ligands can be used in metal affinity chromatography like nitrilotriacetic acid (NTA), carboxymethylated-aspartic acid (CM-Asp), 8-hydroxyquinoline, ortho-phosphoserine, and *N,N,N'* tris(carboxymethyl) ethylenediamine [18, 19]. Columns containing IDA or NTA are used for the purification of histidine-tagged and phosphorylated proteins [20].
2. Histidine-containing proteins can also bind strong in Cu^{2+}-IDA; however, Ni^{2+} offers higher selectivity [17].
3. Sodium chloride (up to 500 mM) and low levels of imidazole (up to 20 mM) in the equilibration buffer reduce the number of proteins that bind nonspecifically to the resin [21, 22].

4. Affinity immobilization of enzymes is reversible and oriented, allowing the immobilized enzyme to adopt the correct folding, leading to better analyte selection [16].
5. Proteins can be eluted from the adsorbent by altering the buffer conditions, as for example: by changing salt concentration, by adding Lewis acids (H^+ and Zn^{2+}), or by adding a competitive agent that has a similar structure to the amino acid that participates in metal binding [17, 21, 22]. Imidazole is a Lewis base with similar structure to His. It is preferred for the elution of 6His-tag proteins [17, 21]. The strength of the metal–protein bond is variable; therefore, gradient elution with buffer containing different concentration of imidazole should be carried out [22, 23].
6. Prepare the substrates fresh each day. Depending on the concentration of the GST in the sample, the assay should be achieved using different volumes of the eluted fraction; each time adjustment of the reaction buffer in cuvette should be taken into account, so the final volume should be retained equal to 1 mL.
7. The extinction coefficient of GSH-CDNB conjugate at 340 nm in 1 mL cuvette is 9.6 mM^{-1} cm^{-1}. Remember to multiply with the dilution factor where it is needed. In case that the ligand-GST complex does not display any activity, the analysis of the eluted fractions can be achieved by SDS-PAGE.

Acknowledgments

E.G.C. thanks IKY Scholarship Programs that is co-financed by Greece and the European Union (European Social Fund- ESF) through the Operational Programme «Human Resources Development, Education and Lifelong Learning» in the context of the project "Reinforcement of Postdoctoral Researchers" (MIS-5001552), implemented by the State Scholarships Foundation (IKY).

References

1. Chronopoulou EG, Ataya F, Labrou NE (2018) A microplate-based platform with immobilized human glutathione transferase A1-1 for high-throughput screening of plant-origin inhibitors. Curr Pharm Biotechnol 19:925–931
2. Wubshet SG, Brighente IM, Moaddel R et al (2015) Magnetic ligand fishing as a targeting tool for HPLC-HRMS-SPE-NMR: α-glucosidase inhibitory ligands and alkylresorcinol glycosides from Eugenia catharinae. J Nat Prod 78:2657–2665
3. Vanzolini KL, Ainsworth S, Bruyneel B et al (2018) Rapid ligand fishing for identification of acetylcholinesterase-binding peptides in snake venom reveals new properties of dendrotoxins. Toxicon 152:1–8
4. Dixon DP, Edwards R (2018) Protein-ligand fishing in planta for biologically active natural products using glutathione transferases. Front

Plant Sci 9:1659. https://doi.org/10.3389/fpls.2018.01659

5. Zhuo R, Liu H, Liu N et al (2016) Ligand fishing: a remarkable strategy for discovering bioactive compounds from complex mixture of natural products. Molecules 21:1516. https://doi.org/10.3390/molecules21111516
6. Wu GF, Jiang XL, Gong YZ et al (2019) Ligand fishing of anti-neurodegenerative components from *Lonicera japonica* using magnetic nanoparticles immobilised with monoamine oxidase B. J Sep Sci 42:1289–1298
7. Van Breemen RB, Nikolic D, Bolton JL (1998) Metabolic screening using on-line ultrafiltration mass spectrometry. Drug Metab Dispos 26:85–90
8. Li Y, Xu J, Chen Y et al (2015) Screening of inhibitors of glycogen synthase kinase-3beta from traditional Chinese medicines using enzyme-immobilized magnetic beads combined with high-performance liquid chromatography. J Chromatogr A 1425:8–16
9. Kool J, Jonker N, Irth H et al (2011) Studying protein–protein affinity and immobilized ligand–protein affinity interactions using MS-based methods. Anal Bioanal Chem 401:1109–1125
10. Asanomi Y, Yamaguchi H, Miyazaki M et al (2011) Enzyme-immobilized microfluidic process reactors. Molecules 16:6041–6059
11. Zhang Y, Shi S, Chen X et al (2014) Functionalized magnetic nanoparticles coupled with mass spectrometry for screening and identification of cyclooxygenase-1 inhibitors from natural products. J Chromatogr B 960:126–132
12. Tao Y, Zhang Y, Cheng Y et al (2013) Rapid screening and identification of alpha-glucosidase inhibitors from Mulberry leaves using enzyme-immobilized magnetic beads coupled with HPLC/MS and NMR. Biomed Chromatogr 27:148–155
13. Allocati N, Masulli M, Di Ilio C et al (2018) Glutathione transferases: substrates, inhibitors and pro-drugs in cancer and neurodegenerative diseases. Oncogene 7:8. https://doi.org/10.1038/s41389-017-0025-3
14. Kim HS, Hage DS (2006) Immobilization methods for affinity chromatography. In: Hage DS (ed) Handbook of affinity chromatography, 2nd edn. Taylor & Francis Group, Boca Raton
15. Gaberc-Porekar V, Menart V (2001) Perspectives of immobilized-metal affinity chromatography. J Biochem Biophys Methods 49:335–360
16. Ueda EK, Gout PW, Morganti L (2003) Current and prospective applications of metal ion-protein binding. J Chromatogr A 988:1–23
17. Arnold FH (1991) Metal-affinity separations: a new dimension in protein processing. Biotechnology (N Y) 9:151–156
18. Gutiérrez R, Martín del Valle EM, Galán MA (2007) Immobilized metal-ion affinity chromatography: status and trends. Sep Purif Rev 36:71e111. https://doi.org/10.1080/15422110601166007
19. Prasanna RR, Vijayalakshmi MA (2010) Immobilized metal-ion affinity systems for recovery and structure function studies of proteins at molecular, supramolecular, and cellular levels. Pure Appl Chem 82:39e55. https://doi.org/10.1016/j.jbiotec.2008.07.614
20. Hage DS, Anguizola JA, Li R et al (2013) Affinity chromatography. In: Liquid chromatography: applications. Elsevier Inc, pp 1–23. https://doi.org/10.1016/B978-0-12-415806-1.00001-2
21. Bornhorst JA, Falke JJ (2000) Purification of proteins using polyhistidine affinity tags. Methods Enzymol 326:245–254
22. Wong JW, Albright RL, Wang N-HL (1991) Immobilized metal ion affinity chromatography (IMAC) chemistry and bioseparation applications. Sep Purif Rev 20:49–106. https://doi.org/10.1080/03602549108021408
23. Kågedal L (1998) Immobilized metal ion affinity chromatography. In: Janson J-C, Ryden L (eds) Protein purification: principles, high-resolution methods, and applications, 2nd edn. Wiley-VCH, New York

Chapter 16

HMG-CoA Reductase as Target for Drug Development

Baskaran Gunasekaran and Mohd Yunus Shukor

Abstract

The main strategy for lowering blood cholesterol levels is through the inhibition of the NADPH-dependent HMG-CoA reductase (3-hydroxy-3-methyl-glutaryl-CoA reductase). The enzyme catalyses the reduction of HMG-CoA to mevalonate and this process is inhibited by statins that form the bulk of the therapeutic agents to treat high cholesterol since the 1970s. Newer drugs that are safer than statins are constantly being developed. The inhibition of candidate drugs to HMG-CoA reductase remains the mainstay of drug development research. The determination of the enzyme activity is important for the correct assessment of potency of the enzyme as well as determining the inhibition of potential therapeutic agents from the plant and microbial extracts. Also, this chapter covers the use of the popular four-parameter logistics model that can yield accurate estimation of the IC_{50} values of therapeutic agents and their 95% confidence intervals.

Key words HMG-CoA reductase, Drug development, Enzyme activity, IC_{50}, Four-parameter logistics model

1 Introduction

Atherosclerosis, which is caused by hypercholesterolemia, is a major cause of heart diseases such as myocardial infarction. An elevated level of plasma cholesterol, particularly LDL and triglyceride, are mainly responsible for hypercholesterolemia which can also lead to other diseases such as obesity, diabetes, and cancer [1–3]. HMG-CoA reductase is the rate-limiting enzyme in cholesterol biosynthesis that catalyzes the conversion of HMG-CoA to mevalonate. The inhibition of HMG-CoA reductase effectively lowers the level of cholesterol in humans and most animals by the activation of sterol-regulatory element-binding protein-2 (SREBP-2) which upregulates the HMG-CoA reductase and LDL receptor that lead to the reduction of cholesterol levels [4].

The most well-known HMG-CoA reductase inhibitors are statins. Nevertheless, long-term consumption of statins has been reported to lead to severe adverse effects such as muscle and liver damage, intracerebral hemorrhage (ICH), rhabdomyolysis,

Nikolaos E. Labrou (ed.), *Targeting Enzymes for Pharmaceutical Development: Methods and Protocols*, Methods in Molecular Biology, vol. 2089, https://doi.org/10.1007/978-1-0716-0163-1_16, © Springer Science+Business Media, LLC 2020

angioedema, and acute renal failure [5–7]. Due to these issues, more screening activities for HMG-CoA inhibitors using plants especially those that have been reported with medicinal values are being carried out with the hope of finding active compounds that exhibit minimal side effects [1, 8].

The reaction scheme for this enzyme is as follows:

$$\mathrm{HMG-CoA} + 2\mathrm{NADPH} + 2\mathrm{H}^+ \rightarrow \mathrm{mevalonate} + 2\mathrm{NADP}^+ + \mathrm{CoA-SH}$$

The enzyme assay is based on the spectrophotometric measurement of the decrease in absorbance at 340 nm, which represents the oxidation of NADPH by the catalytic subunit of HMGR in the presence of the substrate HMG-CoA (*see* **Note 1**). The K_m values for human HMG-CoA reductase using HMG-CoA and NADPH as substrates are determined to be 0.07 mM and 0.021 mM, respectively [9]. The substrate concentration range for conducting the assay should be at least five times the K_m values for the assay to be conducted under saturation conditions or V_{max}.

2 Materials

2.1 Equipment

1. UV–Visible spectrophotometer with thermostatically controlled cell holder (e.g., Beckman Coulter or Shimadzu).
2. Pipettes and repeat pipettor.

2.2 Reagents Supplies

1. 1 mL standard quartz cuvette.
2. Buffer A. Dissolve 2.4 g NaH_2PO_4 (molecular weight is 120) in about 180 mL of deionized water in a volumetric flask. Adjust to pH 7.43 at room temperature with 1 M NaOH or 1 M HCl. Make up the volume to 200 mL with deionized water. The pH will be 7.4 at 37 °C.
3. 0.1 M EDTA (molecular weight is 292.2) stock solution. Dissolve 292 mg of ethylenediaminetetraacetic acid in 5 mL of water. Add 1 M KOH until the pH reached 7.0 and topped up to 10 mL with deionized water (*see* **Note 2**).
4. BSA stock solution (1% w/v), dissolve 60 mg of BSA in 6 mL of deionized water (*see* **Note 3**).
5. KCL stock solution (molecular weight is 74.54) stock solution (120 mM). Dissolve 373 mg in 10 mL deionized water.
6. NADPH (tetrasodium salt, molecular weight is 833.4) stock solution (5 mM), dissolve 20.84 mg NADPH tetrasodium salt in 5 mL of a solution of 5 g $KHCO_3$ in 1 L water (50 mM). Aliquot 1 mL and store at −20 °C.

7. HMG-CoA (molecular weight is 911.661) stock solution (7.5 mM). Dissolve 13.67 mg HMG-CoA in 1 mL water and adjust the pH to 5.5 with 1 M HCl and dilute to 2 mL with water.
8. DTT (molecular weight is 154.253) stock solution (125 mM). Dissolve 192.8 mg in 10 mL of deionized water. Aliquot 1 mL and store at −20 °C.

3 Methods

3.1 HMG-CoA Reductase Enzyme Assay

The HMG-CoA reductase enzyme (Sigma-Aldrich, St. Louis, MO, USA) is used under conditions recommended by the manufacturer. The protein is the catalytic domain of the human enzyme expressed in E. coli in the form of a recombinant GST fusion protein. The concentration of the enzyme is between 0.5 and 0.7 mg protein/mL and the enzyme has specific activities between 2 and 8 units per mg depending on lot number.

1. In a quartz cuvette, add in sequence 730, 50, 40, 100, 50, and 20 μL of buffer A, EDTA, DTT, KCL, NADPH, and HGG-CoA, respectively, from the prepared stock solutions. The final concentrations of EDTA, DTT, KCL, NADPH, and HGG-CoA are 0.4, 0.4, 1.0, 5.0, and 120 mM, respectively.
2. Incubate at 37 °C for 2 min.
3. Initiate the enzymatic reaction (time 0) by adding 5 μL of HMG-CoA reductase and 5 μL of DMSO (control) and mix by carefully and slowly drawing and ejecting the reaction mixture so not to introduce air bubbles into the mixture with the aid of a pipette.
4. Monitor rate of NADPH oxidizes every 20 s for up to 300–720 s at 340 nm (*see* **Note 4**).

3.2 Data Analysis

1. As NADPH is slowly oxidized especially at pH below 7.0, subtraction of readings from the control (minus enzyme) is important.
2. The enzyme activity is calculated according to the following equation:

$$\text{HMG} - \text{CoA Reductase activity} = \left(\frac{\Delta 340\text{nm}/\min(\text{sample}) - \Delta 340\text{nm}/\min(\text{blank})}{(12.44)\times V \times P \times (1\ \text{cm})}\right)$$

where 12.44 = extinction coefficient at 340 nm for NADPH which is 6.22 $\text{mM}^{-1}\ \text{cm}^{-1}$. Since 2 molecules of NADPH are consumed, a double of the value of the extinction coefficient (12.44) is used instead V = volume of the enzyme used (5 μL or 0.005 mL)0.6 = median concentration of the enzyme used (between 0.5 and 0.7 mg protein/mL).

3. Example of calculation

"The oxidation of NADPH in the absence of enzyme after 10 min of incubation was 0.005. In the presence of HMG-CoA reductase, the absolute change of absorbance after 10 min was 0.45. The total reaction mixture was 1 mL, and the cuvette path length is 1 cm."

Thus, the specific activity of the enzyme is:

=(0.45 − 0.005)/(12.44 mM−1 cm−1 × 10 min × 0.005 mL × 0.6 mg/mL × 1 cm)

=1205.79 mM NADPH converted to NADP+/min/mg protein

=1205.79 mmol NADPH converted to NADP+/min/mg per liter assay mixture

=1205.79 μmol NADPH converted to NADP+/min/mg per liter assay mixture

In a 1 mL assay mixture

=1.206 μmol NADPH converted to NADP+/min/mg or 1.206 unit/mg enzymes (*see* **Note 5**).

3.3 HMG-CoA Reductase Enzyme Inhibition

Inhibitory effect of standard inhibitors such as statins or potential active compounds from plant or microbial extracts need to be dissolved in DMSO and tested at various concentrations in the reaction mixture.

1. The calculation for inhibitor activity in the form of percent inhibition is as follows:

$$\%\text{Inhibition} = \left(\frac{\text{Activity (enzyme)} - \text{activity (inhibitor)}}{\text{(Acivity (enzyme)}}\right)$$

2. Determination of IC_{50} value of inhibitors

The four-parameter logistics equation is normally used in fitting the nonlinear HMG-CoA inhibition curve based on the least square fitting [10, 11]. The equation is described as follows:

$$y = \text{Bottom} + \frac{(\text{Top} - \text{Bottom})}{1 + 10^{(\text{Log IC}_{50} - x) * \text{Hillslope}}}$$

where y represents the percent inhibition obtained, x is the concentration of inhibiting agent (log unit), IC_{50} is the concentration of inhibitor resulting in 50% inhibition of signal response, and the Hillslope is the slope-like parameter (Hill coefficient). Software that can be used to find the parameters of the four-parameter model include, for example, PRISM and SigmaPlot (*see* **Notes 6** and **7**).

4 Notes

1. The enzyme, once thawed should be stored on ice and best aliquoted and stored at −70 °C as prolonged storage on ice reduces the enzyme activity.
2. EDTA dissolved under alkaline conditions, and once dissolved should be neutralized.
3. BSA added into the reaction mixture acts as a stability reagent by becoming a sacrificial target for contaminating proteases and also prevents nonspecific binding of HMG-CoA reductase to the walls of the cuvette which can reduce the potency of the enzyme.
4. Plant, bacterial, or fungal extracts may contain active NADPH oxidase, which needs to be tested as a false positive control by eliminating the enzyme from the reaction mixture. Should this activity be detected, proper subtraction of the absorbance values is needed.
5. The assay is meant for a 1 mL reaction mixture, of which most lab will be able to conduct the assay. The assay can be modified to 0.2 mL for a microplate assay using quartz microplate and read on a microplate reader.
6. Nonlinear regression is the most preferred method to fit the nonlinear inhibition curve and finding the parameters and their confidence interval as linearization disrupts the error structure. Data points of more than seven are recommended.
7. The lower the number of points, the wider the confidence interval values for the parameters. Increasing the number of replicates has the opposite effect.

Acknowledgments

This work was supported by the Research University Grant GP-1PS/2013/9399800 entitled "Anti Hypercholesterolemia Properties of *Adatodha vasica*: In vitro and In vivo."

References

1. Salvamani S, Gunasekaran B, Shaharuddin NA et al (2014) Antiartherosclerotic effects of plant flavonoids. Biomed Res Int 2014Article ID 480258, 11 p. https://doi.org/10.1155/2014/480258
2. Baskaran G, Salvamani S, Azlan A et al (2015) Hypocholesterolemic and antiatherosclerotic potential of *Basella alba* leaf extract in hypercholesterolemia-induced rabbits. Evid Based Complement Alternat Med 2015:Article ID 751714, 7 p. https://doi.org/10.1155/2015/751714
3. Whayne TF, Saha SP (2019) Genetic risk, adherence to a healthy lifestyle, and ischemic heart disease. Curr Cardiol Rep 21:1. https://doi.org/10.1007/s11886 019 1086 z
4. Miserez AR, Muller PY, Barella L et al (2002) Sterol-regulatory element-binding protein

(SREBP)-2 contributes to polygenic hypercholesterolaemia. Atherosclerosis 164:15–26. https://doi.org/10.1016/S0021-9150(01)00762-6

5. Maron DJ, Fazio S, Linton MF (2000) Current perspectives on statins. Circulation 101:207–213. https://doi.org/10.1161/01.CIR.101.2.207
6. Naz S, Saleem MW, Haider AW (2018) Angioedema; an unreported adverse effect of pitavastatin. J Ayub Med Coll Abbottabad 30:603–604
7. Jung M, Lee S (2018) Effects of statin therapy on the risk of intracerebral hemorrhage in Korean patients with hyperlipidemia. J Human Pharmacol Drug Ther 39:129–139. https://doi.org/10.1002/phar.2211
8. Salvamani S, Gunasekaran B, Shukor MY, et al (2016) Anti-HMG-CoA reductase, antioxidant, and anti-inflammatory activities of *Amaranthus viridis* leaf extract as a potential treatment for hypercholesterolemia, Evid Based Complement Alternat Med .2016:Article ID 8090841, 10 p. https://doi.org/10.1155/2016/8090841
9. Carbonell T, Freire E (2005) Binding thermodynamics of statins to HMG-CoA reductase. Biochemistry 44:11741–11748. https://doi.org/10.1021/bi050905v
10. Motulsky HJ, Ransnas LA (1987) Fitting curves to data using nonlinear regression: a practical and nonmathematical review. FASEB J 1:365–374. https://doi.org/10.1096/fasebj.1.5.3315805
11. Corsini A, Maggi FM, Catapano AL (1995) Pharmacology of competitive inhibitors of HMG-CoA reductase. Pharmacol Res 31:9–27. https://doi.org/10.1016/1043-6618(95)80042-5

Chapter 17

Lipoxygenases as Targets for Drug Development

Xiao-Yuan Mao

Abstract

Lipoxygenases are key enzymes that catalyze the polyunsaturated fatty acids such as arachidic acid, linoleic acid (LA), and others unsaturated fatty acids. They are involved in important functions such as cell structure, metabolism, and signal transduction mechanisms, finally mediating cell death process, especially ferroptosis, a novel type of cell death modality. Our present protocol described a colorimetric assay for measuring lipoxygenase activity as well as a high-performance liquid chromatography/electrospray ionization tandem mass spectrometry method for the quantification of arachidonic acid metabolites.

Key words Lipoxygenases, Arachidic acid, Leukotrienes, Polyunsaturated fatty acids

1 Introduction

Lipoxygenases (LOXs) are key enzymes that catalyze the polyunsaturated fatty acids (PUFAs) such as arachidic acid (AA), linoleic acid (LA), and other unsaturated fatty acids [1]. LOXs are expressed in immune, neural, epithelial, tumor cells, and other cells [2]. By converting the PUFAs into bioactive metabolites, the enzymes can affect the cell structure, metabolism, and signal transduction, finally mediating cell death process, especially ferroptosis, a novel type of cell death modality [3, 4]. LOXs can be found in both animals and plants. Humans have six ALOX genes while the mouse has seven different ALOX genes. LOXs are named based on the numbered carbon where the LOXs oxygenated their PUFA's substrates, for example, 5-LOX [5]. The activation of LOX enzymes requires either molecular oxygen or lipid hydroperoxides to activate the inactive ferrous form to an active ferric form [6]. Moreover, its activation is also regulated by the N-terminal beta-barrel region of polypeptides, where this region has a similar amino acid sequence to the C2-like domain; therefore, Ca^{2+}-mediated activation via interaction with the plasma membrane has been proposed [7]. Activation of LOX enzymes displays different physiological functions, including inflammation, erythropoiesis,

Nikolaos E. Labrou (ed.), *Targeting Enzymes for Pharmaceutical Development: Methods and Protocols*, Methods in Molecular Biology, vol. 2089, https://doi.org/10.1007/978-1-0716-0163-1_17,

epidermal differentiation and skin development, cell proliferation and carcinogenesis, tumorigenesis, and so on [8].

5-LOX is a member of LOXs family, whose family members also include 12-LOX and 15-LOX. 5-LOX is expressed in the cardiovascular system, endothelial cells, and the entire nervous system including neurons and glial cells [9]. The 5-LOX, by inserting molecular oxygen into carbon 5, catalyzes the conversion of arachidonic acid to 5-hydroxy-peroxy-eicosatetrenoic acid (5-HPETE), which can subsequently be metabolized to 5-hydroxy-eicosatetrenoic acid (5-HETE) as well as different leukotrienes such as leukotriene A4 and leukotriene B4 [10]. Since leukotrienes are potent pro-inflammatory molecules, much work has investigated the role that 5-LOX plays in Alzheimer's disease, atherosclerosis, asthma, and cancer [11].

12/15-LOX is a member of LOXs family, which is widely distributed in different organizations. 12-LOX can be divided into three types, platelet-type 12-lipoxygenase (p12LOX), leukocyte-type 12-lipoxygenase (l12LOX), and epidermis-type 12-lipoxygenase (e12LOX) [12]. l12LOX deriving from rats, mouse, cattle, and pigs is highly homologous to human 15-LOX-1 since both enzymes catalyze AA to produce 12-hydroxy dioctatetraenoic acid (12-HETE) and 15-hydroxy eicosatetraenoic acid (15-HETE) [13]. That is how 12/15-LOX are named and its' metabolites have stereoisomer nomenclature S and R. The lipid metabolites produced by 12/15-LOX enzymes reacting with PUFAs are involved in a variety of diseases including metabolic, vascular, and neurological diseases [14].

2 Materials

2.1 Purification and Colorimetric Assay for LOX Activity

1. Ni-NTA (nickel-nitrilotriacetic acid) affinity adsorbent packet in a column.
2. Lysis buffer: 50 mM Tris (pH 8.0), 500 mM NaCl, 20 mM imidazole, and 100 μM phenylmethanesulfonyl fluoride (PMSF).
3. Pre-equilibration buffer for Ni-NTA column: 50 mM Tris (pH 8.0), 500 mM NaCl, and 20 mM imidazole.
4. Wash buffer: 50 mM Tris-HCl (pH 8.0), 500 mM NaCl, and 50 mM imidazole.
5. Elution buffer: 50 mM Tris-HCl (pH 8.0), 500 mM NaCl, and 200 mM imidazole.
6. Disposable ultrafiltration unit (Amicon).
7. Assay buffer: 50 mM Tris–HCl, adjust the pH to 7.5.
8. Prepare 0.1 N KOH aqueous solution.

9. Prepare 0.2 M HCl aqueous solution.
10. Ferrithiocyanate (FTC) reagents: reagent I (4.5 mM $FeSO_4$ dissolved in 0.2 M HCl) and reagent II (3% w/v NH_4SCN, dissolved in methanol). Mix equivalent volumes of reagent I and reagent II (*see* **Note 1**).
11. Arachidonic acid solution for enzyme assay: Prepare arachidonic acid in ethanol (40 mM). Mix 20 μL of diluted arachidonic acid and 20 μL of 0.1 N KOH. Add 360 μL of deionized water to the mixture (*see* **Note 2**).

2.2 LC/MS/MS for the Detection of Metabolites

1. Prepare 1 N citric acid solution.
2. Solution against free radical peroxidation: add 40 mL of 1 N citric acid and 5 mL of butylated hydroxytoluene (BHT) (1% v/v) to samples.

2.3 Extraction of Metabolites

1. PGE2-d4 (Cayman Chemical, no. 314010-500). Adjust the concentration to 100 ng/mL. Stable at least 1 year (stocks should be stored at −20 °C).
2. Extracting solution for samples: mix hexane and ethyl acetate with equal capacity.
3. Buffer for reconstituting samples: mix methanol and 2 mM ammonium (ratio of 70:30). Adjust the pH to 8.0.
4. YMC ODS-AQ 2.0 3100-mm column (S3 mm 120 Å, Waters Corp., Marlborough, MA).
5. Mobile phases: 2 mM ammonium acetate in methanol. Adjust the pH to 8.0.
6. An LC/MS/MS system (e.g., Quattro Ultima tandem mass spectrometer (Micromass, Beverly, MA) equipped with an Agilent HP1100 binary pump HPLC inlet).

3 Methods

3.1 Purification and Colorimetric Assay for LOX Activity [15]

3.1.1 LOX Purification

1. Purification of LOX is performed according to previous investigation [16]. Harvest recombinant cells by centrifugation at 5000 *g* for 30 min and store at −80 °C.
2. Resuspend cell pellets in lysis buffer.
3. Cells are lysed by sonication (300 W; sonication cycle: 10 s on, 10 s off; total time: 30 min). Centrifuge the suspension at 11,000 *g* for 30 min at 4 °C.
4. Load the supernatant on a Ni-NTA column pre-equilibrated with 50 mM Tris (pH 8.0), 500 mM NaCl, and 20 mM imidazole.

5. Wash the column with 50 mM Tris-HCl (pH 8.0), 500 mM NaCl, and 50 mM imidazole.
6. Elute the supernatant with 50 mM Tris-HCl (pH 8.0), 500 mM NaCl, and 200 mM imidazole (*see* **Note 3**).

3.1.2 Ferrithiocyanate (FTC)-Based LOX Activity Assay

LOX FTC colorimetric assays were performed in 96-well flat-bottom plates in a total assay volume of 200 μL.

1. Add suitable amount of recombinant human LOXs or HEK293T cell lysates in assay wells (95 μL assay buffer).
2. Add 5 μL of arachidonic acid solution to a final concentration of 100 μM to initiate the enzymatic reaction.
3. Add 100 μL of FTC reagents after 6 min at room temperature to terminate the enzymatic reaction.
4. Monitor the absorbance at 480 nm at room temperature, using a microplate reader.
5. Measure total protein concentration of HEK293T cell lysates with the BCA (bicinchoninic acid) method.
6. Scan the absorption spectra of FTC reagents with 5-, 12-, and 15-HpETEs under visible light ranging from 300 to 800 nm.
7. Study the effect of 5-LOX cofactors (Ca^{2+}, ATP, dithiothreitol [DTT], glutathione [GSH], l-α-phosphatidylcholine, and EDTA) and transition metal ions (Fe^{3+}, Cu^{2+}, Co^{2+}, Ni^{2+}, and Zn^{2+}) on the FTC assay.

3.2 LC/MS/MS for the Detection of Metabolites [17]

3.2.1 Preparation of Standard Curves

1. Spike the samples with a mixture containing known quantities of the eicosanoid AA metabolites of interest.
2. Add an aliquot (40 mL) of 1 N citric acid and of 5 mL of BHT (1%) to the samples immediately after spiking to prevent free radical peroxidation.
3. Prepare standard curves for each analyte of interest over the concentration range 10–1000 ng/mL.

3.2.2 Extraction of Metabolites (See Note 4)

1. Add an aliquot (20 mL) of PGE2-d4 (100 ng/mL) to each sample as an internal standard.
2. Extract the samples by the addition of 2 mL of hexane/ethyl acetate (1/1, v/v). Mix and vortex for 2 min. Centrifuge the mixture (1800 *g* for 10 min at 4 °C).
3. Remove the upper organic layer and store in a glass test tube on ice.
4. Extract the lower aqueous layer twice (*see* **step 2**).
5. Pool the organic phases and then evaporate it to dryness under a gentle stream of nitrogen at room temperature.

6. Reconstitute the samples in (200 mL) methanol: 2 mM ammonium acetate buffer, pH 8.0 (70:30) prior to analysis by LC/MS/MS.

3.2.3 Reversed-Phase Electrospray Ionization Mass Spectrometry

LC/MS/MS is performed using a Quattro Ultima tandem mass spectrometer (Micromass, Beverly, MA) equipped with an Agilent HP1100 binary pump HPLC inlet. Eicosanoid metabolites are separated using a YMC ODS-AQ 2.0 3100-mm column (S3 mm 120 Å, Waters Corp., Marlborough, MA). Separation is achieved using a rapid linear gradient of 70–90% methanol in 4 min with injection-to-injection time of 10 min. Mobile phases consisted of 2 mM ammonium acetate, pH 8 (A), and methanol (B). Flow rate is adjusted to 250 mL/min with the column maintained at 40 °C. Injection volumes are 25 mL/sample. The mass spectrometer is operated in electrospray negative ion mode with a cone voltage of 100 V, cone gas flow rate of 117 L/h, and desolvation gas flow rate of 998 L/h. The temperature of the source region is adjusted at 120 °C while that for the desolvation region is 400 °C. Fragmentation for all compounds is performed using argon as the collision gas at a collision cell pressure of 3.7×10^{-3} mbar. The collision energy was 19 V.

4 Notes

1. The FTC reagent should be stored at 4 °C, protected from oxygen and light.
2. The arachidonic acid solution is sensitive and should be used within 1 h.
3. The purified enzyme is stable for at least 4 h on ice.
4. All extraction procedures should be performed at minimum light levels to reduce the potential for photodegradation of the eicosanoid metabolites.

References

1. Brash AR (1999) Lipoxygenases: occurrence, functions, catalysis, and acquisition of substrate. J Biol Chem 274:23679–23682
2. Claesson HE (2009) On the biosynthesis and biological role of eoxins and 15-lipoxygenase-1 in airway inflammation and Hodgkin lymphoma. Prostaglandins Other Lipid Mediat 89:120–125
3. Yang WS, Kim KJ, Gaschler MM, Patel M, Shchepinov MS, Stockwell BR (2016) Peroxidation of polyunsaturated fatty acids by lipoxygenases drives ferroptosis. Proc Natl Acad Sci U S A 113:E4966–E4975
4. Kuhn H, Banthiya S, van Leyen K (2015) Mammalian lipoxygenases and their biological relevance. Biochim Biophys Acta 1851:308–330
5. Mashima R, Okuyama T (2015) The role of lipoxygenases in pathophysiology; new insights and future perspectives. Redox Biol 6:297–310
6. Acosta IF, Laparra H, Romero SP, Schmelz E, Hamberg M, Mottinger JP et al (2009) tasselseed1 is a lipoxygenase affecting jasmonic acid signaling in sex determination of maize. Science 323:262–265

7. Werz O (2007) Inhibition of 5-lipoxygenase product synthesis by natural compounds of plant origin. Planta Med 73:1331–1357
8. Haeggstrom JZ, Funk CD (2011) Lipoxygenase and leukotriene pathways: biochemistry, biology, and roles in disease. Chem Rev 111:5866–5898
9. Chu J, Pratico D (2009) The 5-lipoxygenase as a common pathway for pathological brain and vascular aging. Cardiovasc Psychiatry Neurol 2009:174657
10. Gilbert NC, Bartlett SG, Waight MT, Neau DB, Boeglin WE, Brash AR et al (2011) The structure of human 5-lipoxygenase. Science 331:217–219
11. Radmark O, Werz O, Steinhilber D, Samuelsson B (2007) 5-Lipoxygenase: regulation of expression and enzyme activity. Trends Biochem Sci 32:332–341
12. Kuhn H, Thiele BJ (1999) The diversity of the lipoxygenase family. Many sequence data but little information on biological significance. FEBS Lett 449:7–11
13. Berger M, Schwarz K, Thiele H, Reimann I, Huth A, Borngraber S et al (1998) Simultaneous expression of leukocyte-type 12-lipoxygenase and reticulocyte-type 15-lipoxygenase in rabbits. J Mol Biol 278:935–948
14. Spiteller G (2003) Are lipid peroxidation processes induced by changes in the cell wall structure and how are these processes connected with diseases? Med Hypotheses 60:69–83
15. Lu W, Zhao X, Xu Z, Dong N, Zou S, Shen X et al (2013) Development of a new colorimetric assay for lipoxygenase activity. Anal Biochem 441:162–168
16. Su C, Oliw EH (1998) Manganese lipoxygenase. Purification and characterization. J Biol Chem 273:13072–13079
17. Kempen EC, Yang P, Felix E, Madden T, Newman RA (2001) Simultaneous quantification of arachidonic acid metabolites in cultured tumor cells using high-performance liquid chromatography/electrospray ionization tandem mass spectrometry. Anal Biochem 297:183–190

Chapter 18

Cholinesterase as a Target for Drug Development in Alzheimer's Disease

Piyoosh Sharma, Manish Kumar Tripathi, and Sushant Kumar Shrivastava

Abstract

Alzheimer's disease (AD) is an enormous healthcare challenge, and 50 million people are currently suffering from it. There are several pathophysiological mechanisms involved, but cholinesterase inhibitors remained the major target from the last 2–3 decades. Among four available therapeutics (donepezil, rivastigmine, galantamine, and memantine), three of them are cholinesterase inhibitors. Herein, we describe the role of acetylcholine sterase (AChE) and related hypothesis in AD along with the pharmacological and chemical aspects of the available cholinesterase inhibitors. This chapter discusses the development of several congeners and hybrids of available cholinesterase inhibitors along with their binding patterns in enzyme active sites.

Key words Acetylcholinesterase, Butyrylcholinesterase, Acetylcholine, Butyrylcholine, Catalytic active site, Peripheral anionic site, Oxyanion hole, Acyl binding pocket, Anionic subsite, Amyloid beta, Choline acetyltransferase, Donepezil, Galantamine, Rivastigmine, Tacrine, *N*-Benzylpiperidine

1 Introduction

Alzheimer's disease (AD) is one of the utmost common neurological disorders causing mental deterioration [1]. It is a neurological disorder that affects memory, thinking, and other abilities of a person. It is a common form of dementia most prevalent among the age group of 65 or above. Around 50 million people in the worldwide are affected by AD, and these numbers are predicted to be increased up to 152 million by 2050 [2]. Therefore, AD is a huge, enormous, economic, and societal challenge for healthcare and scientific community [3]. The disease is named after Dr. Alois Alzheimer, who was a German psychiatrist and neuropathologist. Firstly in 1906, he examined the changes in brain tissues of a 50-year-old patient died due to unusual mental illness [4]. Persons infected with the disease are broadly categorized under two groups, namely late-onset type, in which symptoms firstly appear in mid-60 years, and early onset type, in which persons between

Nikolaos E. Labrou (ed.), *Targeting Enzymes for Pharmaceutical Development: Methods and Protocols*, Methods in Molecular Biology, vol. 2089, https://doi.org/10.1007/978-1-0716-0163-1_18, © Springer Science+Business Media, LLC 2020

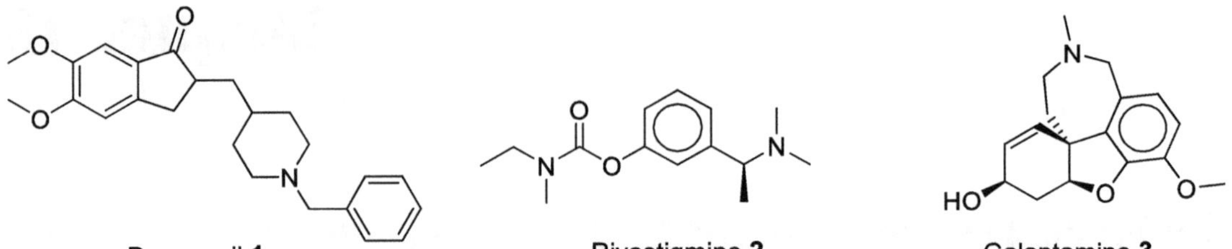

Donepezil **1** Rivastigmine **2** Galantamine **3**

Fig. 1 Structures of USFDA approved cholinesterase inhibitors

mid-30 and mid-60 age groups are included and is a very rare form of the disease [5]. The progression of AD can be categorized into three basic stages: (1) preclinical (no signs or symptoms), (2) mild cognitive impairment, and (3) dementia. The symptoms of AD include loss in person's memory that disrupts daily life, challenges in planning and solving the problems, confusion with time and place, problems with the word in speaking and writing, changes in the mood and personality, and withdrawal from social activities [6].

The pathological hallmarks of the disease which were found in the brain of individuals of AD patients are extracellular amyloid plaques [7], tau protein and intracellular neurofibrillary tangles [8], oxidative stress [9], decline in acetylcholine (ACh) levels [10], etc. Enzymes, namely glycogen synthase kinase-3 (GSK3) [11], lipooxygenases [12], secretases [13], and more specifically cholinesterases (ChEs) [14] are found to be involved in the pathological progression of AD. Despite recent scientific developments [15–23], there is no therapy available till date to prevent the progression of AD. The present therapeutic regimens only provide symptomatic cure [24]. The exact cause of the AD is still unknown, but some key findings revealed that overexpression of acetylcholinesterase (AChE) and extracellular accumulation of "mysterious" β-amyloid (Aβ) plaques are the major hallmarks of this disease [25]. Current pharmacotherapy includes the cholinergic receptor agonist or ChE inhibitors (donepezil **1**, rivastigmine **2**, and galantamine **3**) and NMDA receptor antagonist (memantine) (Fig. 1).

Current pharmacotherapy is slow in the aspect of cognitive impairment and benefits are often marginal. Non-pharmacological therapy like cognitive training and exercise also help in improving the cognitive functions in the patients [26]. Thus, extensive research is needed to develop and discover novel scaffold for this challenging and dreadful disease.

2 Cholinergic Hypothesis

The exact molecular mechanism involved in AD progression is still unclear due to the complexity of the disease. Many theories were proposed to understand the pathogenesis of AD including the cholinergic hypothesis, Aβ, tau, excitotoxic, oxidative stress, apolipoprotein E (ApoE), GSK-3, and CREB signaling pathways [27]. Still, many efforts have been made by the researchers to develop anti-AD drugs based on the suggested hypothesis.

The cholinergic hypothesis was proposed to explain the etiology of AD for the treatment of mild to moderate AD and came into existence in 1970 after the observation of a postmortem brain of patients. Further subsequent discoveries in which the reduced activity of choline acetyltransferase (CAT) was reported, in the septal nuclei and basal forebrain of patients [28]. CAT is responsible for the synthesis of neurotransmitter ACh in the brain, which has a significant role in the neuromodulation of memory, learning, and other cognitive functions [29]. The specific pathway involved in the synthesis of ACh is shown in Fig. 2. Firstly, acetyl CoA produced in mitochondria by the process of glycolysis transfer its acetyl group to the choline (Ch) in the presence of the enzyme CAT for the synthesis of ACh. The biosynthesized ACh is transported by VAChT (vesicular ACh transporter) and facilitates its storage into synaptic vesicles. The increase in the concentration of Ca^{2+} in the axon bulb is responsible for the release of ACh into the synaptic cleft, which binds to cholinergic receptors at the postsynaptic membrane of the dendrite. Cholinesterase enzyme, namely AChE catalyzed hydrolytic breakdown of ACh into the choline and acetate [30]. Further, the uptake of this choline in the presynaptic neuron and the process is repeated for the neurotransmission.

From decades, cholinesterase hypothesis is widely approved for its role in dementia and neurological disorders. This hypothesis states that during AD, cholinergic neurons are severely affected in the basal forebrain of the patient, and it can be detected histopathologically through the loss of neurons and neurochemically by the deficit of marker enzymes involved in ACh synthesis and degradation [31]. These observations led to the rationale for the development of cholinesterase inhibitors having clinical efficacy against [32]. Thus, the effective treatment strategy was to increase the cholinergic function and elevate the level of ACh via inhibiting the AChE and butyrylcholinesterase (BChE) (Fig. 3).

3 AChE and BChE: Structural Differences

Cholinesterases are the ubiquitous class of enzymes divided into two subfamilies, namely AChE (EC 3.1.1.7) and BChE (EC 3.1.1.8) according to their substrate and inhibitor specificities. Evolution of these enzymes occurs as a result of gene duplication method during the early vertebrate evolution. In mammals, gene with chromosome location 7q23 encodes AChE and its multiple forms arise due to mRNA slicing process. Whereas, BChE is encoded by the gene with chromosome location 3q23 [33]. The

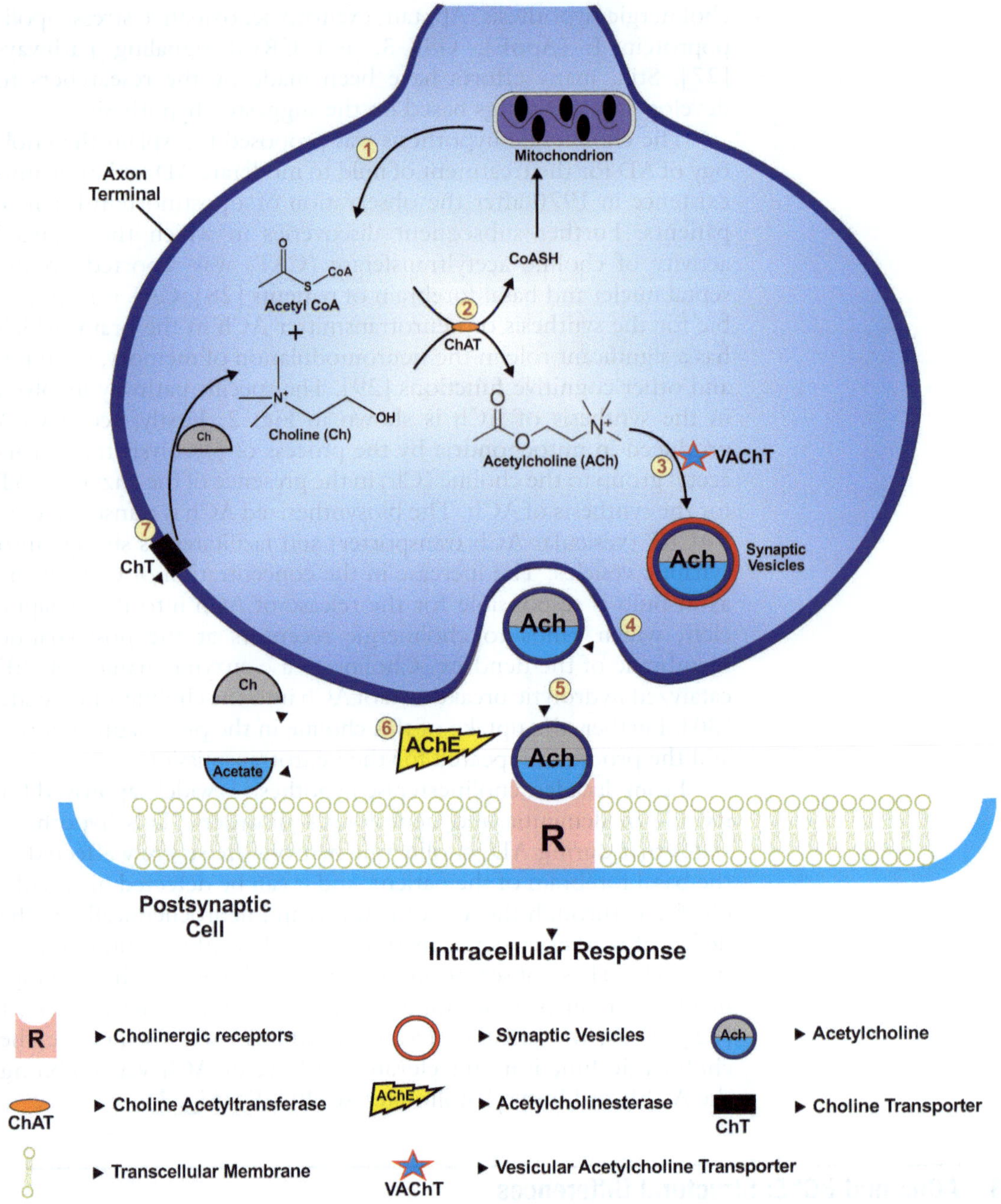

Fig. 2 Synthesis of ACh and cholinergic transmission. (*1*) Acetyl CoA produced via glycolysis in mitochondria; (*2*) transfer of the acetyl group from acetyl CoA to the choline (Ch) via CAT to synthesize ACh; (*3*) VAChT (vesicular ACh transporter) facilitates the storage of ACh into the synaptic vesicles; (*4*) increase in concentration of Ca^{2+} ions in the axon bulb facilitates the release of ACh in the synaptic cleft; (*5*) binding of released ACh to the cholinergic receptors at postsynaptic membrane of dendrites; (*6*) catalytic breakdown of ACh into choline and acetate via AChE; (*7*) uptake of the released choline. (Adapted with permission from Sharma et al. [27])

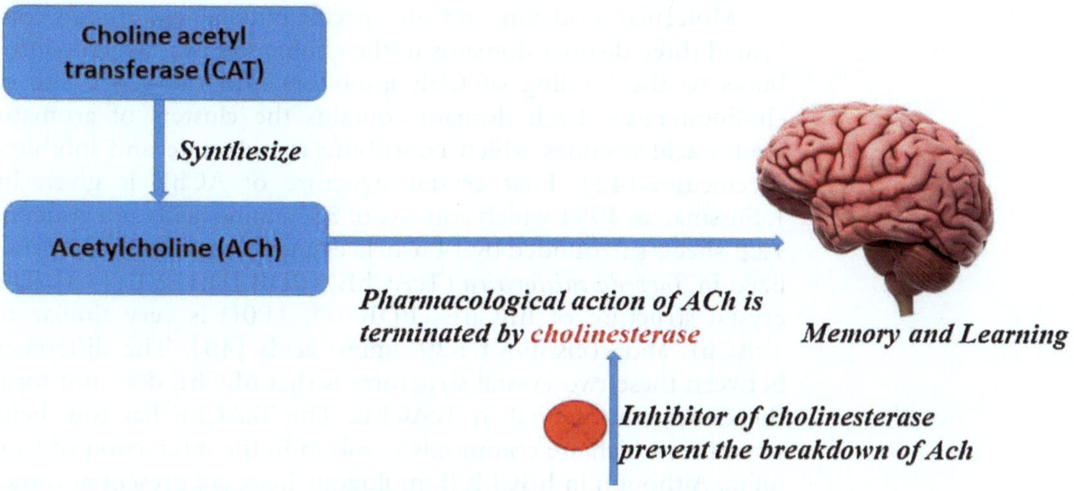

Fig. 3 Cholinesterase hypothesis of memory and learning

difference between these two cholinesterases is due to their preferences for the substrate they hydrolyze. AChE hydrolyzes ACh whereas BChE hydrolyzes butyrylcholine. AChE favors hydrolysis of small substrate ACh into the acetic acid and choline. Whereas, BChE adapts the bulkier substrate like benzoyl- or butyrylcholine for their catalytic decomposition [34]. BChE also significantly catalyzes other esters, namely cocaine, acetylsalicylic acid, heroin, physostigmine, and organophosphates [35]. In addition, both these enzymes are also found to be involved in cell differentiation and development activities [36]. It was found that the level of BChE increases in AD patients with enhancing the amyloid plaque formation [37]. Both the cholinesterases (AChE and BChE) belong to α/β-fold protein family due to the presence of central β-sheet surrounded by the α-helices. These structures were also observed in other proteins such as thyroglobulin (precursor of hormone) and in the cell adhesion molecules (such as neuroligin) [38].

AChE exists in the two molecular forms, namely "asymmetric form" localized in the neuromuscular junction and "globular form" which is anchored in the membrane of hydrophobic domain. Membrane-bound globular tetrameric (G4) form is found mostly in the brain as compared to monomeric (G1) form. During AD, the ratio of these two molecular forms G1 and G4 were disturbed in hippocampal and cortex regions due to selective loss of G4 form [39]. Within species, the overall tertiary structure of cholinesterase enzymes was found to be similar with ~50% amino acids identity [40]. AChE is found to be less glycosylated as compared to BChE, which affects the stability and the pharmacokinetic properties and not their catalytic properties [41].

Molecular modeling and site-specific mutagenesis studies confirmed three distinct domains in the cholinesterases, which contributes to the binding of ChE inhibitors into the active site of cholinesterases. Each domain contains the clusters of aromatic amino acid residues which contribute to substrate and inhibitor specificities [42]. First crystal structure of AChE is given by J. Sussman in 1991 which consists of 537 amino acids and contains 12 β-sheets surrounded by 14 α-helices and was successfully crystallized in *Torpedo californica* (TcAChE) (PDB ID:1EVE) [43]. The crystal structure of hBChE (PDB ID: 1P0I) is very similar to TcAChE and consists of 529 amino acids [40]. The difference between these two crystal structures is that hBChE does not form the dimer as observed in TcAChE. The TcAChE has four-helix structure, which are commonly involved in the interaction of subunits. Although in hBChE homologous helix are present at dimer interface (residues 362–375 and 514–529), but their orientation is not antiparallel like TcAChE but form a 45° angle.

The outer gorge of AChE is termed as peripheral anionic site (PAS) [44]. The entrance of PAS is lined by numerous aromatic amino acid residues. In BChE, lining of the gorge has lesser aromatic amino acid residues (six in number as compared to 14 in AChE). This PAS region is lined with aromatic amino acid residues, namely Tyr70, Tyr72, Tyr121, Tyr279, and Tyr334 for TcAChE and Asp70, Tyr332 for hBChE responsible for the π interactions with inhibitors. PAS enhances the catalytic efficiency by trapping the substrate or inhibitor toward the active site. This region is also found to be involved in the promotion of Aβ aggregation. Thus, inhibitors targeting the PAS and active site also prevents the Aβ production and deposition along with the AChE inhibition activity [45].

The inner gorge portion of the active site in cholinesterase consists of four subsites, namely, acyl binding pocket, anionic subsite, oxyanion hole, and catalytic active site (CAS). CAS also called as esteratic site of enzyme consists of amino acids, Ser200, Glu327, and His400 in TcAChE and Ser198, Glu325, His438 in hBChE, where hydrolysis of substrate occurs. The CAS residues help in designing inhibitors and understanding the molecular mechanisms of ACh binding to the muscarinic and nicotinic receptors.

Oxyanion hole is the crucial determinant in active site geometry, which accommodates the acetyl group of ACh [46]. The oxyanion hole is responsible for the transition state stabilization of the enzyme-substrate complex. The crystal structure of AChE revealed that three-pronged oxyanion hole was created by peptidic NH group of two glycines Gly118 and Gly119 and one Ala201 residues present in TcAChE. In hBChE, oxyanion hole is similar to TcAChE and consists of highly conserved Gly116, Gly117, and Ala199 [47].

The acyl binding pocket in cholinesterase is responsible for substrate specificity. In TcAChE, acyl binding pocket consists of aromatic residues (Trp233, Phe288, Phe290, and Phe331) and during catalysis, acyl group of choline esters are accommodated in this area (Moralev, 2001). But in hBChE, two of these phenyl residues were replaced with valine and leucine to create larger pocket (500 Å^3 in BChE versus 300 Å^3 in AChE) [48]. This larger pocket (Trp231, Val288, Leu286, and Phe329) allows BChE to hydrolyze bulkier substrate and ligands compared to AChE, resulting in its lower specificity.

In cholinesterases, there also exists an anionic subsite, which accommodates the quaternary amine of ACh by aromatic residue tryptophan with cation–π interactions. The substrate is guided inside the active gorge via interaction with a phenylalanine residue. In TcAChE, this gorge consists of Trp84, Glu199, and Phe330 residues. In hBChE, Phe330 residue is replaced by Ala328. Lack of this phenylalanine residue in BChE influences the affinity for some inhibitors.

4 Development of Cholinesterase Inhibitors

4.1 Donepezil and Benzylpiperidine Analogs

Donepezil, an AChEI is one of the most commonly prescribed drugs for the treatment of mild to moderate AD. But like other USFDA approved drugs, donepezil also provides symptomatic relief only and not useful in halting the progression of AD [49, 50]. Donepezil also found to be effective in inhibiting Aβ aggregation [51] and increasing the expression of nicotinic receptors in the cortex [52]. These pathophysiological mechanisms observed to be beneficial in preventing the AD disease progression to some extent. Donepezil also found to be neuroprotective as it inhibited specifically the R-form of AChE and not the S-form [43].

Donepezil arguably is the most successful AChE inhibitor available to treat AD. It has oral bioavailability of about 100% with a longer half-life ($t_{1/2}$ = 70–80 h) [53]. It has comparably fewer side effects than other available therapeutics and also not showed dose-dependent hepatotoxicity. The discovery of donepezil was started with the development of *N*-benzylpiperazine derivatives showed micromolar inhibitory potential. Further, with the replacement of the piperidine nucleus instead of piperazine ring dramatically increased the potency [54]. Finally, among several substitutions at another terminal, 4,5-dimethoxyindanone nucleus showed significant inhibitory potential and devoid of any side effect. Donepezil is extensively investigated molecule, and its detailed SAR study has been conducted. The ring expansion of indanone nucleus decreased the activity drastically. The 4,5-dimethoxy substituents showed maximum inhibitory potential, while carbonyl group in indanone is also essential for activity and its replacement with indanol or indene nucleus, remarkably decreased the activity. Several bridging groups were also evaluated to connect the benzylpiperidine nucleus with indanone ring. This bridge is essential for AChE inhibitory potential, and direct connection between both the nucleuses decreased the potency dramatically. The order of potency with varied linker length is in the following order: 3 > 1 > 5 > 2 > 4. Further, piperidine nucleus is also an essential part, and its replacement with piperazine significantly decreased the inhibitory potency of AChE [55].

Donepezil has very high inhibitory potency against AChE in the nanomolar range. It binds to AChE in a unique orientation extending through to the PAS residues to the deep into the gorge and interacts with the CAS (PDB Code: 4EY7) [43]. The 5,6-dimethoxyindanone nucleus aligned toward PAS region, while benzyl piperidine ring extended into CAS. At PAS, phenyl part of indanone nucleus involved in a π–π stacking interaction with Trp286, while interacted hydrophobically with aromatic amino acid residues (Tyr72, Tyr124, and Tyr341) and through charged interaction with Asp74. At acyl binding pocket, the oxygen atom of indanone ring involved in H-bonding interaction with Phe295, while Phe297 interacted hydrophobically. At anionic subsite, benzyl ring exhibited π–π stacking interaction with Trp86 residue. The –NH group of piperidine ring involved in π–cation interactions with the Trp86 and Phe338 residues, while Glu202 showed charged interaction. At oxyanion hole, donepezil also interacted with Gly120 and Gly121 residues. Further, in the deep gorge of AChE, benzylpiperidine nucleus aligned toward CAS and interacted with CAS residues Ser203 and His447 through polar interactions.

The superior pharmacological profile of donepezil has prompted several researchers to develop its congeners, and extensive research efforts have been made to explore and develop the *N*-benzylpiperidine analogs with the modification particularly made at the terminal indanone nucleus. Initially, Andreani and coworkers modified the donepezil by replacing indanone ring with indole nucleus and introducing the double bond connecting the basic benzylpiperidine nucleus and indole ring. The results suggested very lower inhibitory potential of compounds against AChE compared to standard donepezil and tacrine. The reason for decreased potency was cited as loss of interaction of compounds with Trp279 residue. Another reason mentioned was the rigidity of the molecule that might have hindered the penetration of compounds into AChE gorge [56]. Contreras and coworkers have studied the minaprine derivatives and found that most active compound of the series has *N*-benzylpiperidine ring as a pharmacophoric feature for the AChE inhibition [57]. Further, Omran and coworkers have designed series of donepezil hybrids with modification in central amide functionality along with terminal indanone portion. The most potent compound (**4**) of the series showed AChE inhibition with an $IC_{50} = 0.06$ μM potency [58].

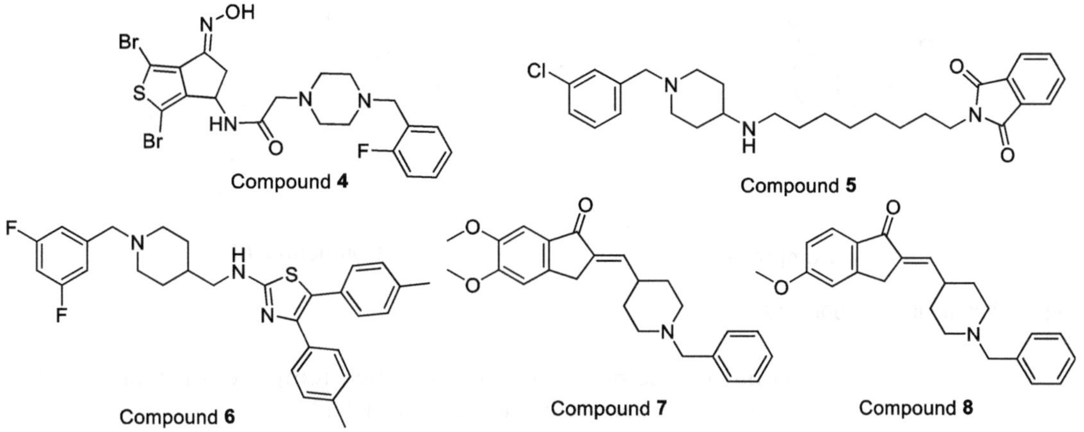

Fig. 4 Donepezil congeners and *N*-benzylpiperidine analogs

Malawska and coworkers have designed 28 novel donepezil-based hybrids containing *N*-benzylpiperidine nucleus combined with a phthalimide or indole moieties. The most active compound of the series (2-(8-(1-(3-chlorobenzyl)piperidin-4-ylamino)octyl) isoindoline-1,3-dione) (**5**) showed selective BChE inhibition (IC_{50} = 0.72 μM) along with Aβ anti-aggregation activity. The compound also has significant BBB permeability [59]. Shidore et al. have designed a series of hybrid structures connecting *N*-benzylpiperidine nucleus with diarylthiazole moiety. The most potent lead molecule of the series *N*-[(1-(3,5-difluorobenzyl) piperidin-4-yl)methyl]-4,5-bis(p- tolyl)thiazol-2-ylamine (**6**) showed significant AChE (IC_{50} = 0.30 μM) and BChE (IC_{50} = 1.84 μM) along with AChE-induced Aβ inhibition, anti-oxidant, and anti-apoptotic activities [60]. Further, Costanzo et al. have developed donepezil analogs with dual inhibitory potential against AChE and BACE-1. The most promising candidates of the designed analogs, (E)-2-((1-benzylpiperidin-4-yl)methylene)-5,6-dimethoxy-2,3-dihydro-1H-inden-1-one (7) and (E)-2-((1-benzylpiperidin-4-yl)methylene)-5-methoxy-2,3-dihydro-1H-inden-1-one (**8**) exhibited significant inhibition of AChE (IC_{50} = 0.058 μM, 0.043 μM) and BACE-1 (IC_{50} = 0.697 μM, 0.333 μM), respectively (Fig. 4) [61]. Recently, to improve the binding of donepezil on BACE-1 along with AChE, two rigid donepezil analogs (**9** and **10**) were designed (Fig. 5). Both compounds have the presence of a double bond to connect the *N*-benzylpiperidine with indanone nucleus. Among both these compounds, the 2-methoxy group was removed from one of them. This rigidification of the donepezil resulted in a likely entropy-enthalpy compensation with solvation effects contributing primarily to AChE binding affinity. Molecular docking studies also revealed the better binding affinity of these compounds to BACE-1 active pocket. Overall, the study has suggested that these rigid molecules

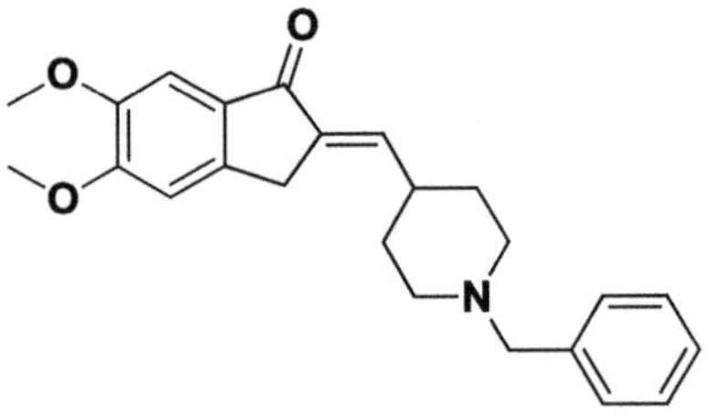

Compound **9**

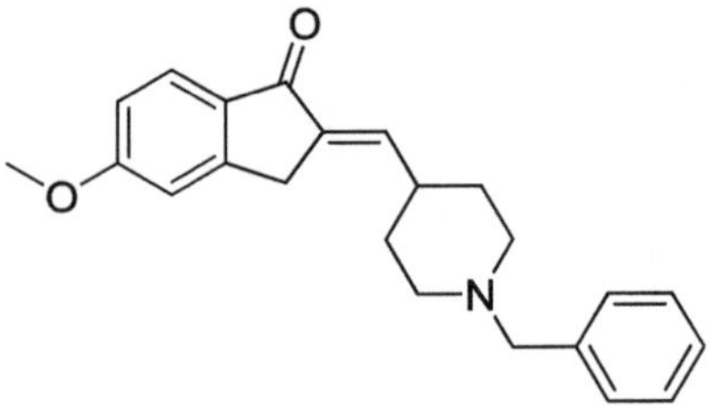

Compound **10**

Fig. 5 Rigid analogs of donepezil

could be the new structural designed template for dual inhibition strategy against AChE and BACE-1 [62].

4.2 Galantamine and Derivatives

Galantamine ((4aS,6R,8aS)-3-methoxy-11-methyl-4a,5,9,10,11,12-hexahydro-6H-benzo[2,3]-benzofuro[4,3-cd]azepin-6-ol) is a selective, reversible, and competitive inhibitor of AChE [63]. Firstly, it was isolated as a natural alkaloid from a plant called Caucasian snowdrop (*Galanthus woronowii*) belonged to family Amaryllidaceae [64]. Additional to AChE inhibitory activity, galantamine also acts on nicotinic receptor as an allosteric modulator [65]. Its binding to the AChE causes conformational changes in the structure of the receptor, which potentiates the nicotinic receptor and enhances its postsynaptic response. Galantamine is absorbed 100% following its oral administration and has a large volume of distribution. It is well-tolerated drug having only a few gastric side effects [66]. Galantamine is marketed since 2001 under the brand name of Reminyl®. Galantamine is also reported for increasing GABA and glutamate release in hippocampal slices [63].

The co-crystallized structure of human AChE with (−)-galantamine (PDB ID: 4EY6) showed active binding site interactions with all the important amino acid residues. At PAS, (−)-galantamine interacted hydrophobically with Tyr72, Tyr124, Trp286, and Tyr341 and through charged interaction with Asp74. At CAS, it involved in polar interactions with Ser203 and His447 along with additional H-bonding interaction with Ser203 residue. At anionic subsite and acyl binding pocket, (−)-galantamine involved in hydrophobic interactions with Trp86, Glu202, Phe338, Trp236, Phe295, and Phe297 residues. (−)-Galantamine also involved in glycine type of interactions with Gly120, Gly121, and Gly122 residues of oxyanion hole [67].

Several studies were conducted to modify and substitute galantamine for developing its congeners and analogs (Fig. 6). In an investigation, optically pure open D-ring analogs were synthesized, and among them, ethyl (2-(3-hydroxy-9-(hydroxymethyl)-6-methoxy-3,4,4a,9b-tetrahydrodibenzo[*b,d*]furan-1-yl)ethyl)(methyl)carbamate (**11**) was found to be a most potent inhibitor of AChE. SAR studies also revealed that substitution of *N*-atom in

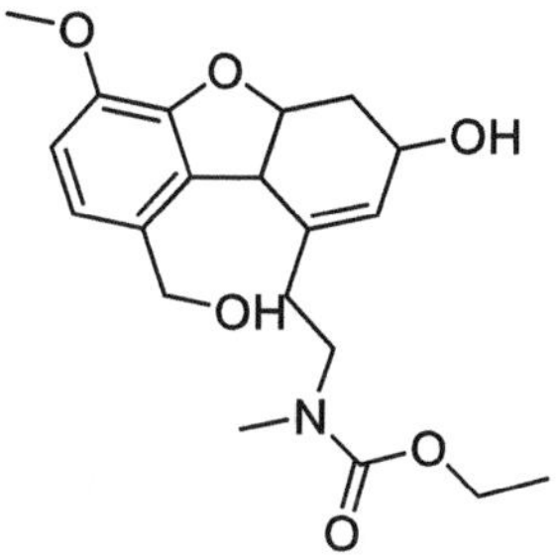

Compound **11**

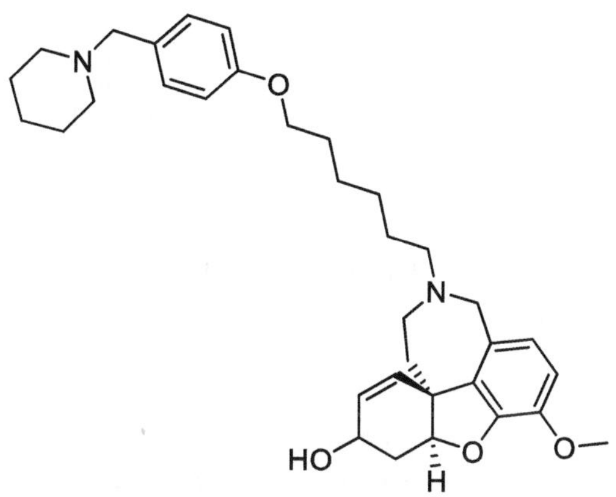

Compound **12**

Fig. 6 Galantamine analogs

galantamine might be favorable for AChE inhibitory activity due to enhanced interaction with PAS residues [68]. Several *N*-substituted benzylamino moieties were explored as a major pharmacophoric group for AChE inhibitory activity [69]. Also, alkyl chain lengths were also modified and compound (4aS,8aS)-3-methoxy-11-(6-(4-(piperidin-1-ylmethyl)phenoxy)hexyl)-4a,5,9,10,11,12-hexahydro-6H-benzo[2,3]benzofuro[4,3-cd]azepin-6-ol (**12**) was found to be most active [70].

4.3 Rivastigmine and Analogs

Rivastigmine ((S)-3-[1-(dimethylamino)ethyl]phenyl-*N*-ethyl-*N*-methylcarbamate) is a reversible cholinesterase inhibitor prescribed under the brand name Exelon® since 2000 for symptomatic treatment of moderate to severe AD [71]. Rivastigmine showed inhibition toward both cholinesterases (AChE and BChE) with more potency against BChE [71]. Thereby, it is most commonly used at later stages of AD, when the concentration of BChE is higher than AChE.

Chemically, it is a carbamate ester, which causes carbamoylation of the active site residues in cholinesterase enzyme to impair its function. The carbamylated complex of rivastigmine with cholinesterase enzyme prevents enzyme-catalyzed hydrolysis of ACh for several hours instead of elimination of the drug from plasma. Thereby, the half-life of rivastigmine is only 1 h, but its effect lasts for about 10 h due to the slow regeneration of the original baseline AChE activity [72]. The most common side effects of rivastigmine are similar to donepezil and galantamine with fewer gastrotoxic reactions. Rivastigmine is rapidly and completely absorbed following oral administration. It rapidly crosses the BBB and selective AChE inhibitor in CNS over the peripheral nervous system. It is a more potent inhibitor of AChE in the prefrontal cortex and hippo campal regions compared to other brain regions [73].

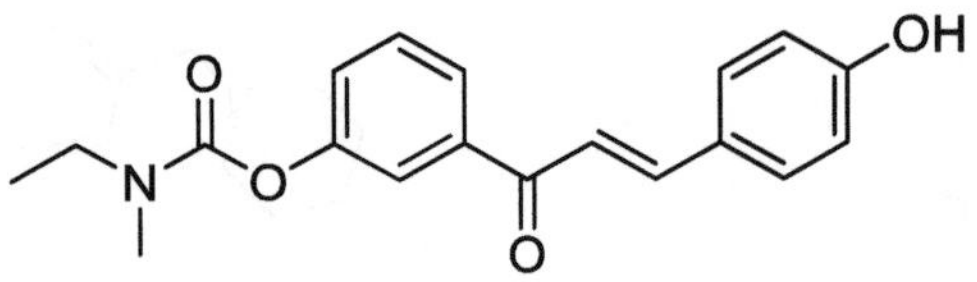

Compound **13**

Fig. 7 Rivastigmine analogs

Variously modified congeners of rivastigmine were developed and studied. The substitution of methyl chain instead of ethyl resulted into 14–23 fold decline in AChE inhibitory potential. The replacement of methyl at carbamoyl *N* with 1-phenylethyl substituent also caused significant loss of AChE and BChE inhibition [74]. Recently, Wang et al. have developed chalcone-rivastigmine-based hybrids for finding potent cholinesterase inhibitor (E)-3-(3-(4-Hydroxyphenyl)acryloyl)phenyl ethyl(methyl)carbamate (**13**, Fig. 7) with a comparable pharmacological profile to rivastigmine [75].

4.4 Tacrine and Related Hybrids

Tacrine was the first approved reversible cholinesterase inhibitor prescribed for the treatment of AD in 1993 under the brand name of Cognex® [76]. Tacrine bound to the α-anionic site of cholinesterase and was also among the first reported dual cholinesterase inhibitor with more selectivity toward BChE over AChE [77]. Tacrine is aminoacridine derivative and was used of palliative treatment of mild to moderate AD before being abandoned. The crystal structure information from the complex of Tetronarce Californica AChE-tacrine (PDB ID: 1ACJ) [78] revealed the hydrophobic π-π stacking of tacrine with Trp84 and Phe330 residues. The ring nitrogen of tacrine also bonded with CAS residue His440 through H-bonding. The use of tacrine has been largely abandoned due to severe hepatotoxicity and other side effects such as nausea, vomiting, dizziness, diarrheas, seizures, and syncope [79]. Tacrine also enhances the production of ROS and depletion of glutathione levels in liver cells, and that might be the reason of its hepatotoxicity [80, 81].

Several tacrine analogs were designed (Fig. 8) to decrease its toxicity, which includes 1-hydoxytacrine (**14**), 7-methoxytacrine (**15**), and bis(7)-tacrine (**16**). 1-Hydroxytacrine was developed to impart more water solubility, enhance the glucuronidation, and subsequent elimination, but this compound was found to be somewhat less potent than tacrine to inhibit AChE. 7-Methoxytacrine was also found to be less toxic along with the more inhibitory potential. The dimeric structure of tacrine, bis-(7)tacrine has improved the pharmacological profile over tacrine and having lesser side effects also. This dimer is a heptamethylene-linked dimer

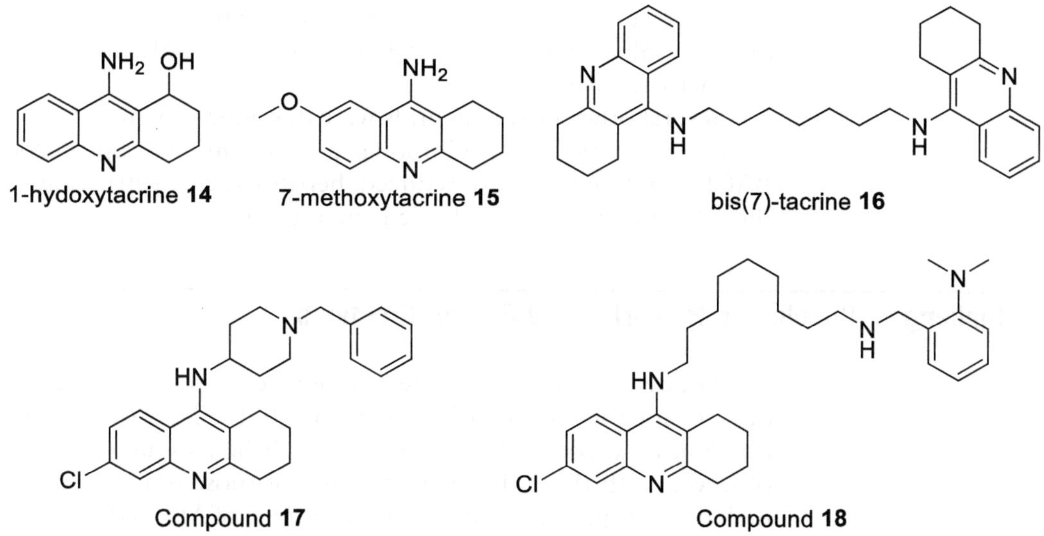

Fig. 8 Tacrine analogs and hybrids

having more selectivity against AChE over BChE [82]. Further, replacement of heptamethylene dimer with cystamine (disulfide bridge) dimer improved the cholinesterase inhibition along with additional AChE-induced Aβ aggregation inhibitory potential [83]. Recanatini and coworkers have examined a series of 9-aminoacridines and found that addition of halogens, methyl, or nitro substituents at position 7 leads to increased AChE inhibitory potential of compounds. They have identified two features: (1) negative steric feature at position 7 and (2) favorable hydrophobic groups at position 6 to improve the potency of compounds against AChE [84].

Tacrine hybrids were designed and developed widely to provide new molecules with improved inhibitory potential against AChE/BChE, decreased hepatotoxicity, and impart additional pharmacological properties. Oxoisoaporphine-tacrine hybrids were designed to improve binding toward PAS-AChE along with CAS region, which leads to the improved ability of compounds for Aβ aggregation inhibition [85]. Mao and coworkers have designed tacrine-based hybrids (**17**) with selective and potent AChE inhibitory property ($IC_{50} = 0.55$ nM) along with inhibition of Aβ aggregation [86]. Jin et al. have developed several tacrine-donepezil hybrids and found that *N*-(1-benzylpiperidin- 4-yl)-6-chloro-1,2,3,4-tetrahydroacridin-9-amine (**18**) inhibited AChE and BChE in nanomolar concentration [87]. Tacrine-trolox hybrids were designed by Kuca and coworkers to provide multifunctional roles in AD treatment [88]. These designed hybrids exhibited significant cholinesterase inhibition along with potential antioxidant abilities. Tacrine-cinnamate and tacrine-cinnamylidene acetate hybrids also found to produce substantial anti-Aβ aggregatory, anti-AChE, and antioxidant activities [89].

Several other tacrine hybrids were developed to impart cholinesterase inhibition and additional antioxidant, neuroprotection, BACE-1 inhibition, and reduced hepatotoxicity [90]. Some of these hybrids are discussed briefly in Table 1.

5 Challenges, Concluding Remarks, and Future Directions

Cholinesterase inhibitors were explored enormously in last 2–3 decades for their substantial role in AD but failed miserably with no clinical outcome. There are several challenges and limitations observed with these therapeutics. One of those limitations is the high doses requirement in an advanced stage of AD, and it provides symptomatic relief only without affecting the progression of AD [91]. ACHEIs not provide high CNS selectivity and cause several side reactions such as gastrotoxicity and hepatotoxicity [92]. There is also a question about the action of a cholinesterase inhibitor on nonspecific cholinergic receptors (nicotinic or muscarinic). Only specific M1 muscarinic receptor agonists are found to have a significant role in AD, and action on other nonspecific receptors might lead to unwanted side reactions [93]. This limitation arises the serious questions for its use in AD. There is also evidence that nicotinic cholinergic receptors have a role in both, memory improvement and impairment [94]. Therefore, it is required to develop specific and selective molecules, and more detailed studies need to be conducted. Another negative aspect with cholinesterase therapeutics is the high cost involved in treatment with available US FDA approved drugs. Although failure and several negative outcomes with these therapeutics, the truth lies with the fact that there is no other alternative available than cholinesterase inhibitors for the effective treatment of AD. But there is a need to focus on long-term goals and development of the molecules using several computational tools and repurposing approach. The research efforts should be more focused on the clinical outcome.

Table 1
Tacrine-based hybrids and their functional roles

S. No.	Tacrine hybrids	Lead structure(s)	Functions	Ref.
1.	7-Methoxytacrine-p-anisidine hybrids	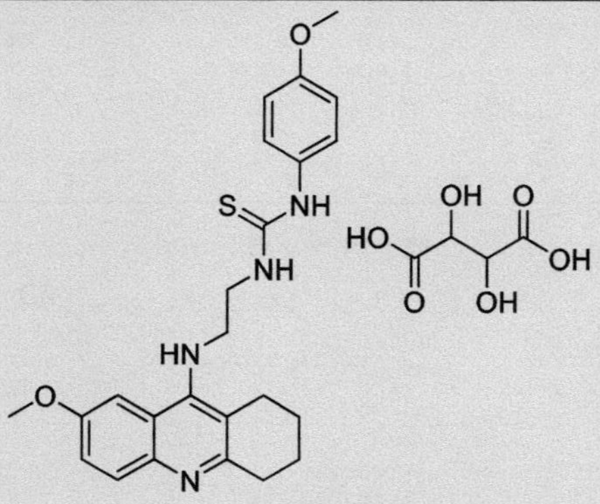 Compound **19**	Higher BChE selectivity (AChE: $IC_{50} = 43.6$ µM, BChE: $IC_{50} = 1.03$ µM)	[95]
		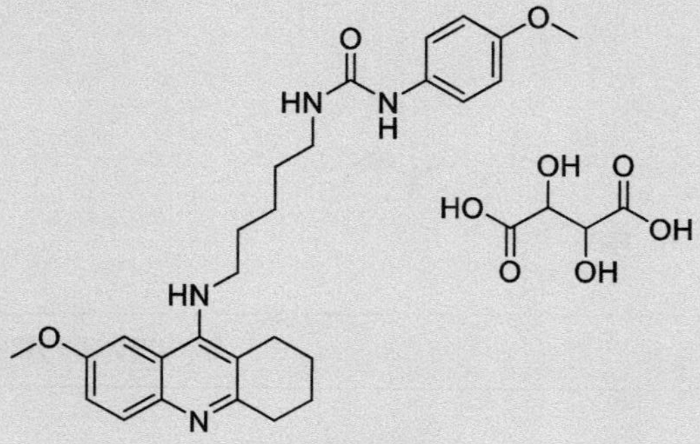 Compound **20**	Higher selectivity toward AChE (AChE: $IC_{50} = 1.35$ µM, BChE: $IC_{50} = 10.9$ µM)	

(continued)

Table 1
(continued)

S. No.	Tacrine hybrids	Lead structure(s)	Functions	Ref.
2.	Tacrine-trolox hybrids	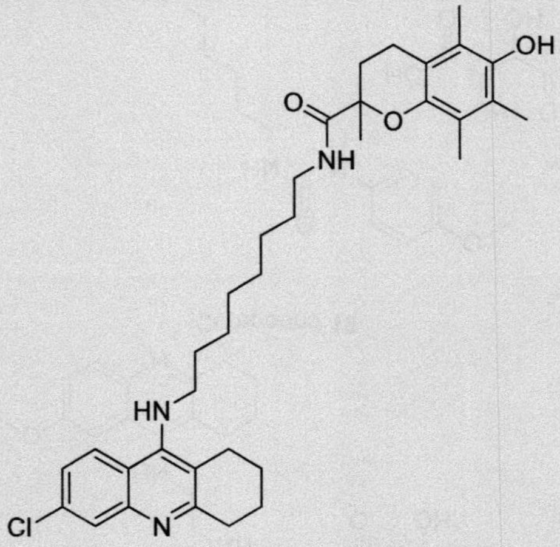 Compound **21**	AChE ($IC_{50} = 0.08$ μM), BChE ($IC_{50} = 0.54$ μM), high brain permeability, neuroprotective, and antioxidant	[88]

(continued)

Table 1
(continued)

S. No.	Tacrine hybrids	Lead structure(s)	Functions	Ref.
3.	Tacrine hybrids with cinnamate and cinnamylidene acetate	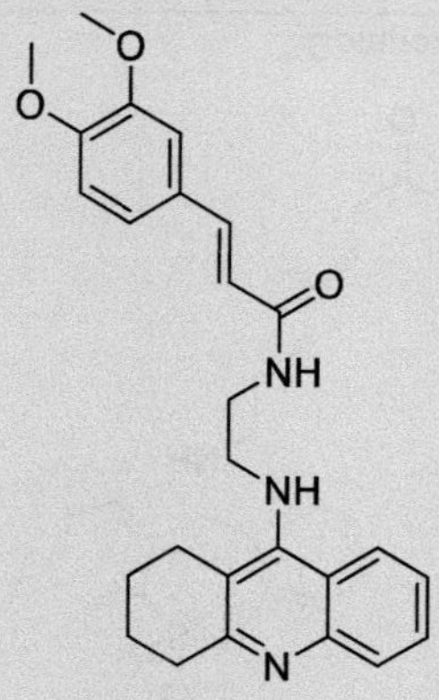 Compound **22**	AChE inhibition ($IC_{50} = 0.09$ μM), antioxidant, anti-Aβ aggregation (19.6%)	[89]
		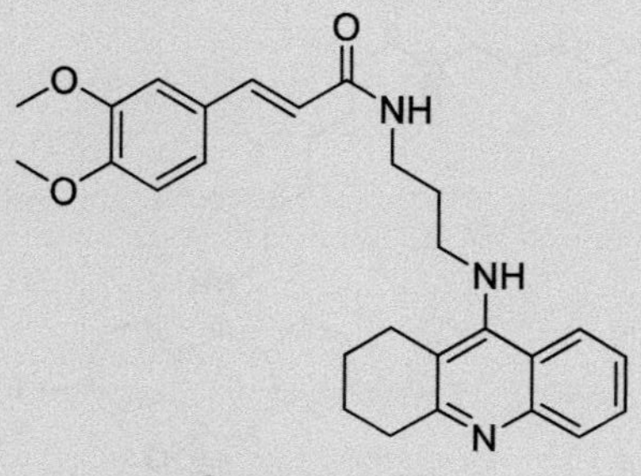 Compound **23**	AChE inhibition ($IC_{50} = 0.09$ μM), antioxidant, anti-Aβ aggregation (56.5%)	

(continued)

Table 1
(continued)

S. No.	Tacrine hybrids	Lead structure(s)	Functions	Ref.
4.	Tacrine-tianeptine hybrids	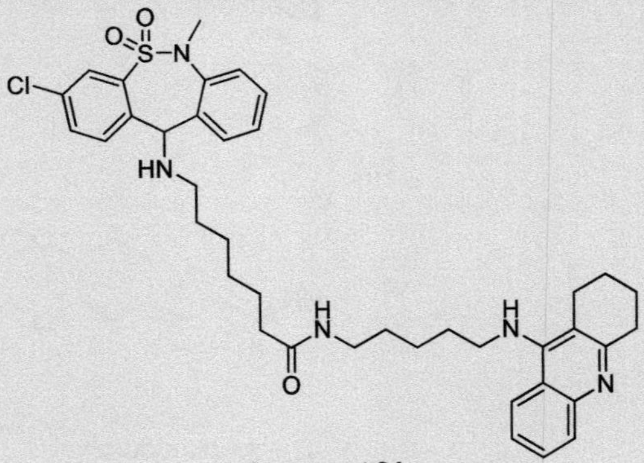 Compound **24**	AChE ($IC_{50} = 108.04$ nM), BChE ($IC_{50} = 6.97$ nM), and neuroprotective	[96]
		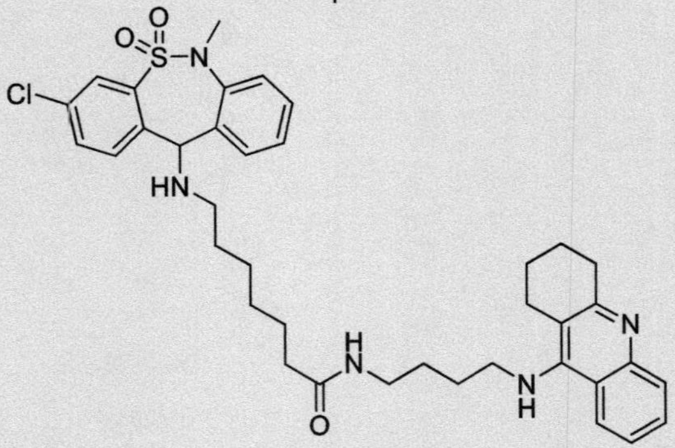 Compound **25**	AChE ($IC_{50} = 156.9$ nM), BChE ($IC_{50} = 3.59$ nM), and neuroprotective	

(continued)

Table 1
(continued)

S. No.	Tacrine hybrids	Lead structure(s)	Functions	Ref.
5.	Tacrine-benzofuran hybrids	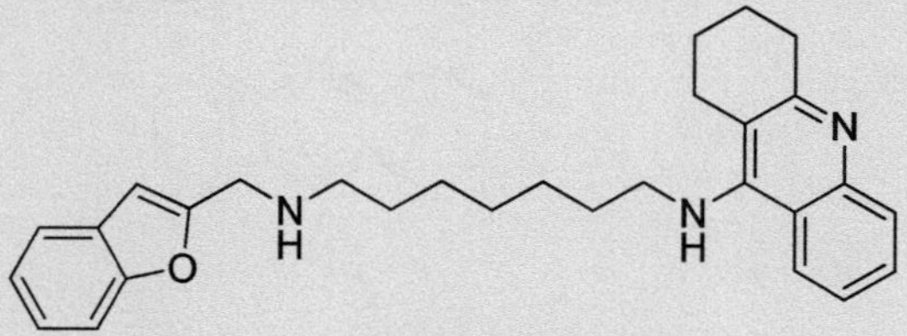 Compound **26**	AChE (IC_{50} = 0.86 nM), BChE (IC_{50} = 2.18 nM), BACE-1 (IC_{50} = 1.35 μM), anti-Aβ aggregation (61.3%), neuroprotective, in vivo enhancement of cognition and memory functions	[97]
6.	Indolotacrine analogs	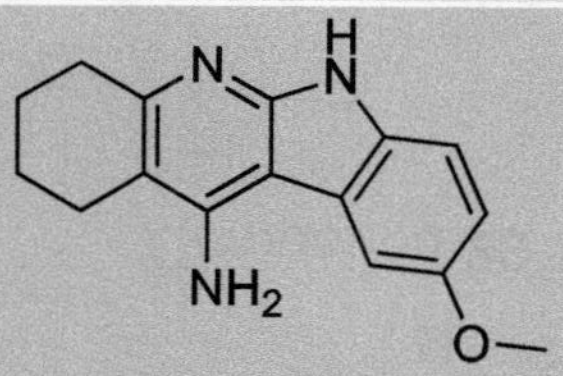 Compound **27**	AChE (IC_{50} = 1.5 μM), BChE (IC_{50} = 2.4 μM), MAO-A (IC_{50} = 0.49 μM), MAO-B (IC_{50} = 53.9 μM), and good brain permeability	[98]
7.	Tacrine-cysteine hybrids	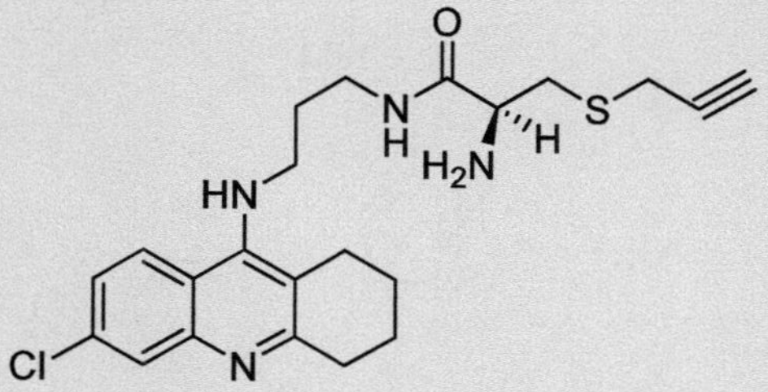 Compound **28**	AChE inhibition (IC_{50} = 0.97 μM), antioxidant, and anti-Aβ aggregation	[99]

(continued)

Table 1 (continued)

S. No.	Tacrine hybrids	Lead structure(s)	Functions	Ref.
8.	Tacrine-1,2,3-triazole hybrids	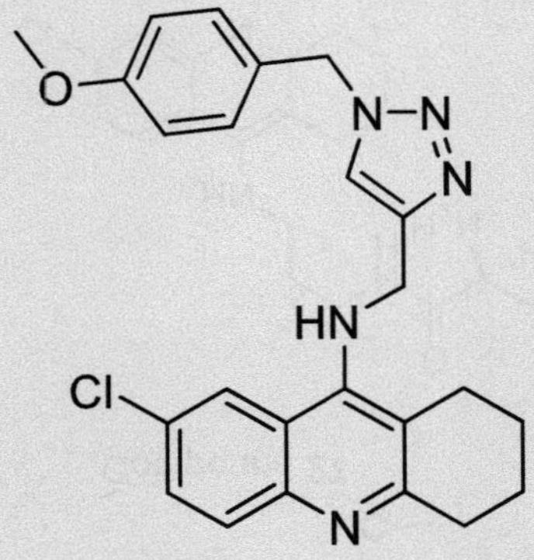 Compound **29**	Cholinesterase inhibition (AChE: $IC_{50} = 0.521$ μM, BChE: $IC_{50} = 1.853$ μM); in vivo improvement in cognitive dysfunctions	[100]
		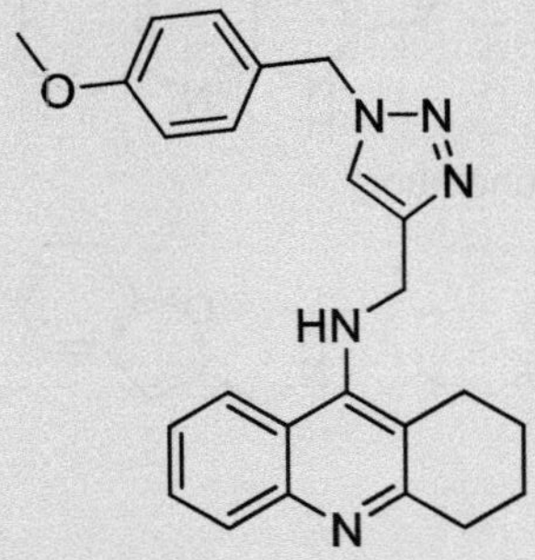 Compound **30**	Cholinesterase inhibition (AChE: $IC_{50} = 2.000$ μM, BChE: $IC_{50} = 0.055$ μM); in vivo improvement in cognitive dysfunctions	

(continued)

Table 1
(continued)

S. No.	Tacrine hybrids	Lead structure(s)	Functions	Ref.
9.	Tacrine-resveratrol fused hybrids	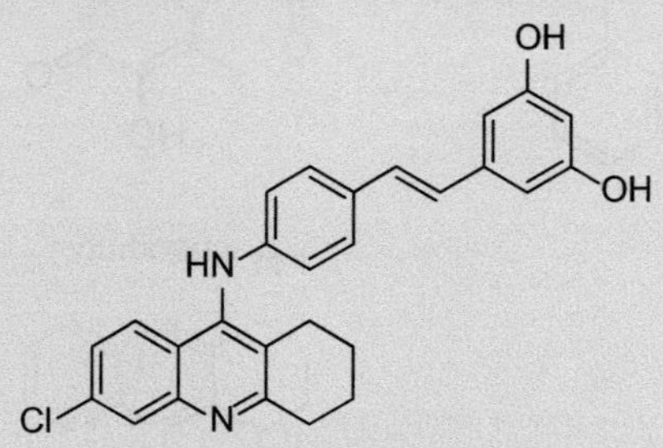 Compound **31**	AChE inhibition (AChE: $IC_{50} = 8.8\ \mu M$); anti-Aβ aggregation, and anti- inflammatory and immuno-modulatory properties in neuronal and glial AD cell models	[101]
10.	Tacrine-scutellarin hybrids	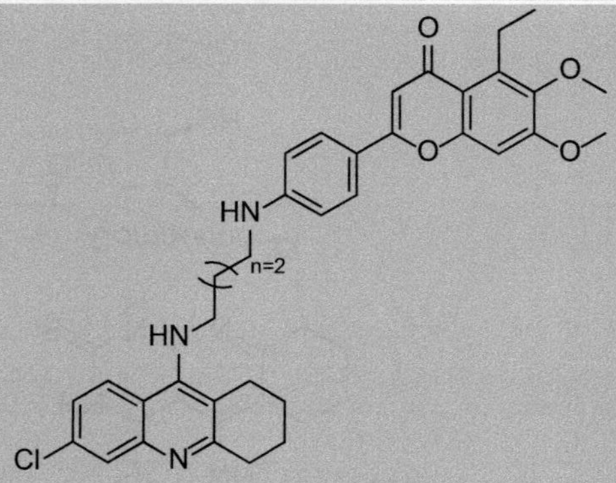 Compound **32**	AChE ($IC_{50} = 1.63$ nM), BChE ($IC_{50} = 1210$ nM), and antioxidant	[102]

(continued)

Table 1 (continued)

S. No.	Tacrine hybrids	Lead structure(s)	Functions	Ref.
11.	Tacripyrimidines	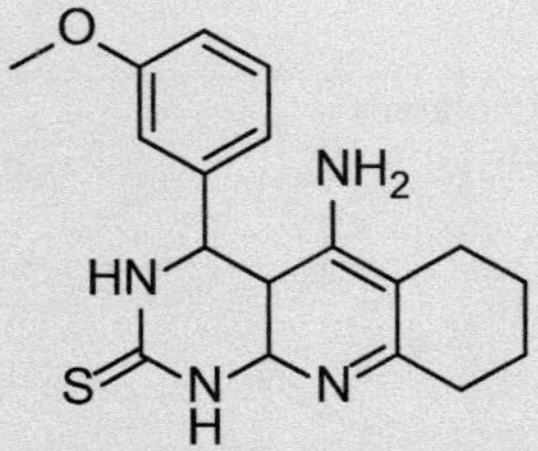 Compound 33	AChE ($IC_{50} = 3.05$ μM), BChE ($IC_{50} = 3.19$ μM), calcium channel blocker (30.40%)	[103]
12.	Tacrine hybrids with carbohydrate derivatives	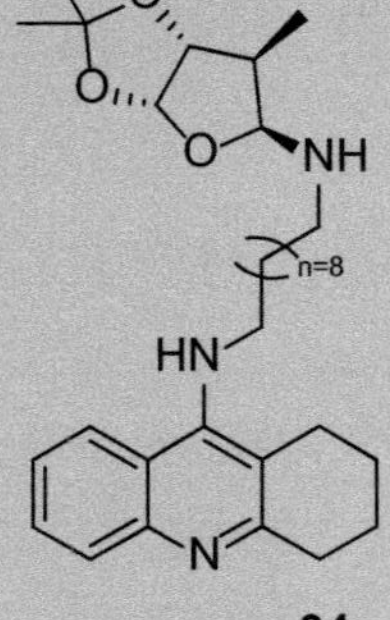 Compound 34	AChE inhibition ($IC_{50} = 2.2$ nM). BChE inhibition ($IC_{50} = 4.93$ nM)	[104]
13.	Tacrine-deferiprone hybrids	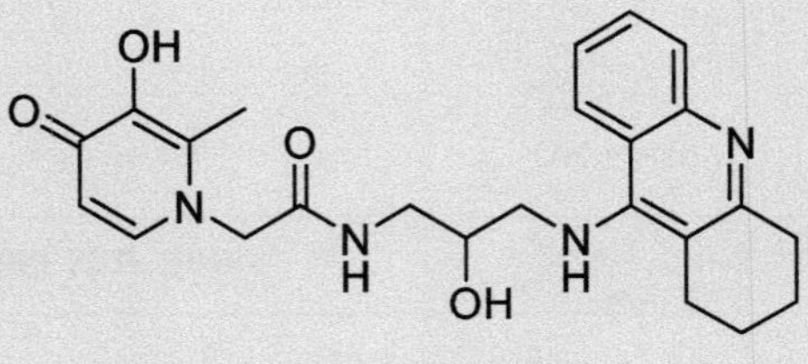 Compound 35	AChE inhibition ($IC_{50} = 0.96$ μM), antioxidant, and anti-Aβ aggregation	[105]

(continued)

Table 1
(continued)

S. No.	Tacrine hybrids	Lead structure(s)	Functions	Ref.
14.	IAA-tacrine hybrids	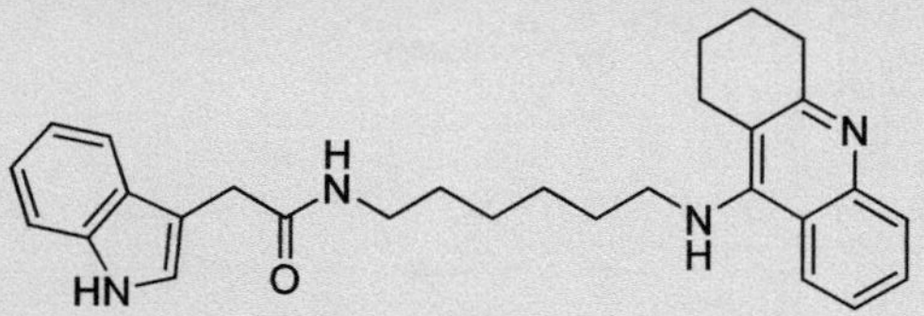 Compound **36**	Cholinesterase inhibition (AChE: $IC_{50} = 0.173$ μM, BChE: $IC_{50} = 0.066$ μM)	[106]
15.	Tacrine-selegiline hybrids	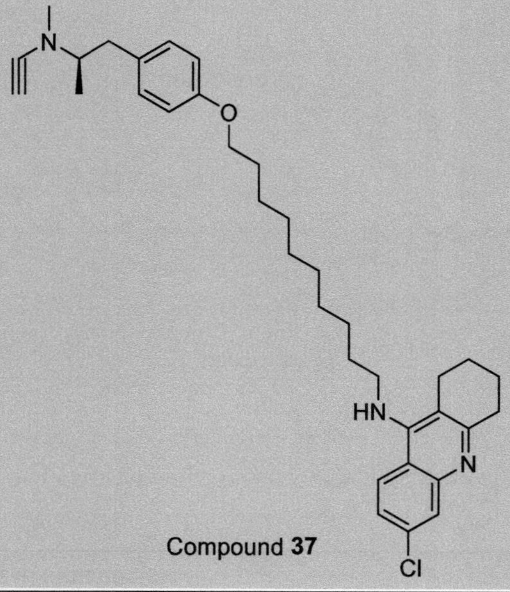 Compound **37**	AChE ($IC_{50} = 23.2$ nM), BChE ($IC_{50} = 2.03$ nM) and MAO (MAO-A: $IC_{50} = 0.31$ μM, MAO-B: $IC_{50} = 0.35$ μM) inhibitory potential	[107]

(continued)

Table 1
(continued)

S. No.	Tacrine hybrids	Lead structure(s)	Functions	Ref.
16.	Tacrine-ferulic acid hybrids	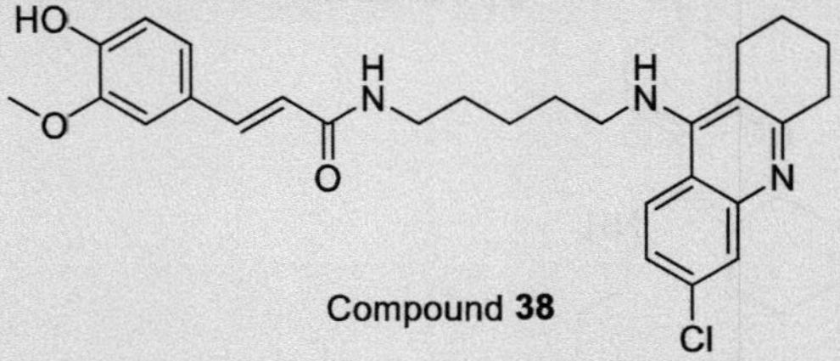 Compound **38**	AChE inhibition (AChE: $IC_{50} = 3.2$ nM), and antioxidant	[108]
17.	Tacrine-lipoic acid hybrids	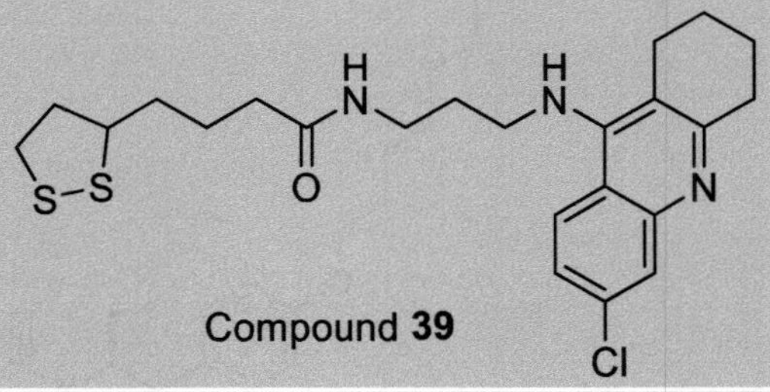 Compound **39**	AChE inhibition (AChE: $IC_{50} = 0.48$ μM), and antioxidant	[109]
18.	Tacrine-carbazole hybrids	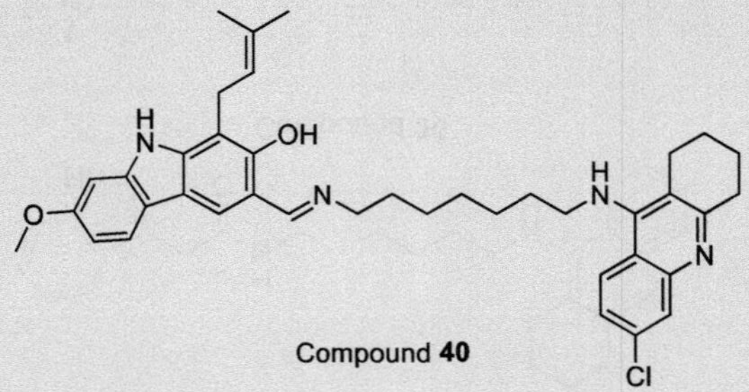 Compound **40**	AChE inhibition (AChE: $IC_{50} = 0.48$ μM); antioxidant, and ABTS radical scavenging activity	[110]

(continued)

Table 1 (continued)

S. No.	Tacrine hybrids	Lead structure(s)	Functions	Ref.
19.	Tacrine-chromene hybrids	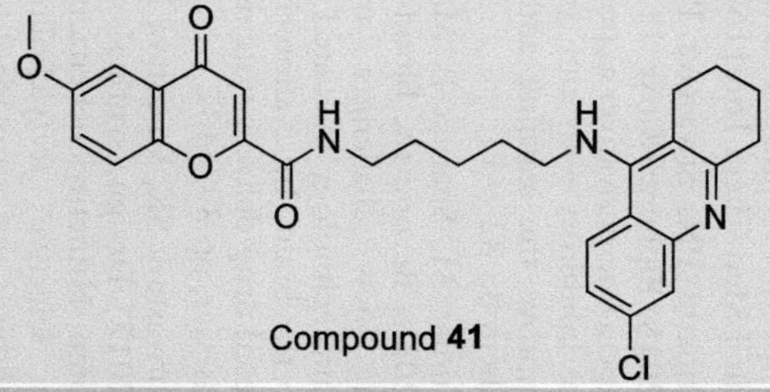 Compound **41**	BChE inhibition (BChE: $IC_{50} = 35$ pM), antioxidant, and anti-Aβ aggregation	[111]
20.	Tacrine-melatonin hybrids	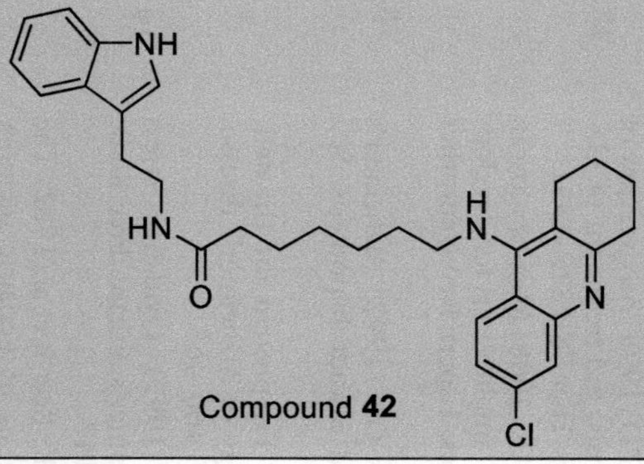 Compound **42**	AChE inhibition (AChE: $IC_{50} = 0.2$ nM); antioxidant, and neuroprotective	[112]

Acknowledgments

The authors are thankful to Department of Health Research (DHR), Ministry of Health and Family Welfare (MHFW), Government of India, New Delhi for providing Young Scientist grant to Mr. Piyoosh Sharma in newer areas of Drug Chemistry (25011/215-HRD/2016-HR).

References

1. Whitehouse PJ, Price DL, Struble RG et al (1982) Alzheimer's disease and senile dementia: loss of neurons in the basal forebrain. Science 215:1237–1239
2. Patterson C (2018) In: World Alzheimer report 2018. The state of the art of dementia research: new frontiers. London, pp 32–36
3. Cuijpers Y, Van Lente H (2015) Early diagnostics and Alzheimer's disease: beyond 'cure' and 'care'. Technol Forecast Soc Change 93:54–67
4. Hippius H, Neundörfer G (2003) The discovery of Alzheimer's disease. Dialogues Clin Neurosci 5:101
5. Forsyth E, Ritzline PD (1998) An overview of the etiology, diagnosis, and treatment of Alzheimer disease. Phys Ther 78:1325–1331
6. Gold CA, Budson AE (2008) Memory loss in Alzheimer's disease: implications for development of therapeutics. Expert Rev Neurother 8:1879–1891
7. Hardy JA, Higgins GA (1992) Alzheimer's disease: the amyloid cascade hypothesis. Science 256:184
8. Goedert M (1993) Tau protein and the neurofibrillary pathology of Alzheimer's disease. Trends Neurosci 16:460–465
9. Markesbery WR (1997) Oxidative stress hypothesis in Alzheimer's disease. Free Radic Biol Med 23:134–147
10. Bartus RT, Dean RR, Beer B et al (1982) The cholinergic hypothesis of geriatric memory dysfunction. Science 217:408–414
11. Hooper C, Killick R, Lovestone S (2008) The GSK3 hypothesis of Alzheimer's disease. J Neurochem 104:1433–1439
12. Qu T, Manev R, Manev H (2001) 5-Lipoxygenase (5-LOX) promoter polymorphism in patients with early-onset and late-onset Alzheimer's disease. J Neuropsychiatry Clin Neurosci 13:304–305
13. Cole SL, Vassar R (2007) The Alzheimer's disease β-secretase enzyme, BACE1. Mol Neurodegener 2:22
14. Talesa VN (2001) Acetylcholinesterase in Alzheimer's disease. Mech Ageing Dev 122:1961–1969
15. Shrivastava SK, Sinha SK, Srivastava P et al (2019) Design and development of novel p-aminobenzoic acid derivatives as potential cholinesterase inhibitors for the treatment of Alzheimer's disease. Bioorg Chem 82:211–223
16. Srivastava P, Tripathi PN, Sharma P et al (2019) Design and development of some phenyl benzoxazole derivatives as a potent acetylcholinesterase inhibitor with antioxidant property to enhance learning and memory. Eur J Med Chem 163:116–135
17. Tripathi PN, Srivastava P, Sharma P et al (2018) Biphenyl-3-oxo-1, 2, 4-triazine linked piperazine derivatives as potential cholinesterase inhibitors with anti-oxidant property to improve the learning and memory. Bioorg Chem 85:82–96
18. Kumar M, Sharma P, Maheshwari R et al (2018) Beyond the blood–brain barrier: facing new challenges and prospects of nanotechnology-mediated targeted delivery to the brain. In: Nanotechnology-based targeted drug delivery systems for brain tumors. Elsevier, pp 397–437
19. Shrivastava SK, Srivastava P, Upendra T et al (2017) Design, synthesis and evaluation of some N-methylenebenzenamine derivatives as selective acetylcholinesterasc (AChE) inhibitor and antioxidant to enhance learning and memory. Biorg Med Chem 25:1471–1480
20. Sinha SK, Shrivastava SK (2013) Synthesis, evaluation and molecular dynamics study of some new 4-aminopyridine semicarbazones as an antiamnesic and cognition enhancing agents. Biorg Med Chem 21:5451–5460
21. Sinha SK, Shrivastava SK (2013) Design, synthesis and evaluation of some new 4-aminopyridine derivatives in learning and memory. Bioorg Med Chem Lett 23:2984–2989

22. Sinha SK, Shrivastava SK (2012) Synthesis and evaluation of some new 4-aminopyridine derivatives as a potent antiamnesic and cognition enhancing drugs. Med Chem Res 21:4395–4402
23. Tripathi PN, Srivastava P, Sharma P et al (2019) Design and development of novel N-(pyrimidin-2-yl)-1, 3, 4-oxadiazole hybrids to treat cognitive dysfunctions. Biorg Med Chem 27:1327–1340
24. Rosini M, Simoni E, Minarini A et al (2014) Multi-target design strategies in the context of Alzheimer's disease: acetylcholinesterase inhibition and NMDA receptor antagonism as the driving forces. Neurochem Res 39:1914–1923
25. Rosini M, Simoni E, Bartolini M et al (2008) Inhibition of acetylcholinesterase, β-amyloid aggregation, and NMDA receptors in Alzheimer's disease: a promising direction for the multi-target-directed ligands gold rush. J Med Chem 51:4381–4384
26. Hogan DB (2007) Progress update: pharmacological treatment of Alzheimer's disease. Neuropsychiatr Dis Treat 3:569
27. Sharma P, Srivastava P, Seth A et al (2019) Comprehensive review of mechanisms of pathogenesis involved in Alzheimer's disease and potential therapeutic strategies. Prog Neurobiol 174:53–89
28. Davies P, Maloney A (1976) Selective loss of central cholinergic neurons in Alzheimer's disease. Lancet 308:1403
29. Thompson P, Wright D, Counsell CE et al (2012) Statistical analysis, trial design and duration in Alzheimer's disease clinical trials: a review. Int Psychogeriatr 24:689–697
30. Hebb C (1972) Biosynthesis of acetylcholine in nervous tissue. Physiol Rev 52:918–957
31. Bartus RT (2000) On neurodegenerative diseases, models, and treatment strategies: lessons learned and lessons forgotten a generation following the cholinergic hypothesis. Exp Neurol 163:495–529
32. Lanctôt KL, Herrmann N, Yau KK et al (2003) Efficacy and safety of cholinesterase inhibitors in Alzheimer's disease: a meta-analysis. Can Med Assoc J 169:557–564
33. Taylor P (1998) Development of acetylcholinesterase inhibitors in the therapy of Alzheimer's disease. Neurology 51:S30–S35
34. Pezzementi L, Nachon F, Chatonnet A (2011) Evolution of acetylcholinesterase and butyrylcholinesterase in the vertebrates: an atypical butyrylcholinesterase from the Medaka Oryzias latipes. PLoS One 6:e17396
35. Darvesh S, Hopkins DA, Geula C (2003) Neurobiology of butyrylcholinesterase. Nat Rev Neurosci 4:131
36. Behra M, Cousin X, Bertrand C et al (2002) Acetylcholinesterase is required for neuronal and muscular development in the zebrafish embryo. Nat Neurosci 5:111
37. Darvesh S, Cash MK, Reid GA et al (2012) Butyrylcholinesterase is associated with β-amyloid plaques in the transgenic APPSWE/PSEN1dE9 mouse model of Alzheimer disease. J Neuropathol Exp Neurol 71:2–14
38. Cousin X, Hotelier T, Giles K et al (1998) aCHEdb: the database system for ESTHER, the α/β fold family of proteins and the Cholinesterase gene server. Nucleic Acids Res 26:226–228
39. Holzgrabe U, Kapková P, Alptüzün V et al (2007) Targeting acetylcholinesterase to treat neurodegeneration. Expert Opin Ther Targets 11:161–179
40. Nicolet Y, Lockridge O, Masson P et al (2003) Crystal structure of human butyrylcholinesterase and of its complexes with substrate and products. J Biol Chem 278:41141–41147
41. Moral-Naranjo MT, Cabezas-Herrera J, Campoy FJ et al (1997) Glycosylation of cholinesterase forms in brain from normal and dystrophic Lama2dy mice. Neurosci Lett 226:45–48
42. Saxena A, Redman AM, Jiang X et al (1997) Differences in active site gorge dimensions of cholinesterases revealed by binding of inhibitors to human butyrylcholinesterase. Biochemistry 36:14642–14651
43. Kryger G, Silman I, Sussman JL (1999) Structure of acetylcholinesterase complexed with E2020 (Aricept®): implications for the design of new anti-Alzheimer drugs. Structure 7:297–307
44. Almeida JSD, Cavalcante SFDA, Dolezal R et al (2019) Molecular modeling studies on the interactions of aflatoxin B1 and its metabolites with the peripheral anionic site (PAS) of human acetylcholinesterase. J Biomol Struct Dyn 37(8):2041–2048
45. Inestrosa NC, Alvarez A, Perez CA et al (1996) Acetylcholinesterase accelerates assembly of amyloid-β-peptides into Alzheimer's fibrils: possible role of the peripheral site of the enzyme. Neuron 16:881–891
46. Zhang Y, Kua J, Mccammon JA (2002) Role of the catalytic triad and oxyanion hole in acetylcholinesterase catalysis: an ab initio

QM/MM study. J Am Chem Soc 124:10572–10577
47. Ordentlich A, Barak D, Kronman C et al (1998) Functional characteristics of the oxyanion hole in human acetylcholinesterase. J Biol Chem 273:19509–19517
48. Bajda M, Więckowska A, Hebda M et al (2013) Structure-based search for new inhibitors of cholinesterases. Int J Mol Sci 14:5608–5632
49. Whitehead A, Perdomo C, Pratt RD et al (2004) Donepezil for the symptomatic treatment of patients with mild to moderate Alzheimer's disease: a meta-analysis of individual patient data from randomised controlled trials. Int J Geriatr Psychiatry 19:624–633
50. Sharma P, Tripathi A, Tripathi PN et al (2019) Design and development of multitarget-directed N-benzylpiperidine analogs as potential candidates for the treatment of Alzheimer's disease. Eur J Med Chem 167:510–524
51. Camps P, Formosa X, Galdeano C et al (2008) Novel donepezil-based inhibitors of acetyl- and butyrylcholinesterase and acetylcholinesterase-induced β-amyloid aggregation. J Med Chem 51:3588–3598
52. Kume T, Sugimoto M, Takada Y et al (2005) Up-regulation of nicotinic acetylcholine receptors by central-type acetylcholinesterase inhibitors in rat cortical neurons. Eur J Pharmacol 527:77–85
53. Tiseo P, Rogers S, Friedhoff L (1998) Pharmacokinetic and pharmacodynamic profile of donepezil HCl following evening administration. Br J Clin Pharmacol 46:13
54. Sugimoto H, Ogura H, Arai Y et al (2002) Research and development of donepezil hydrochloride, a new type of acetylcholinesterase inhibitor. Jpn J Pharmacol 89:7–20
55. Sugimoto H, Yamanish Y, Iimura Y et al (2000) Donepezil hydrochloride (E2020) and other acetylcholinesterase inhibitors. Curr Med Chem 7:303–339
56. Andreani A, Cavalli A, Granaiola M et al (2001) Synthesis and screening for Antiacetylcholinesterase activity of (1-benzyl-4-oxopiperidin-3-ylidene) methylindoles and-pyrroles related to donepezil. J Med Chem 44:4011–4014
57. Contreras J-M, Rival YM, Chayer S et al (1999) Aminopyridazines as acetylcholinesterase inhibitors. J Med Chem 42:730–741
58. Omran Z, Cailly T, Lescot E et al (2005) Synthesis and biological evaluation as AChE inhibitors of new indanones and thiaindanones related to donepezil. Eur J Med Chem 40:1222–1245
59. Więckowska A, Więckowski K, Bajda M et al (2015) Synthesis of new N-benzylpiperidine derivatives as cholinesterase inhibitors with β-amyloid anti-aggregation properties and beneficial effects on memory in vivo. Biorg Med Chem 23:2445–2457
60. Shidore M, Machhi J, Shingala K et al (2016) Benzylpiperidine-linked diarylthiazoles as potential anti-Alzheimer's agents: synthesis and biological evaluation. J Med Chem 59:5823–5846
61. Costanzo P, Cariati L, Desiderio D et al (2016) Design, synthesis, and evaluation of donepezil-like compounds as AChE and BACE-1 inhibitors. ACS Med Chem Lett 7:470–475
62. Caliandro R, Pesaresi A, Cariati L et al (2018) Kinetic and structural studies on the interactions of *Torpedo californica* acetylcholinesterase with two donepezil-like rigid analogues. J Enzyme Inhib Med Chem 33:794–803
63. Villarroya M, García AG, Marco-Contelles J et al (2007) An update on the pharmacology of galantamine. Expert Opin Investig Drugs 16:1987–1998
64. Heinrich M, Teoh HL (2004) Galanthamine from snowdrop—the development of a modern drug against Alzheimer's disease from local Caucasian knowledge. J Ethnopharmacol 92:147–162
65. Maelicke A, Samochocki M, Jostock R et al (2001) Allosteric sensitization of nicotinic receptors by galantamine, a new treatment strategy for Alzheimer's disease. Biol Psychiatry 49:279–288
66. Farlow MR (2003) Clinical pharmacokinetics of galantamine. Clin Pharmacokinet 42:1383–1392
67. Cheung J, Rudolph MJ, Burshteyn F et al (2012) Structures of human acetylcholinesterase in complex with pharmacologically important ligands. J Med Chem 55:10282–10286
68. Fang L, Fang X, Gou S et al (2014) Design, synthesis and biological evaluation of D-ring opened galantamine analogs as multifunctional anti-Alzheimer agents. Eur J Med Chem 76:376–386
69. Guzior N, Wieckowska A, Panek D et al (2015) Recent development of multifunctional agents as potential drug candidates for the treatment of Alzheimer's disease. Curr Med Chem 22:373–404
70. Atanasova M, Yordanov N, Dimitrov I et al (2015) Molecular docking study on galantamine derivatives as cholinesterase inhibitors. Mol Inform 34:394–403

71. Polinsky RJ (1998) Clinical pharmacology of rivastigmine: a new-generation acetylcholinesterase inhibitor for the treatment of Alzheimer's disease. Clin Ther 20:634–647

72. Gottwald MD, Rozanski RI (1999) Rivastigmine, a brain-region selective acetylcholinesterase inhibitor for treating Alzheimer's disease: review and current status. Expert Opin Investig Drugs 8:1673–1682

73. Jann MW, Shirley KL, Small GW (2002) Clinical pharmacokinetics and pharmacodynamics of cholinesterase inhibitors. Clin Pharmacokinet 41:719–739

74. Bolognesi ML, Bartolini M, Cavalli A et al (2004) Design, synthesis, and biological evaluation of conformationally restricted rivastigmine analogues. J Med Chem 47:5945–5952

75. Wang L, Wang Y, Tian Y et al (2017) Design, synthesis, biological evaluation, and molecular modeling studies of chalcone-rivastigmine hybrids as cholinesterase inhibitors. Biorg Med Chem 25:360–371

76. Crismon ML (1994) Tacrine: first drug approved for Alzheimer's disease. Ann Pharmacother 28:744–751

77. Pohanka M (2011) Cholinesterases, a target of pharmacology and toxicology. Biomed Pap Med Fac Palacky Univ Olomouc 155 (3):219–229

78. Harel M, Schalk I, Ehret-Sabatier L et al (1993) Quaternary ligand binding to aromatic residues in the active-site gorge of acetylcholinesterase. Proc Natl Acad Sci U S A 90:9031–9035

79. Blackard WG Jr, Sood GK, Crowe DR et al (1998) Tacrine: a cause of fatal hepatotoxicity? J Clin Gastroenterol 26:57–59

80. Pessayre D, Mansouri A, Haouzi D et al (1999) Hepatotoxicity due to mitochondrial dysfunction. Cell Biol Toxicol 15:367–373

81. Lagadic-Gossmann D, Rissel M, Le Bot M et al (1998) Toxic effects of tacrine on primary hepatocytes and liver epithelial cells in culture. Cell Biol Toxicol 14:361–373

82. Tumiatti V, Minarini A, Bolognesi M et al (2010) Tacrine derivatives and Alzheimer's disease. Curr Med Chem 17:1825–1838

83. Minarini A, Milelli A, Tumiatti V et al (2012) Cystamine-tacrine dimer: a new multi-target-directed ligand as potential therapeutic agent for Alzheimer's disease treatment. Neuropharmacology 62:997–1003

84. Recanatini M, Cavalli A, Belluti F et al (2000) SAR of 9-amino-1,2,3,4-tetrahydroacridine-based acetylcholinesterase inhibitors: synthesis, enzyme inhibitory activity, QSAR, and structure-based CoMFA of tacrine analogues. J Med Chem 43:2007–2018

85. Tang H, Zhao L-Z, Zhao H-T et al (2011) Hybrids of oxoisoaporphine-tacrine congeners: novel acetylcholinesterase and acetylcholinesterase-induced β-amyloid aggregation inhibitors. Eur J Med Chem 46:4970–4979

86. Mao F, Huang L, Luo Z et al (2012) O-hydroxyl-or o-amino benzylamine-tacrine hybrids: multifunctional biometals chelators, antioxidants, and inhibitors of cholinesterase activity and amyloid-β aggregation. Biorg Med Chem 20:5884–5892

87. Jin H, Nguyen T, Go M (2014) Acetylcholinesterase and butyrylcholinesterase inhibitory properties of functionalized tetrahydroacridines and related analogs. Med Chem (Los Angeles) 4(10):688–696

88. Nepovimova E, Korabecny J, Dolezal R et al (2015) Tacrine–trolox hybrids: a novel class of centrally active, nonhepatotoxic multi-target-directed ligands exerting anticholinesterase and antioxidant activities with low in vivo toxicity. J Med Chem 58:8985–9003

89. Quintanova C, Keri RS, Marques SM et al (2015) Design, synthesis and bioevaluation of tacrine hybrids with cinnamate and cinnamylidene acetate derivatives as potential anti-Alzheimer drugs. MedChemComm 6:1969–1977

90. Sameem B, Saeedi M, Mahdavi M et al (2017) A review on tacrine-based scaffolds as multi-target drugs (MTDLs) for Alzheimer's disease. Eur J Med Chem 128:332–345

91. Small G, Bullock R (2011) Defining optimal treatment with cholinesterase inhibitors in Alzheimer's disease. Alzheimers Dement 7:177–184

92. Gauthier S (2001) Cholinergic adverse effects of cholinesterase inhibitors in Alzheimer's disease. Drugs Aging 18:853–862

93. Jiang S, Li Y, Zhang C et al (2014) M1 muscarinic acetylcholine receptor in Alzheimer's disease. Neurosci Bull 30:295–307

94. Maelicke A, Albuquerque EX (2000) Allosteric modulation of nicotinic acetylcholine receptors as a treatment strategy for Alzheimer's disease. Eur J Pharmacol 393:165–170

95. Korabecny J, Andrs M, Nepovimova E et al (2015) 7-Methoxytacrine-p-anisidine hybrids as novel dual binding site acetylcholinesterase inhibitors for Alzheimer's disease treatment. Molecules 20:22084–22101

96. Ceschi MA, Da Costa JS, Lopes JPB et al (2016) Novel series of tacrine-tianeptine hybrids: synthesis, cholinesterase inhibitory

activity, S100B secretion and a molecular modeling approach. Eur J Med Chem 121:758–772

97. Zha X, Lamba D, Zhang L et al (2015) Novel tacrine–benzofuran hybrids as potent multitarget-directed ligands for the treatment of Alzheimer's disease: design, synthesis, biological evaluation, and X-ray crystallography. J Med Chem 59:114–131
98. Benek O, Soukup O, Pasdiorova M et al (2016) Design, synthesis and in vitro evaluation of indolotacrine analogues as multitarget-directed ligands for the treatment of Alzheimer's disease. ChemMedChem 11:1264–1269
99. Keri RS, Quintanova C, Chaves S et al (2016) New tacrine hybrids with natural-based cysteine derivatives as multitargeted drugs for potential treatment of Alzheimer's disease. Chem Biol Drug Des 87:101–111
100. Najafi Z, Mahdavi M, Saeedi M et al (2017) Novel tacrine-1,2,3-triazole hybrids: in vitro, in vivo biological evaluation and docking study of cholinesterase inhibitors. Eur J Med Chem 125:1200–1212
101. Jeřábek J, Uliassi E, Guidotti L et al (2017) Tacrine-resveratrol fused hybrids as multi-target-directed ligands against Alzheimer's disease. Eur J Med Chem 127:250–262
102. Spilovska K, Korabecny J, Sepsova V et al (2017) Novel tacrine-scutellarin hybrids as multipotent anti-Alzheimer's agents: design, synthesis and biological evaluation. Molecules 22:1006
103. Chioua M, Buzzi E, Moraleda I et al (2018) Tacripyrimidines, the first tacrine-dihydropyrimidine hybrids, as multi-target-directed ligands for Alzheimer's disease. Eur J Med Chem 155:839
104. Lopes JPB, Silva L, Da Costa Franarin G et al (2018) Design, synthesis, cholinesterase inhibition and molecular modelling study of novel tacrine hybrids with carbohydrate derivatives. Biorg Med Chem 26:5566–5577
105. Chand K, Candeias E, Cardoso SM et al (2018) Tacrine–deferiprone hybrids as multi-target-directed metal chelators against Alzheimer's disease: a two-in-one drug. Metallomics 10:1460–1475
106. Cheng Z-Q, Zhu K-K, Zhang J et al (2019) Molecular-docking-guided design and synthesis of new IAA-tacrine hybrids as multifunctional AChE/BChE inhibitors. Bioorg Chem 83:277–288
107. Lu C, Zhou Q, Yan J et al (2013) A novel series of tacrine–selegiline hybrids with cholinesterase and monoamine oxidase inhibition activities for the treatment of Alzheimer's disease. Eur J Med Chem 62:745–753
108. Fang L, Appenroth D, Decker M et al (2008) Synthesis and biological evaluation of NO-donor-tacrine hybrids as hepatoprotective anti-Alzheimer drug candidates. J Med Chem 51:713–716
109. Rosini M, Andrisano V, Bartolini M et al (2005) Rational approach to discover multipotent anti-Alzheimer drugs. J Med Chem 48:360–363
110. Thiratmatrakul S, Yenjai C, Waiwut P et al (2014) Synthesis, biological evaluation and molecular modeling study of novel tacrine-–carbazole hybrids as potential multifunctional agents for the treatment of Alzheimer's disease. Eur J Med Chem 75:21–30
111. FernáNdez-Bachiller MI, PéRez CN, Monjas L et al (2012) New tacrine–4-oxo-4 H-chromene hybrids as multifunctional agents for the treatment of Alzheimer's disease, with cholinergic, antioxidant, and β-amyloid-reducing properties. J Med Chem 55:1303–1317
112. Rodríguez-Franco MI, Fernández-Bachiller MI, Pérez C et al (2006) Novel tacrine–melatonin hybrids as dual-acting drugs for Alzheimer disease, with improved acetylcholinesterase inhibitory and antioxidant properties. J Med Chem 49:459–462

Correction To: Enzyme–Ligand Interaction Monitored by Synchrotron Radiation Circular Dichroism

Rohanah Hussain, Charlotte S. Hughes, and Giuliano Siligardi

Correction to:
Chapter 6 in: Nikolaos E. Labrou (ed.), *Targeting Enzymes for Pharmaceutical Development: Methods and Protocols*, Methods in Molecular Biology, vol. 2089, https://doi.org/10.1007/978-1-0716-0163-1_6

The original version of the chapter "Enzyme–Ligand Interaction Monitored by Synchrotron Radiation Circular Dichroism" was previously published non-open access. It has now been changed to open access under a CC BY 4.0 license and the copyright holder has been updated to 'The Author(s).' The book has also been updated with the change.

The updated online version of this chapter can be found at
https://doi.org/10.1007/978-1-0716-0163-1_6

Nikolaos E. Labrou (ed.), *Targeting Enzymes for Pharmaceutical Development: Methods and Protocols*, Methods in Molecular Biology, vol. 2089, https://doi.org/10.1007/978-1-0716-0163-1_19, © Springer Science+Business Media, LLC 2020

Index

Nikolaos E. Labrou (ed.), *Targeting Enzymes for Pharmaceutical Development: Methods and Protocols*, Methods in Molecular Biology, vol. 2089, https://doi.org/10.1007/978-1-0716-0163-1, © Springer Science+Business Media, LLC 2020

MIX
Papier aus verantwortungsvollen Quellen
Paper from responsible sources
FSC® C105338

If you have any concerns about our products,
you can contact us on
ProductSafety@springernature.com

In case Publisher is established outside the EU,
the EU authorized representative is:
Springer Nature Customer Service Center GmbH
Europaplatz 3, 69115 Heidelberg, Germany

Printed by Libri Plureos GmbH
in Hamburg, Germany